SOCIAL GEOGRAPHY

SOCIAL GEOGRAPHY

A READER

Edited by
Chris Hamnett
Professor of Human Geography,
King's College London

A member of the Hodder Headline Group
LONDON • NEW YORK • SYDNEY • AUCKLAND

First published in Great Britain in 1996 by
Arnold, a member of the Hodder Headline Group
338 Euston Road, London NW1 3BH

Copublished in the US, Central and South America by
John Wiley & Sons, Inc.,
605 Third Avenue,
New York, NY 10158–0012

British Library Cataloguing in Publication Data
A catalogue entry for this book is available from the British Library

Library of Congress Cataloging-in-Publication Data
A catalog record for this book is available from the Library of Congress

ISBN 0 340 66282 4 (Pb)
ISBN 0 470 23639 6 (Wiley)

ISBN 0 340 66281 6 (Hb)
ISBN 0 470 23640 X (Wiley)

Composition by J&L Composition Ltd, Filey, North Yorkshire
Printed and bound in Great Britain by J. W. Arrowsmith, Bristol

CONTENTS

ACKNOWLEDGEMENTS

The editor and publishers would like to thank the following for permission to use copyright material in this book:

Blackwell Publishers for C. Katz, 'Sow What You Know: The Struggle for Social Reproduction in Rural Sudan', *Annals of the Association of American Geographers* 81 (1991), pp. 488–514, D. Hindson, M. Byerley and M. Morris, 'From Violence to Reconstruction: The Making, Disintegration and Remaking of an Apartheid City', *Antipode* 26 (1994), pp. 323–50, A. Merrifield, 'Situated Knowledge through Exploration: Reflections on Bunge's "Geographical Expeditions"', *Antipode* 27 (1995), pp. 49–70, D. Harvey and L. Chatterjee, 'Absolute Rent and the Structuring of Space by Governmental and Financial Institutions', *Antipode* 6 (1973), pp. 22–36, and D. Harvey, 'Social Justice, Postmodernism and the City', *International Journal of Urban and Regional Research* 16 (1992), pp. 588–601; The University of Chicago Press for E. W. Burgess, 'The Growth of the City: An Introduction to a Research Project' from R. E. Park, E. W. Burgess and R. D. McKenzie (eds), *The City* (1925), pp. 47–62; Elsevier Science Ltd for K. H. Halfacree, 'Locality and Social Representation: Space, Discourse and Alternative Definitions of the Rural', *Journal of Rural Studies* 9 (1993), pp. 23–37; Oxford University Press for R. Black, 'Livelihoods under Stress: A Case Study of Refugee Vulnerability in Greece', *Journal of Refugee Studies* 7 (1994), pp. 360–77; Pion Ltd for G. Pratt and S. Hanson, 'Gender, Class and Space', *Society and Space* 6 (1988), pp. 15–35, L. Peake, '"Race" and Sexuality: Challenging the Patriarchal Structuring of Urban Social Space', *Society and Space* 11 (1993), pp. 415–32, K. J. Anderson, 'Cultural Hegemony and the Race-Definition Process in Chinatown, Vancouver, 1880–1980', *Society and Space* 6 (1988), pp. 127–49, G. Rose, 'Locality, Politics and Culture: Poplar in the 1920s', *Society and Space* 6 (1988), pp. 151–68, and A. Charlesworth, 'Contesting Places of Memory: The Case of Auschwitz', *Society and Space* 12 (1994), pp. 579–93; Royal Geographical Society-Institute of British Geographers for D. Ley, 'Social Geography and the Taken-for-Granted World', *Transactions of the Institute of British Geographers* 2 (1977), pp. 498–512, H. P. M. Winchester, 'Women and Children Last: The Poverty and Marginalization of One-parent Families', *Transactions of the*

Institute of Geographers 15 (1990), pp. 70–86, and Susan J. Smith, 'Society, Space and Citizenship: A Human Geography for the "New Times"?' *Transactions of the Institute of British Geographers* 14 (1989), 144–56.

INTRODUCTION

An introduction to a book of this nature generally has to do several things. First, it has to define its focus or area of study. Second, it has to examine the history and development of the field, and third it has to provide some kind of theoretical and organizing structure for the papers selected for inclusion. In this introduction I will briefly attempt to deal with these issues. The question of definition is not an easy one as Emrys Jones (1975) has pointed out:

> If they are not too facile to be useful, definitions will always be a reflection of the time, the outlook or the philosophy of the investigator, the state of empirical knowledge of the field, and the scale of investigation. Rarely do all these strands come together to satisfy a large number of people and to easily identify an academic discipline (p. 1).

This is very true, and given the revolutions in thinking and approach which have characterized human geography in the last 25 years, the task is made all the more difficult. What is clear, however, is that some attempt is necessary, lest social geography is defined solely in an atheoretical, eclectic fashion. Unfortunately, social geography has sometimes been characterized by a 'seemingly inherent ambiguity' (Eyles, 1974), and what Pacione (1987) termed its 'eclecticism of study topics'. As Buttimer (1968) pointed out, 'no generally accepted definition of social geography exists. The variety of literature which has appeared under the title social geography is astounding'. Similarly, according to Knox (1987), social geography 'encompasses an eclectic mixture of ideas, theories and empirical research' (p. 1), and Cater and Jones (1989) have suggested that social geography is suffering from an 'identity crisis'. I believe that these criticisms considerably overstate the nature of the problems facing contemporary social geography and derive, in part, from the rapidity of intellectual change over the last 25 years which have led to a sometimes bewildering shift of emphasis and approach.

Much of the early work in social geography was concerned with the study of differences in ways of life in different parts of the world, generally in predominantly agrarian and rural societies, and attempts to explain these in terms of their relation to the differences in environment, climate, etc. Although the determinism of Ratzel's *Anthropogeographie* was replaced by 'more elastic concepts of possibilism' in the French school of social geography (Buttimer, 1968), the emphasis was on the relationship of different patterns of living or *genres de vie*, with nature and the ways in which human groups 'secured the material necessities of life within a functional social order'. But, as Buttimer (1968) rightly points out, this type of social geography was 'admirably suited to the study of European – particularly French – rural society of the early twentieth century', and she notes that 'most of the early studies in social geography were regional in character, and their excellence consisted more in their artistic cohesion and integrative descriptions than in their analytical or theoretical expertise' (p. 137). Similar comments can be made about Fleure and Estyn Evans on their work on the character or 'personality' of a region. While of historical interest, these issues are not of great relevance to the way most people in western, urban, industrial societies live today. As Marshall McLuhan (1964) perceptively commented, 'we drive forward into the future, looking into the rear view mirror of the past'. And as Pahl (1971) pointed out, 'Since the idea that "geographers start from the soil, not from society" was until recently widely held by most geographers, and is indeed still held by some, it is easy to understand why social geography has been slow to develop' (p. 125).

Some early work tended to equate social geography with human geography *tout court*. I consider this conflation very confusing and unhelpful given that the terms human and physical geography are now widely accepted, and that human geography is generally seen to subsume within it all aspects of human activity – economic, political, social and cultural. Fitzgerald (1946) suggested that social geography was concerned 'with spatial arrangements or patterns over the world of phenomena which are of social, as distinct from political or economic significance to man'. This parallels the definition of social geography made by Ray Pahl (1970b). It is, he said, 'the processes and patterns involved in an understanding of socially defined populations in their spatial setting'. This has the merit of being inclusive, but it has been criticized by Jones (1975) on the grounds that the term 'spatial setting' can be construed as 'nothing more than a descriptive point of reference system' (p. 3), and he argues that 'much more emphasis must be given to the way(s) in which the use and perception of this space enters as an element in the patterns and even plays a part in the processes'. In addition, the meaning of the term 'socially defined' populations can be queried on the grounds that there is an element of tautology involved. Eyles (1974) has also suggested that 'the implication seems to be that social geography is human geography without an economic component' (p. 29). If this were correct, I would accept Eyles's criticism that it is untenable, but to argue that social geography should be primarily concerned with social activity and behaviour does not necessarily imply that it has no economic component.

The field embraced by social geography is potentially vast. If we exclude the sphere of production and economic activity, which is widely considered to be the province of economic geography, and politics which is widely considered to be the province of political geography, there is a good case that social and cultural geography between them embrace most, if not all, of the remainder of the geography of human activity. But in some ways this simply displaces the definitional problem one remove. Economic geography studies the geography of the economy, political geography studies the geography of political activity, cultural geography examines the geography of culture, and social geography studies the 'social'. But what is 'the social'? Does it entail society as a whole, in the sense of everything that is not economic, political or cultural, and does it include the social structure and 'social activity', whatever that might mean?

The approach I take is that social geography is primarily concerned with the study of the geography of social structures, social activities and social groups across a wide range of human societies. As such, social geography embraces those areas of human activity which are not directly concerned with the production and distribution of goods and services (the economic). There are, of course, certain overlaps. Certain areas of economic geography embrace the analysis of consumption, which is also important for studies of housing, health care and education which are traditionally seen as the province of social geography. I do not see this as a problem. Economic, social and political activity cannot and should not be divided into separate, hermetically sealed compartments. To attempt to do so is to try artificially to compartmentalize human life and activity. What is appropriate, however, is to identify particular foci for intellectual study and analysis, and traditionally social geography has focused on those areas of social life and activity which are concerned with social reproduction rather than economic production or politics and organized political activity. But, here again, there are some areas of overlap between political and social geography concerning, for example, local social conflicts over facilities or turf, riots and other disturbances. It is inappropriate to attempt to legislate which sphere of activity these fall under. What is important is that they are adequately studied and analysed, whether by social or political geographers or both.

This focus on the geography of social structures, activities, social groups and their attitudes and perceptions, is deliberately intended to be as wide ranging as possible, because it seems excessively limiting and restrictive to argue that social geography should be concerned only with specific aspects of social activity. Nor do I think it desirable to suggest that social geography should be limited to a specific theoretical or methodological perspective, structuralist, phenomenological or structurationist, though I would suggest that it should endeavour to link analyses of structure and human agency wherever possible. Thus, at its broadest, social geography can be argued to comprise the study of the geography of the major social divisions such as race, class, gender and ethnicity, along with household and family structures. In addition, it includes the study of key areas of social reproduction such as the geography of health, housing and education, along with many other aspects of social activity such as crime, policing, festivals, carnivals. While I think it is important for social geographers to focus on questions of social inequality and the distribution of and access to scarce resources and facilities, it would be unduly restrictive to argue that social geography should be limited to such questions. This would be to legislate politically what is, and what is not, an acceptable area of study and it would exclude many interesting and important areas of study.

One of the most challenging questions faced by modern social geographers is whether, and if so how, to differentiate social from cultural geography. This is a relatively recent issue, given the communality of interest between the two areas. Until the early 1980s, cultural geography was still strongly influenced by the legacy of Carl Sauer and his followers in the Berkeley School. As Cosgrove and Jackson (1987) note: 'their concerns were dominantly rural and antiquarian, narrowly focused on *physical artifacts* (log cabins, fences and field boundaries). Although rarely explicitly stated, their work rested on the ethnological assumption that distinctive geographical areas (landscapes) could be identified and described by mapping visible elements of material culture produced by unitary cultural groups' (p. 96).

This approach was challenged in the late 1970s and early 1980s by Blaut (1979), Cosgrove (1983, 1984) and others who sought to introduce a radical cultural geography which aimed, amongst other things, to link symbolic landscapes to social formations. Although the landscape theme is still important within cultural geography, the import of ideas from linguistics and semiotics has meant that the symbolic qualities of landscape are now seen as far more important, and the influence of cultural theorists such as Raymond Williams, Stuart Hall, Clifford Geertz (1973, 1983) and others has been very important in reshaping and widening cultural geography. As Cosgrove and Jackson (1987) stated, the 'new' cultural geography is interested in 'the contingent nature of culture, in dominant ideologies and in forms of resistance to them'. They add that it also asserts the centrality of culture in human affairs:

> Culture is not a residual category, the surface variation left unaccounted for by more powerful economic analyses; it is the very medium through which social change is experienced, contested and constituted (p. 95).

The revival of the 'new' cultural geography in Anglo-America has been very influential in the revival of social geography. In many respects, the two have proceeded hand-in-hand, and it is difficult to separate the two areas of work. Jones (1975) notes that, though the central themes are easily distinguishable:

> it is sometimes very difficult to say where cultural geography ends and social geography begins – and probably undesirable . . . But on the whole, the social geographer rarely gets enmeshed in the palimpsest of behavioural patterns which give historical depth to the cultural landscape (p. 3).

The distinction I have drawn for the sake of selecting papers for this reader is between concrete social structures and material practices on the one hand and systems of meaning and belief on the other. Geertz (1973) suggests that cultural analysis is the study of webs of significance, the symbolic aspects of social action. I have focused more on material practices and less, but not exclusively so, on questions concerning symbolic meanings, signs and signifiers. It is impossible to draw a firm line between the two, and the notion of a social world which exists independently of cultural meanings and beliefs is difficult, if not impossible, to sustain.

I am making a distinction here between work which deals directly with the geography of social action, social groups or behaviour, and work which deals with the ways in which such action, groups and behaviours, and their representations have been interpreted by cultural geographers and others. Thus, I have *not* included such contemporary cultural geography as Daniels and Rycroft's (1993) study of Alan Sillitoe's Nottingham novels, Gillian Rose's (1994) analysis of representation, discourse and the cultural politics of place, Gruffudd's (1994) analysis of historiography, rurality and national identity in interwar Wales, Brian Graham's (1994) analysis of Estyn Evans's work on Ireland and Irish identity, Nuala Johnson's (1994) study of the role of public statuary in the celebration of the 1798 rebellion in Ireland and the construction of a heroic past. Nor am I including Lily Kong's (1995) work on music and cultural politics in Singapore or Cohen's (1995) work on music and the sensuous production of place. The justification in all cases is that such studies are more concerned with representations, and the analysis of representations, than they are with concrete material practices or activity.

To attempt to draw a line between the analysis of social structure and activity *per se* and analysis of the representation of such activity in literature, music and landscape is not easy. And, if we accept Geertz's argument that culture consists of the webs of signification, it is difficult to analyse social structure and action independently of the meanings ascribed to them by human actors. This point is powerfully made by Katz (Chapter 12, this volume) in the context of her study of social reproduction in rural Sudan. She argues that how people act within their environment is crucially dependent on how they come to know it. In other words, social action is dependent on cultural understanding. I would strongly endorse this view and the importance Katz accords to human agency in the reproduction of social structures.

The distinction I wish to draw, however, is between studies which *primarily* focus on social structures, attitudes and actions (and their representations), such as Susan Smith's (1993) work on crowd behaviour, procession rituals and the carnivalesque in the context of the Borders area of rural Scotland, or Peter Jackson's (1988, 1992) studies of Caribbean carnivals in Toronto and London, Rose's (Chapter 11, this volume) study of locality, politics and culture in Poplar in the 1920s or Valentine's (1993) work on lesbian timespace strategies and sexual identity, and those studies which focus *primarily* on the representation of human activity and the analysis of such representations in literature, art, the built environment and so on. In making this distinction, I am not saying that questions of representation, discourse and identity are unimportant. They are extremely important, and I have tried, where possible, to give them coverage. The distinction is thus primarily one of emphasis. I am also aware that I am excluding a good deal of exciting recent work which straddles this (artificial) boundary but, given the constraints on space and difficulties of selection, it is necessary to limit the coverage of the reader. A parallel reader in this series on cultural geography is being edited by Deryck Holdsworth.

The question of antecedents is almost as difficult as that of coverage. While it would have been misguided to attempt to edit a reader which dealt in detail with the history and development of social geography, it is necessary to give some idea of the development of the area, and to include some of the papers which have influenced contemporary work. I have not, however, attempted to go back and trace the early development of social geography in the work of Vidal de la Blache (1926), Barrow (1923), Visher (1929), Firey (1945), Gilbert and Steel (1945) and others. Instead, I have focused on developments since 1970 which mark a turning point in the development of human geography in general, with the shift from descriptive geography and spatial science toward marxist-influenced and interpretative work.

Trying to provide a coherent structure for such a reader is very difficult, not least because some of the most interesting current research ranges over such a wide area. The task is rather akin to shuffling a pack of cards and deciding whether it is best to organize them by suit, colour, face value or rank. Each arrangement can be justified, and each has some merit. The solution I have chosen is to group a number of papers into an initial 'frameworks' section, highlighting the importance of different approaches, followed by a section on gender, class, race and space, a section on locality, politics and culture, and a final section titled towards a progressive human geography. But groupings of this sort always have some arbitrary element to them, and readers should not assume that this is the only, or necessarily, the most appropriate one. Finally, I should note that I have left the original language of the papers untouched.

SECTION ONE
FRAMEWORKS OF ANALYSIS

Editor's introduction

The early origins of contemporary social geography can be traced back to two main bodies of literature, the first rural and the second urban. The rural tradition, exemplified in the work of Vidal de la Blache (1926) and the French school of human geography, was primarily concerned with the relationships between social groups and their environments, and de la Blache introduced the concepts of *genre de vie* and *milieu* (Buttimer, 1968, 1971; Jones, 1975; Ley, Chapter 3, this volume). Each region and place, says Ley, was 'considered holistically as an intimate amalgam of environment and decision-making men, as an object with a subject' (p. 52). The approach was primarily interpretative, but Brunhes and others subsequently stressed the analysis of 'landscape facts' as objective subjects for analysis, a tradition later developed by Carl Sauer's landscape school. As Ley points out:

> The material facts had sprung loose from their everyday human world, artifact replaced attitude, and land-use morphology triumphed over regional personality. The transition from a science of man in place to a science of phenomena prepared the way for a scientism which ultimately abstracted place to a geometry of space, and reduced man to a pallid . . . figure' (p. 52).

The other principal tradition was an urban one, grounded in the attempts of a number of American sociologists to make sense of the massive growth of cities, and associated immigration, in the late nineteenth and early twentieth centuries. The two main figures associated with the Chicago school are Ernest Burgess (1919) and Robert Park (1916), and their edited book *The City* (Park and Burgess, 1925) on the work of the Chicago school was published in 1925. The theoretician was Park. Originally a journalist, Park went to Chicago University in 1914 to research for a Ph.D. and, influenced by the writings of Darwin and the plant ecologists, he developed a theory of the city based on an analogy between plant communities and human communities. As Robson (1969) summarizes it:

> The starting point of Park's ecological theory was Darwin's concept of the web of life: the intimate interrelationship between organism(s) . . . and

> between organism and the environment. Since man is an organic creature, Park argued that he is subject to the general laws of the organic world. This laid the basis for his use of biological analogy (pp. 9–10).

Robson goes on to note that, as Park recognized, the analogy is imperfect as human beings are subject to other impulses than the basic need for survival. Thus, in order to accommodate both the social and sub-social factors, Park organized his conceptual framework of human ecology around a distinction between *community* and *society*. Community, in Park's view, was built on what he saw as 'sub-social' or biotic competition between anonymous individuals in a struggle for existence. At this level, the various processes recognized by plant ecologists could be translated into human terms. Most important of these, says Robson, was the concept of *competition*, whereby human beings competed for limited space and access to the most desirable locations. This competition was reflected in land values and led to the segregation of types of person in terms of their ability to compete economically. A second process was that of *dominance*, which is one of the fundamental concepts of plant ecologists. Within different types of plant associations, one species exerts a dominant influence controlling the environmental conditions which encourage or discourage other types of species. In the city, the Central Business District forms the dominant element within the urban areas, and within certain areas particular activities exert dominance. The slum area was seen as the area of minimum choice, and so it collected a population which was homogeneous in terms of its 'economic competence', even though in ethnic terms it might be very heterogeneous. Finally, Robson points out that dominance was very strongly connected to the concepts of *invasion* and *succession*, and the analogy with the plant world is close. These concepts were applied to the city, in relation to the invasion of residential areas by commercial activities or the invasion of higher-status residential areas by lower-status groups, or one type of ethnic area by another ethnic group (pp. 10–11).

Robson (1969) points out that other important concepts were developed from the ecological analogy. The first, related to the process of dominance, was the concept of the gradient, seen in the land value gradient, and reflected in a number of social gradients such as income, crime and mental illness, which were influenced by the sifting and sorting effects of land values on land use. Second, says Robson, was the concept of the 'natural area' which was linked to the process of segregation. Some viewed natural areas as being delineated by land use patterns, while others saw them primarily as social and cultural features, comprising homogeneous areas of race, language and income. The presumed high degree of internal homogeneity within natural areas was seen to make them act as social sub-systems within the city as a whole.

The underlying concepts were brought together in an overall spatial model by Burgess's well-known concentric zone theory of city growth which, with its five component rings, was, says Robson:

> the logical spatial expression of the ecological principles of central dominance, segregation, invasion and succession. Development outwards was accompanied by the differentiation of the successive rings of new urban growth and the invasion by different elements of older inner rings of the city (p. 13).

Burgess himself introduced his essay (Chapter 1, this volume) by arguing that 'The outstanding fact of modern society is the growth of great cities' (p. 26), and he went on to state that this is seen most clearly in the United States, where all the manifestations of modern urban life such as the skyscraper, the subway, the department store and the daily newspaper are to be found. He argued that while there had been studies of the growth of the urban population, a more significant feature of great cities was their tendency to physical expansion, and he focused on this in his essay. Unfortunately, although he examined 'expansion as a process', he argued that:

> The typical processes of the expansion of the city can best be illustrated . . . by a series of concentric circles, which may be numbered to designate both the successive zones of urban extension and the types of areas differentiated in the process of expansion [p. 27] . . . In the expansion of the city a process of distribution takes place which sifts and sorts and relocates individuals and groups by residence and occupation (p. 30).

Although he stated that 'it hardly needs to be added that neither Chicago nor any other city fits perfectly into this ideal scheme', generations of urban social geographers and urban sociologists occupied themselves looking for similar patterns in a variety of other cities. The game of 'hunt the Chicago model' was firmly established, and reached its apogee in the late 1960s and early 1970s, when a combination of detailed census data availability, the introduction of computers and a positivist scientific methodology allowed geographers to engage in quantitative analysis of urban residential structure (Robson, 1969). In the excitement of the hunt, the conditions prevailing in Chicago and other cities at the time, large-scale immigration, an unconstrained free market and rapid urbanization tended to be overlooked, even though Burgess had pointed to the importance of immigration and population growth.

Park and Burgess also touched on a number of other issues which were taken up by later generations of urban theorists. Paralleling Simmel's (1903) work 'The metropolis and mental life', written in turn-of-the-century Berlin, Burgess pointed to the importance of mobility, stimulation and segmentation of social relationships, an issue subsequently taken up by Louis Wirth (1939) in his classic essay

'Urbanism as a way of life'. All of the social theorists at this time were concerned with the question of what Park and Burgess termed 'social disorganization', that is to say, crime, vice, delinquency, promiscuity, child abandonment, etc., all of which were seen to be concentrated in areas of high mobility in the zone of transition. This disorganization was seen to be a particular feature of inner urban cities, and it generated a long interest in the social ecology of crime and mental illness (Castle and Gittus, 1957; Wallis and Maliphant, 1967; Giggs, 1973; Herbert, 1976). See Hamnett (1979) for a critique of this tradition of research.

Meanwhile, on the rural front researchers such as Williams (1956, 1963) and Littlejohn (1963) were examining the sociology of village life. Frankenberg (1966) summarized much of this work in his study of communities in Britain. But, with the gradual breakdown of traditional rural society and rural ways of life (Friedland, 1982), and the expansion of urban areas and commuters into previously agricultural areas, questions began to be asked by Pahl (1965, 1966) about the nature of the rural–urban distinction.

Urban social geography: from patterns to processes

Traditional urban social geography was based on the work of Burgess and Hoyt and consisted of attempts to delineate the urban residential structure in terms of some mixture of concentric zones and sectors (Jones, 1960). For the most part, these were essentially descriptive, and as Johnston (1966) noted, 'there has been no movement towards the elucidation of general principles' (p. 23). One of the first people to attempt to put the study of urban residential geography on a more rigorous footing was Ron Johnston, who published a series of papers in the 1960s which tried to identify some general principles, culminating in a theoretical paper (Johnston, 1973) which attempted to formulate a general model of intra-urban residential patterns in different societies, linking this to stages of social modernization and development of the division of labour and class structures.

For most of the 1960s and early 1970s, urban social geography was strongly dominated by two main approaches: first, the quantitative analyses of urban social structure utilizing census data to produce urban factorial ecologies and social area analyses. These analyses of urban social patterns utilized the new techniques such as factor analysis and principal components analysis to pull out the main dimensions of urban residential differentiation. At first this work seemed promising, but as knowledge accumulated it became clear that there were three principal dimensions to urban residential structure – class, race and household structure – which were common to most cities, and this work began to grind to a halt around 1970 as declining returns set in (see Hamnett, 1972 and Johnston, 1972b for reviews). Once the patterns had been mapped there was the question of where to go next.

Second, there was a developed body of work on urban migration, linked to residential choice and preference approaches to the subject. These focused on people's knowledge of residential space, their preferences for different attributes and types of area, and the choices they made. Typical examples of this work are Adams (1969), Adams and Gilder (1972), Brown and Moore (1970), Brown and Holmes (1971) and Wolpert (1965, 1967). The assumption of this work was that migration was very important in modifying the social characteristics of different residential areas.

The crucial shortcoming of this work, with the advantage of hindsight, was that it focused on choice and preference and the residential search process, with no understanding of the context within which choice and preference are exercised, the nature of the housing market and its institutional structure, or the unequal distribution of choice and preference and the economic, social and political constraints experienced by different sections of the community. The existence of a highly differentiated income structure, of institutional red-lining of parts of the city for mortgages and racial and ethnic prejudice did not figure in these analyses. Consequently, though it limped on into the mid 1970s, there was less and less interest in studies of choice, preference and search behaviour. As with the studies of residential differentiation, declining returns had begun to set in (Bassett and Short, 1980).

Access and inequality: the development of urban managerialism

It is perhaps interesting to speculate what would have happened to urban social geography in the absence of any radical shift but, in the event, it was transformed in the early 1970s by three parallel developments. The first was the emergence of what was termed the 'managerialist' approach, pioneered by Ray Pahl (1970a). This argued that urban sociology and urban geography had largely ignored the extent to which the distribution of, and access to, scarce resources in the urban system were frequently allocated by an influential group of urban managers such as estate agents, building society managers, local authority housing managers and health professionals. Pahl (1970a) argued that, to understand the distribution of resources in the urban system, it was important to understand the principles by which such resources were allocated and managed. He also argued that there are major spatial inequalities in the distribution of, and access to, scarce urban resources. He summarized the main propositions in his argument as follows:

> (a) There are fundamental *spatial* constraints on access to scarce resources and facilities. Such constraints are generally expressed in time/cost distance.
> (b) There are fundamental *social* constraints on access to scarce urban facilities. These reflect the distribution of power in society and are illustrated

by: bureaucratic rules and procedures, social gatekeepers who help to distribute and control urban resources.
(c) Populations in different localities differ in their access and opportunities to gain the scarce resources and facilities, holding their economic position or their position in the occupational structure constant. The situation which is structured out of (a) and (b) may be called a socio-spatial or socio-ecological system. Populations limited in this access to scarce urban resources and facilities are the *dependent* variable; those controlling access, the managers of the system, would be the *independent* variable.
(d) Conflict in the urban system is inevitable. The more the resource or facility is valued by the total population in a given locality, or the higher the value and the scarcer the supply in relation to demand, the greater the conflict (p. 215).

Pahl's statement represented an important advance in that it identified both social and spatial constraints on access to scarce urban resources, it directed attention to the role of the urban managers/ social gatekeepers and the rules and procedures they follow, and it argued that conflict was inevitable. It thus linked together questions of distribution, access, allocation and power, and Pahl argued that: 'The basic framework for urban sociology is *the pattern of constraints which operates differentially in given localities*' (1970a, p. 218). His focus was on the spatial distribution of differential life chances and the links between occupational and income inequalities and the urban system.

Pahl argued that there are scarce and desirable resources in any society and, as those with power might be expected to distribute facilities to their own advantage, the spatial structure would then reflect the distribution of power in society. However, he argued that all societies intervene to redress the unequal distribution of power and that the nature and type of intervention is likely to 'vary according to the specific ideologies and historical experience of the society concerned' (Pahl, 1970a). He also argued that 'The spatial structure partly reflects and partly determines the social structure', and that the 'sheer permanence of the built environment means that the distribution of economic rewards which creates a social structure at one period of time becomes fossilized at a later period of time' (p. 187). In arguing this, Pahl foreshadowed some of the subsequent debates in the late 1970s and 1980s about the relationships between society and space and the crucial constitutive role of space in social processes and structures.

The second major challenge to quantitative social geography was pioneered by a group of radical human geographers all loosely linked to the journal of radical geography *Antipode*, founded at Clarke University by Dick Peet in 1969. *Antipode* introduced liberal (in the 1960s and 1970s usage of the term) and marxist perspectives into mainstream human geography for the first time. Their impact, combined with the publication of David Harvey's book *Social Justice and the City* in 1973, was dramatic. Within the space of two or three years, urban social geography was radically transformed as a new generation of marxist- and managerialist-influenced young geographers began to rethink the focus of, and the approaches to the subject. It is difficult for those who did not experience the transformation in urban social geography from the choice and preference and behaviourally based approaches of the late 1960s, and the analyses of census data, to the radical, marxist-influenced work of David Harvey, to understand the dramatic nature of the shift.

One of the first contributions was made by David Harvey (1969) in his essay on the redistribution of real income in an urban system, and subsequently developed in *Social Justice and the City*. Harvey's approach was somewhat technocratic and strongly policy orientated, but the questions he posed were very similar to those raised by Ray Pahl. Harvey argued that we should focus our attention on

> the mechanisms which connect allocational decisions (whether public or private) on such things as transport networks, industrial zoning, location of public facilities, location of households, and so on, with their inevitable distributional effects upon the real income of different groups in the population. These distributional effects are exceedingly important. Yet they are very poorly understood and the mechanisms relating allocation and distribution remain obscure (p. 51).

The work of Pahl, Harvey and what can be termed the Antipode school of radical geography, led to a spate of work on questions of the distribution of and access to urban facilities, particularly as they affected the poor. Typical of this early work is Pierre de Vise's paper (1972) on Cook County Hospital and what he termed Chicago's Apartheid Health system. He argued that the absence of a proper health care insurance system in the USA and the flight of physicians out of the, largely black, inner areas of American cities and into the white suburbs, had led to an acute shortage of basic health care in the inner cities. Thus, the deprived inner city population had to visit the understaffed and under-resourced emergency rooms of the city centre public hospitals as the first, and often the only, point of contact with the health system.

De Vise's paper reflects the style of many of the early *Antipode* papers in the wake of the social unrest surrounding the Vietnam War, the struggle for civil rights and attempts at major social reform made by Presidents Kennedy and Johnston in the 'Great Society Programme'. The authors of these early papers were angry, radical and activist. Their interest in questions of social theory was strongly wedded to social inequality and social reform. Theory was seen as informing practice, not as something divorced from it and, in this respect, the work of the early 1970s social geographers, both in the USA and the UK, was geared to the analysis of current social problems and the formulation of political and policy critiques (Davis and Albaum, 1972).

Although Pahl subsequently rejected managerialism as an approach, and Harvey switched his focus away from what he termed liberal, reformist formulations to radical, marxist ones, their work was enthusiastically taken up by a number of urban geographers and sociologists (see Williams, 1978b, 1982 for a review) and related work on the geography of welfare and social well-being by David Smith (1977) and Paul Knox (1975) amongst others led to a growing interest in questions of the distribution of, and access to, scarce urban resources (Smith and Eyles, 1974). As Knox (1978) points out:

> the physical accessibility of people to jobs, houses, educational and medical facilities, welfare services, shops and public open spaces is central to the study of welfare geography. Indeed, the idea of accessibility as a fundamental source of 'real income' was a major theme in the earlier writings of Harvey and Pahl, whose emphasis was placed on the 'hidden mechanisms' affecting the welfare of different sociospatial groups arising from the location or allocation of resources in an urban system (p. 415).

This approach was reinforced by the appearance in 1970 of Hagerstrand's pioneering paper 'What about people in Regional Science?', which focused attention on the activity spaces of people's daily lives and the structure and type of constraints on people's activities. It led to a spate of work on what came to be known as 'time' geography and time budgets (Carlstein *et al.*, 1978; Pred and Palm, 1974, 1978) and on the constraints facing particular groups such as car-less women living in suburbia.

The third challenge to the analytical frameworks which had prevailed during the 1980s was David Harvey's abandonment of liberal, reformist approaches to questions of social justice, and his switch to marxist analysis. In his paper 'Revolutionary and counter-revolutionary theory in geography and the problem of ghetto formation' (Harvey, 1972), subsequently reproduced in amended form as a chapter in *Social Justice and the City* (Harvey, 1973), he argued that:

> The quantitative revolution has run its course and diminishing marginal returns are apparently setting in as yet another piece of factorial ecology, yet another attempt to measure the distance–decay effect, yet another attempt to identify the range of a good, serve to tell us less and less about anything of great relevance. There is a clear disparity between the sophisticated theoretical and methodological framework which we are using and our ability to say anything really meaningful about events as they unfold around us (p. 6).

Harvey argued that what was needed was revolutionary theory in geography. First he specified what it does not entail:

> It does not entail yet another empirical investigation of the social conditions in the ghettoes. In fact, mapping even more evidence of man's patent inhumanity to man is counter-revolutionary in that it allows the bleeding-heart liberal in us to pretend we are contributing to a solution when in fact we are not. This kind of empiricism is irrelevant (1972, p. 10).

He then raised the crucial question of 'how and why would we bring about a revolution in geographic thought?' He rejected positivism, phenomenology and behaviouralism as inadequate in various ways and identified marxism as the best way forward. Taking the pattern of urban land uses as an example, he argued that the neo-classical economic models of Alonso and Muth all assume that urban land use is determined through a process of competitive bidding for land in which different groups have different resources. Thus, 'a variety of city structures can emerge depending on the preferences of the rich group who can always use their resources to dominate the preferences of poor groups' (p. 8).

Harvey then suggested that we assume that the theory of land rent is true, in the sense used by logical positivists, and predicts that poor groups will live where they can least afford to live; that is, on expensive land in the inner city. But he argues that the only valid policy is to eliminate the conditions which give rise to the truth of the theory. 'In other words we want the von Thünen model of the urban land market to become *not* true. The simplest approach to this is to eliminate the mechanism which gives rise to the truth of the theory. The mechanism in this case is competitive bidding for the use of the land' (p. 8). Harvey argued from this that although most analysts concede the serious nature of urban problems such as ghetto formation, 'few call into question the forces which rule at the heart of our economic system. We thus discuss everything except the . . . characteristics of the capitalist market economy' (1972, p. 10).

The focus of research in urban social geography rapidly shifted away from a concern with the delineation of patterns *per se*, towards the social, economic and political processes which produced the patterns. This was paralleled by a shift of attention away from residential patterns and migration to analysis of the structure and operation of the housing market. This was assisted by another major piece of work by David Harvey (Harvey and Chatterjee, Chapter 2, this volume) on the structure of Baltimore which, in retrospect, can be argued (along with Harvey's (1974) paper in *Regional Studies*) to be one of the most important papers produced in urban geography. Its impact was dramatic and led to a major re-orientation of urban social geography in the subsequent 10 years.

The paper addressed the questions of how the macro and micro features of housing markets related to each other in theory and practice, and how what is termed 'absolute rent' is realized in the housing market of large cities. The starting point was adoption of a methodological position termed *contextual relations* which allowed them to deal with relations between the individual and the wider social totality, and reciprocal structuring this entails. In many ways this was a forerunner of what later became known as the structuration approach (Gregory, 1981). The paper sought to relate economic policy, and the role of housing within it as a Keynesian counter-cyclical regulator, to the local level through the role of financial and governmental institutions.

Harvey and Chatterjee (chapter 2, this volume) pointed to the variety of institutions such as Savings and Loans, mortgage bankers, savings and commercial banks which operate in the housing market and the differential scope and geographic range of their operations. Whereas mortgage, saving and commercial banks are not limited to housing, Federal and State Savings and Loans are confined to housing and are designed to 'promote the thrift of the people locally to finance their own homes'. Harvey and Chatterjee then asked: 'How do all of these financial and governmental complexities relate to the individual?' They pointed out that in the typical micro-economic models of residential differentiation, it is assumed that income is the relevant determinant of housing choice. But they argued that, in fact, 'It is the ability to obtain credit and a mortgage that is, for most people, the immediate determinant' (p. 36). They accepted that this is income related but argued that it is also a function of the policies of financial and governmental institutions, and noted that institutions have distinctive policies to the nature of the housing stock.

They then proceeded to outline the ways in which these factors coalesce in the Baltimore housing market, pointing to the development of a structured relationship which has distinct geographical manifestations in that different parts of the city are dominated by particular types of lenders. Then, in what was perhaps the key passage of the paper, they argued that:

> This geographical structure forms a 'decision environment' in the context of which individual households make housing choices. These choices are likely, by and large, to conform to the structure and to reinforce it. The structure itself is a product of history. Attempts to change the structure can . . . generate considerable social conflict . . . It is this process of transformation of and within a structure that must be the focus for understanding residential differentiation (p. 41).

This is clearly far removed from the previous attempts to examine residential choices and preferences in isolation. Not only is Harvey's approach far more realistic in its understanding of the complex web of influences on housing market structure and operation, but his linkage of structure and individual agency was theoretically far more sophisticated than what had gone before. Not surprisingly, Harvey's work, with its sophisticated analysis of mortgage red-lining, block busting and landlord behaviour, led to a rash of new work. One of the first attacks on the traditional choice- and preference-dominated approaches to urban social geography was mounted by Fred Gray (1975), in a paper entitled 'Non-explanation in urban geography', and Duncan (1976) argued that:

> Previous research and our own experience show that, in the situation of unequal distribution of wealth, prestige, and power, the study of constraints and allocation, rather than preference and demand, provides a more realistic viewpoint from which to understand housing situations and housing markets (p. 10).

When the focus of attention did shift back to migration, it was looking more at the role of allocation and institutional constraints on the migration process, a shift which had been initiated by Murie (1974), and Short (1978) noted that:

> The differentially priced housing market, the unequal distribution of income and the policies of the housing finance institutions all affect the response of households to new housing needs . . . The decisions of individual households are more adequately explained as a form of adaptive behaviour in relation to the nature of the housing system, which in turn is shaped by the nature of the wider society, than by . . . consumer preference arguments (p. 546).

This kind of radical reappraisal of the nature of the processes shaping urban residential patterns was well under way in the late 1970s (Dingemans, 1978; Williams, 1978a; Wolfe *et al.*, 1980; Bassett and Short, 1980) when Neil Smith's (1979) paper on gentrification caused another theoretical shock. Smith launched an attack on demand-based theories of gentrification, and David Ley's 'post-industrial' explanation, arguing that the so-called urban renaissance has been stimulated more by economic than cultural forces. He argued that:

> to explain gentrification according to gentrifiers' actions alone, while ignoring the role of builders, developers, landlords, real estate agents and tenants is excessively narrow. A broader theory of gentrification must take the role of producers as well as consumers into account, and when this is done, it appears that the needs of production – in particular the need to earn a profit – are a more decisive initiative behind gentrification than consumer preference (p. 540).

Smith went on to develop his rent gap theory of gentrification whereby, as a result of suburbanization and capital depreciation in the inner city, property values in the inner city declined and the potential ground rent of a particular site exceeded the capitalized ground rent, creating a 'rent gap'. When the gap is large enough, gentrification or redevelopment will take place. Criticisms of Smith's theory have been made by Hamnett (1984, 1991), Rose (1984), Munt (1987) and others arguing, amongst other things, that Smith neglected the key question of where the potential gentrifiers come from, but Smith's approach has been very influential and, along with Walker's work on suburbanization, helped further to strengthen the marxist analysis of urban space.

Marxism, behaviouralism and phenomenology

The impact of marxism on social geography was not met passively. There was a strong resistance by behavioural and phenomenological geographers. The behavioural approach, which stressed the analysis of human behaviour as a window into attitudes and motivation and found its strongest expression in revealed preference theory (Golledge *et al.*, 1972; Mercer, 1972), had a relatively short life span, as its deficiencies became apparent. Essentially, it was inductive, attempting to work backwards from patterns of behaviour in space to an understanding of the generative preferences and motivations. But this approach lacked both an understanding of the importance of structures which influence behaviour and of human consciousness and understanding.

In 1977 David Ley published an important paper entitled 'Social geography and the taken-for-granted world' (Chaper 3, this volume) in which he argued that social geography lacked both a clear theoretical direction and an appropriate philosophical underpinning. He suggested that the shift away from a concern with spatial patterns to an attempt to understand social processes cannot avoid studying the 'subjective' because, as he put it: 'in the taken-for-granted world of everyday experience, which is the ground of group behaviour and decision-making, every object is always an object for a subject'. Ley argued in favour of phenomenology as an epistemology which takes the everyday world as its starting point, and suggests Alfred Schutz's theory of action as an appropriate underpinning for a social geography 'concerned with the social and cognitive processes which lend meaning to place, and guide the decision-making of both groups and organisations'.

Clearly written, and cogently argued, Ley's paper looks at the development of social geography from the French rural school of Vidal de la Blache to the Chicago school of urban sociology, and traces this through to the emergence of behavioural geography. But he argued that behavioural geography is flawed by a methodological ambiguity, namely the dualism of cognitive process and spatial form, and its underlying positivist psychologism. Instead, he argued that the question of meaning for human beings is of crucial importance, and that only the phenomenological approach can deal with the real intersubjectivity of the life-world in all its complexity. Ley also argued that there is a meaning to space as well as to time: 'Each place should be seen phenomenologically, in its relational context, as an object for a subject. To speak of a place is not to speak of an object alone, but of an image and an intent . . . Thus place always has meaning, it is always "for" its subject' (p. 61). Conversely, Ley also argues that if places are meaningless without subjects, so too, people removed from places are people of uncertain identity. Ley argues that this simple dialectic permits an understanding of many phenomena such as long-range migration:

> The distant metropolis is never perceived in the perfect material terms that the gravity model with its economic determinism would have us believe. The metropolis has a meaning, it is, in Park's words, a state of mind, and it is always this meaning for the subject that precedes action; creative decision-making is not pre-empted by a mechanistic gravity field (p. 61).

Ley went on, anticipating recent debates about multiple identities, subjective realities, dominant meanings, etc., to argue that:

> In contemporary urbanism a place may commonly have a multiple reality; its meaning may change with the intent of the subject, and a plurality of subjects may simultaneously hold a different meaning for the same place. Usually, however, a dominant meaning holds sway . . . In the same way, mundane and taken-for-granted features in an environment can point beyond themselves to local societal values (p. 62).

He gives the example of graffiti which may, he suggests, allow us to see the various 'intangible attitudes which govern everyday relations between groups and space in the American inner city' (p. 62). See Ley and Cybriwsky's (1974) paper on urban graffiti as territorial markers for a discussion. More generally, Ley argued that 'the meaning of a place systematically attracts groups with similar interests and lifestyles; places are selected and retained by reflective decision-makers on the basis of their perceived image and knowledge'. The result, says Ley, is that 'the city becomes a mosaic of social words each supporting a group of similar intent, who in their habitual interaction reinforce the character both of their group and of their place' (p. 62). He also suggests that the personality of place ranges in scale from the nation state to the local area, and argues that to have meaning, and thereby reality in experience, a sense of place, nationality or community, must have a shared meaning for a plurality of subjects. In arguing this, Ley anticipates much recent social and cultural geography on place, identity and meaning. Twenty years on, his paper can be seen to have provided a basis for much subsequent work.

A similar set of arguments to those of Ley were made by Eyles (1981) in his paper 'Why geography cannot be Marxist: towards an understanding of lived experience'. As the title of the paper suggests, Eyles argued that in order to understand lived experience it is necessary to transcend some of the marxist orthodoxies, particularly those of structural marxism (see Duncan and Ley, 1982, for a trenchant critique of structural marxism). Eyles does not dispute marxism's contribution to human geography, but he argues that structural marxism has led to a limited understanding of lived experience in that it has treated such experience as a determinate outcome of structural processes. In other words, structure is seen to determine behaviour. The target of Eyles's attack is not marxism *tout court*, but structural or Althusserian marxism, and he is clearly sympathetic to the approach of Gramsci and E. P. Thompson. Indeed, Eyles concludes that while geography cannot be marxist in the sense of accepting structural marxism as an all-embracing theoretical system, the validity of which is assumed to be given, it must be *marxian*, in that marxism has provided a set of fundamental insights into the nature of the relationships which structure human existence.

The argument about the relationship between social and economic structures and human consciousness and agency was developed during the late 1970s and early 1980s in a series of major papers by Gregory (1981), Duncan and Ley (1982) and others. In his paper 'Human agency and human geography', Gregory introduced human geographers to the idea of 'structuration theory' developed by Tony Giddens (1979). He distinguished between four models of human agency and structure. The first, which he labelled 'reification' sees society as a pre-given set of structures which are external to and constraining upon human agency. This is typified in social theory by Durkheim and some neo-marxist formulations. The second, which Gregory labels as 'voluntarism', sees society as being constituted by intentional action and is typified by the approach of Max Weber. In its extreme form, this was the view taken by the British ex-prime minister, Margaret Thatcher, who stated that there was 'no such thing as society, only individuals and their families'. Where this left such institutions as Parliament, the police, the armed forces, or the Trades Union Congress and the Confederation of British Industry, remained a mystery.

The third approach, 'dialectical reproduction', typified by the work of Peter Berger, views society as forming the individuals who create society in a continuous dialectic. Society is 'an externalisation of man, and man a conscious appropriation of society'. Finally, Gregory identifies the 'structuration' approach of Habermas and Giddens, which sees social systems as 'both the medium and the outcome of the practices that constitute them'. In this approach, which proved influential in human geography for a period, (Sarre, 1986, 1987; Gregson, 1987), individuals are seen as the 'bearers' of social structures. The actions of individuals, in their social contexts, act to reproduce and change social structures, whilst simultaneously constrained by them. A variant of this approach is utilized by Harvey and Chatterjee (Chapter 2, this volume).

Arguably, one of the most important shifts in social geography, and in human geography as a whole, was in the late 1970s with the debate over the relative importance of spatial patterns and social processes. During the mid 1970s, a number of human geographers began actively to challenge what Sayer (1977) has termed the 'spatial fetishism' of the discipline. Sayer defined this as the 'clinging to some asocial, ahistorical purely spatial effect', where 'the spatial effect of social processes can be described by some mysterious mathematical or geometrical laws that have an identical impact, irrespective of the social processes'.

The problem, as Duncan (1979) noted, was that 'while geographers have studied space in isolation from process, sociologists and economists have studied process abstracted from space' (p. 2). It became clear that geography needed to be grounded in an understanding of social processes, a plea made by Pahl in 1970a, but it began to be realized that the links were not simply one way: from social process to spatial patterns. Spatial patterns were not simply the result or outcome of social processes (Hamnett, 1979), the relationships between society and space were two way, if indeed they could be analysed independently. Social processes do not, for the most part, take place on the head of a proverbial pin: they have a spatial dimension to them, and they take place within a spatial context shaped by the past. In other words, social processes have an inherent spatiality (Gregory and Urry, 1985; Massey, 1985).

Representation and discourse

One of the most important recent developments in social-cultural geography has been a realization of the significance of the concepts of representation, narrative and discourse. Previously, it was generally assumed that the world presented itself to us more or less directly for analysis in an unproblematic fashion.

It is difficult to determine exactly when geographers began to take questions of representation seriously. It appears to have taken place gradually through the second half of the 1980s and early 1990s as the work of Said (1979) on Orientalism, Spivak (1988), and the work of the French linguistic theorists such as Derrida slowly percolated into geographical thinking (Barnett and Low, 1996). Derrida (1973, 1978, 1981) and other authors have argued that representation and reality cannot be sharply distinguished in language, that language use rests on representation, and that 'language is the place where reality and representation meet' (Barnett, 1993). This argument is associated with the view that there is nothing beyond the text, that there is no privileged interpretation of texts, there are only various readings, and that it is the task of critical theorists to deconstruct texts. As will be realized, this has led to a number of problems regarding the nature of representation and language and the relationship between the 'signified' and the 'arbitrary nature of the sign'.

Similar arguments have also taken hold in philosophy, with pragmatist antirealists such as Rorty (1980, 1991) arguing that: 'We have to drop the notion of a God's eye point of view, a way the world is apart from our descriptions of it in language' (Rorty, 1982). A pragmatist must insist, says Rorty, 'that there is no such thing as the way the thing is in itself, under no description, apart from any use to which human beings might want to put it' (1991, p. 99). Clearly, this constitutes a strong challenge for objectivism and naturalism of any kind, though see Geras (1995) for a powerful critique of Rorty's position. The debates over the implications of deconstruction have been pursued by Strohmayer and Hannah (1992), Hannah and Strohmayer (1991), and Barnett (1993) and will not be rehearsed here. The key point to stress is that as Duncan and Sharp (1993) note, 'The recent introduction of postmodern epistemologies into geographical work has exposed the problematic representational nature of descriptions of the world and has highlighted the cultural specificity of those representations' (p. 473).

One focus has been on the nature of the representations produced by Westerners, often, but not always, white, male and middle class, regarding those who are the objects of representation, typically non-Western or post-colonial 'Others'. The issue was given its clearest expression by Said (1979), who argued that the representations of the Orient in Western discourse traditionally served to portray the Orient as alien and 'other'. But, more generally, the implication of deconstruction has been a growth of relativism, rejection of what are termed 'grand narratives' such as marxism, an interest in the production of narratives, and in the act of interpretation and reading texts. Little is taken at face value any longer, and everything is said to be the product of specific discourses.

Duncan and Ley (1993a) note that though the task of scholars is to represent the world to others, representation has not been problematized either within geography or within Anglo-American social science, but has been taken for granted. They suggest that within twentieth-century Anglo-American human geography it is possible to identify four major modes or types of representation, two of which operate within the framework of mimetic representation (the belief that we should strive to produce an accurate reflection of the world), and two of which pose various challenges to mimesis. They suggest that the dominant mode of representation in most human geography until the 1950s was that of 'descriptive fieldwork' based upon observation: 'The assumption underlying this position is that trained observation, transcribed into clear prose and unencumbered by abstract theorizing, produces an accurate understanding of the world'. The second mode of mimetic representation was loosely based upon positivist science and was popular within geography from the 1950s on, with the exception of cultural geography where a stress on the interpretation of differences between individual places prevailed. They state that:

> The third type of practice is a postmodernism which represents a radical attack upon the mimetic theory of representation and the search for truth . . . It is anti-foundational in that it explicitly rejects totalizing ambitions of modern social science . . . Such an epistemology, if taken seriously, is inescapably and radically relativist . . . The fourth type of practice is interpretative and its basis is hermeneutics. Unlike the first two positions it acknowledges the role of the interpreter and therefore rules out mimesis in the strict sense of the term. It is precisely the interpersonal and intercultural nature of the hermeneutic method which poses a challenge to mimesis, since a 'perfect' copy of the world is clearly not possible if the interpreter is present in that textual copy (p. 3).

The implications of this debate are gradually percolating into many areas of human geography. Thus, White and Jackson (1995) have recently pointed to the need for population geographers to consider the implications of social construction theories and the politics of position and to examine critically the nature and utility of pre-given, 'objective' data categories such as 'household head', race, ethnicity, occupation and the like. This has had both beneficial and detrimental consequences. Categories, theories and standpoints are the subject of growing critical examination, but some debates have begun to turn in on themselves, and have become more interested in the deconstruction of existing work, than in the production of knowledge of the world outside the text (which is sometimes asserted not to exist independently of the text).

The paper I have chosen to highlight the importance of representation and discourse is Halfacree's (1993) study of 'Locality and social representation: space, discourse and alternative definitions of the rural' (Chapter 4, this volume). This paper is very useful in a number of ways. First, it discusses the question of representation and discourse; second, it examines the relationship between society and space which has been an important debate within human geography in the 1980s; and third, it raises the distinction between notions of urban and rural which has been important within social geography, although as Halfacree shows it is a highly problematic distinction (see also Saunders, 1985; Sayer, 1985 for a fuller discussion of the debates over the definition of the 'urban'). Halfacree outlines three broad approaches to the definition of the rural. The first is that of descriptive definition such as Cloke's index of rurality, which concentrates on the observable or measurable. Halfacree argues that descriptive methods only describe the rural (or the urban), they do not define it. The second is the socio-cultural approach, which examines the way in which peoples' characteristics vary with the type of environment in which they live. This mirrors the approach of Wirth (1939) to 'urbanism as a way of life'. Halfacree points out that researchers soon realized that there was no simple dichotomy between 'rural' and 'urban' areas, and the idea of a rural–urban continuum was devised. This was criticized by Pahl (1966, 1970a) and it was argued that sociological characteristics of a place could not be read off from its relative location on a continuum (Newby, 1986).

Halfacree argues that the main theoretical criticism of both the descriptive and the socio-cultural definitions of the rural is that they have an 'erroneous conceptualization of the relationship between space and society' (p. 73) and he advances the view that space is neither absolute (it does not possess causal powers) nor relative. He argues instead that space is both produced, and serves as a means of creating further space (it is a resource). He outlines a useful review of the literature on the history of conceptualizing space and the relations between society and space (Gregory and Urry, 1985). He then goes on to argue that if we are to develop a conception of rural space within social science then two criteria have to be satisfied: first, that there are clear structures operating which are unambiguously associated with the local level and, second, that these structures allow us to distinguish between urban and rural environments. Hoggart (1990) is not convinced by these arguments and suggests that we should do away with the 'rural' which he sees as a 'chaotic conception' lacking any explanatory power. Instead, he argues that we should look at extra-rural structures which both mould, and are moulded by, local circumstances. But Halfacree argues that this 'does *not* mean that we have to assign "the rural" to the dustbin of research history. This is because there is . . . another means of defining the rural, which directs our attention to the realm of discourse and . . . social representations'. He goes on to argue that:

> There is an alternative way of defining rurality which, initially, does not require us to abstract causal structures operating at the rural scale. This alternative comes about because 'the rural' and its synonyms are *words and concepts understood and used by people in everyday talk* (p. 78, emphasis in original).

What Halfacree does in the paper is to argue that we should investigate the status of the rural in discourse since the term rural is a symbolic shorthand which we use to describe something. It can be argued that this is simply to reintroduce the concept of the rural through the back door of discourse, but nonetheless it represents an interesting and novel argument. Halfacree then proceeds to outline the theory of 'social representations' as developed in the work of Moscovici (1976, 1981) and Potter and Wetherell (1987). He argues that, in contrast to orthodox social psychology, social representation theory attaches great importance to what Giddens terms 'practical consciousness'. Halfacree argues that people use social representations in order to deal with the complexity of the social world, and he stresses the key social dimension of representations: their importance for communications requires an agreed code of meanings. Potter and Wetherell (1987) question the overly cognitive status of social representations, argue that the concept underplays the role of language and put forward the rival concept of 'interpretative repertoire': 'a lexicon or register of terms and metaphors drawn upon to characterise and evaluate actions and events'. As Halfacree rightly points out, both Moscovici and Potter and Wetherell tend to neglect the socio-political aspects involved in the production and reproduction of social representations, but he deploys the concept of social representation to argue that our attempts at defining the rural can be seen as 'academic discourses', and he suggests that it is crucial to link these to lay discourses. There is, he says, a growing realization of the need to study 'people's accounts of the form, content and context of their daily lives', and he suggests that rural social scientists have been as guilty as anyone else in neglecting these ordinary views of the world which form the crucial raw material for social scientific accounts.

Halfacree argues from this that if we are to accept the status of the rural as a social representation, it is necessary to view lay discourses as interpretative repertoires 'derived from a disembodied but none the less real social representation of the rural' which, however distorted and partial they may be, produce very real effects. To construct a representational definition of the rural needs a hermeneutic and interpretive methodology to analyse the intellectual shorthand which allows spatial metaphors and place images to convey a complex set of associations, however imprecise they may prove to be. Finally, Halfacree suggests that analysis of the social representations of the rural requires us to examine the 'sign' and its meanings and referents. Pratt (1991) sees the sign ('rurality') as being increasingly detached from its signification (social representations of the rural are becoming more diverse) and he suggests the sign and its significant are becoming divorced from their referent (the rural locality), leading to cross-cutting discourses about rurality. Halfacree asks to what extent this may mean that the symbolic is taking precedence over the material, and he poses the question of whether 'severing of the social representations of the rural from any material referent also makes the former a site of social struggle within discourse, as promoters of competing representations strive for hegemony' (pp. 84–5). This question emerges in a number of different contexts – see for example Mills's (1988) work on the postmodern landscape of gentrification in Vancouver, and Crilley's (1993) work on Docklands and the extent to which developers and others are active in promoting images of lifestyles as marketing tools for selling places (Kearns and Philo, 1993).

1 Ernest W. Burgess,
'The Growth of the City: An Introduction to a Research Project'

Excerpts from Robert E. Park, Ernest W. Burgess and Roderick D. McKenzie, *The City*, Chapter II. Chicago and London: The University of Chicago Press (1925)

The outstanding fact of modern society is the growth of great cities. Nowhere else have the enormous changes which the machine industry has made in our social life registered themselves with such obviousness as in the cities. In the United States the transition from a rural to an urban civilization, though beginning later than in Europe, has taken place, if not more rapidly and completely, at any rate more logically in its most characteristic forms.

All the manifestations of modern life which are peculiarly urban – the skyscraper, the subway, the department store, the daily newspaper, and social work – are characteristically American. The more subtle changes in our social life, which in their cruder manifestations are termed 'social problems', problems that alarm and bewilder us, as divorce, delinquency, and social unrest, are to be found in their most acute forms in our largest American cities. The profound and 'subversive' forces which have wrought these changes are measured in the physical growth and expansion of cities. That is the significance of the comparative statistics of Weber, Bücher, and other students.

These statistical studies, although dealing mainly with the effects of urban growth, brought out into clear relief certain distinctive characteristics of urban as compared with rural populations. The larger proportion of women to men in the cities than in the open country, the greater percentage of youth and middle-aged, the higher ratio of the foreign-born, the increased heterogeneity of occupation increase with the growth of the city and profoundly alter its social structure. These variations in the composition of population are indicative of all the changes going on in the social organization of the community. In fact, these changes are a part of the growth of the city and suggest the nature of the processes of growth.

The only aspect of growth adequately described by Bücher and Weber was the rather obvious process of the *aggregation* of urban population. Even more significant than the increasing density of urban population is its correlative tendency to overflow, and so to extend over wider areas, and to incorporate these areas into a larger communal life. This paper, therefore, will treat first of the expansion of the city, and then of the less-known processes of urban metabolism and mobility which are closely related to expansion.

Expansion as physical growth

The expansion of the city from the standpoint of the city plan, zoning, and regional surveys is thought of almost wholly in terms of its physical growth.

Traction studies have dealt with the development of transportation in its relation to the distribution of population throughout the city. The surveys made by the Bell Telephone Company and other public utilities have attempted to forecast the direction and the rate of growth of the city in order to anticipate the future demands for the extension of their services. In England, where more than one-half of the inhabitants live in cities having a population of 100,000 and over, the lively appreciation of the bearing of urban expansion on social organization is thus expressed by C. B. Fawcett:

> One of the most important and striking developments in the growth of the urban populations of the more advanced peoples of the world during the last few decades has been the appearance of a number of vast urban aggregates, or conurbations, far larger and more numerous than the great cities of any preceding age . . .
>
> These great aggregates of town dwellers are a new feature in the distribution of man over the earth. At the present day there are from thirty to forty of them, each containing more than a million people, whereas only a hundred years ago there were, outside the great centers of population on the waterways of China, not more than two or three. Such aggregations of people are phenomena of great geographical and social importance; they give rise to new problems in the organization of the life and well-being of their inhabitants and in their varied activities. Few of them have yet developed a social consciousness at all proportionate to their magnitude, or fully realized themselves as definite groupings of people with many common interests, emotions and thoughts.[1]

In Europe and America the tendency of the great city to expand has been recognized in the term 'the metropolitan area of the city', which far overruns its political limits, and in the case of New York and Chicago, even state lines. The metropolitan area may be taken to include urban territory that is physically contiguous, but it is coming to be defined by that facility of transportation that enables a business man to live in a suburb of Chicago and to work in the Loop, and his wife to shop at Marshall Field's and attend grand opera in the Auditorium.

Expansion as a process

No study of expansion as a process has yet been made, although the materials for such a study and intimations of different aspects of the process are contained in city planning, zoning, and regional surveys. The typical processes of the expansion of the city can best be illustrated, perhaps, by a series of concentric circles, which may be numbered to designate both the successive zones of urban extension and the types of areas differentiated in the process of expansion (Figure 1.1).

This figure represents an ideal construction of the tendencies of any town or city to expand radially from its central business district – on the map 'The Loop' (I). Encircling the downtown area there is normally an area in transition, which is being invaded by business and light manufacture (II). A third area (III) is inhabited by the workers in industries who have escaped from the area of deterioration (II) but who desire to live within easy access of their

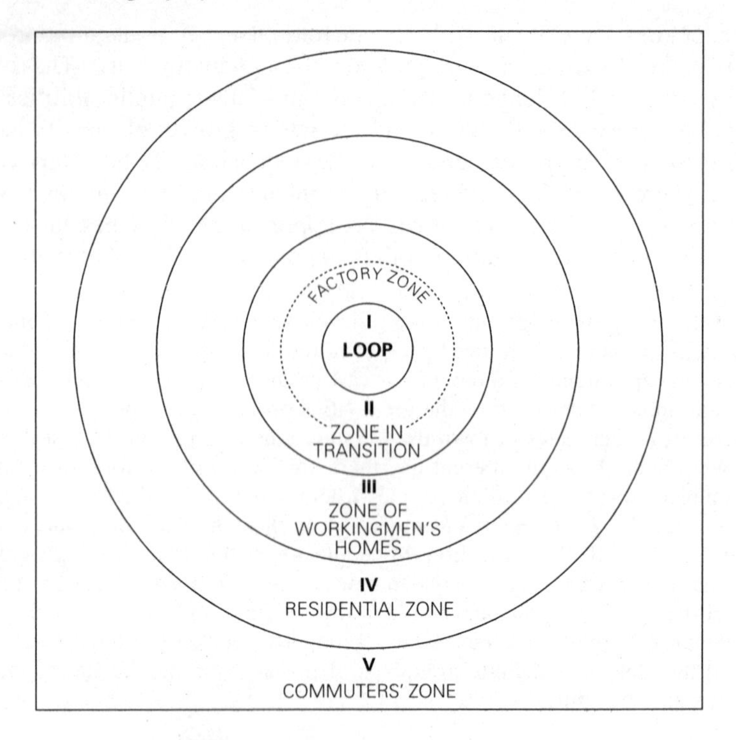

Figure 1.1 The growth of the city

work. Beyond this zone is the 'residential area' (IV) of high-class apartment buildings or of exclusive 'restricted' districts of single family dwellings. Still farther, out beyond the city limits, is the commuters' zone – suburban areas, or satellite cities – within a thirty- to sixty-minute ride of the central business district.

This figure brings out clearly the main fact of expansion, namely, the tendency of each inner zone to extend its area by the invasion of the next outer zone. This aspect of expansion may be called *succession*, a process which has been studied in detail in plant ecology. If this chart is applied to Chicago, all four of these zones were in its early history included in the circumference of the inner zone, the present business district. The present boundaries of the area of deterioration were not many years ago those of the zone now inhabited by independent wage-earners, and within the memories of thousands of Chicagoans contained the residences of the 'best families'. It hardly needs to be added that neither Chicago nor any other city fits perfectly into this ideal scheme. Complications are introduced by the lake front, the Chicago River, railroad lines, historical factors in the location of industry, the relative degree of the resistance of communities to invasion, etc.

Besides extension and succession, the general process of expansion in urban growth involves the antagonistic and yet complementary processes of concentration and decentralization. In all cities there is the natural tendency for local and outside transportation to converge in the central business district. In the down-town section of every large city we expect to find the department stores, the skyscraper office buildings, the railroad stations, the great hotels, the theaters, the art museum, and the city hall. Quite naturally, almost inevitably, the economic, cultural, and political life centers here. The relation of centralization to the other processes of city life may be roughly gauged by the fact that over half a million people daily enter and leave Chicago's 'Loop'. More recently sub-business centers have grown up in outlying zones. These 'satellite loops' do not, it seems, represent the 'hoped for' revival of the neighborhood, but rather a telescoping of several local communities into a larger economic unity. The Chicago of yesterday, an agglomeration of country towns and immigrant colonies, is undergoing a process of reorganization into a centralized decentralized system of local communities coalescing into sub-business areas visibly or invisibly dominated by the central business district.

Expansion, as we have seen, deals with the physical growth of the city, and with the extension of the technical services that have made city life not only livable, but comfortable, even luxurious. Yet the processes of expansion, and especially the rate of expansion, may be studied not only in the physical growth and business development, but also in the consequent changes in the social organization and in personality types. How far is the growth of the city, in its physical and technical aspects, matched by a natural but adequate readjustment in the social organization? What, for a city, is a normal rate of expansion, a rate of expansion with which controlled changes in the social organization might successfully keep pace?

Social organization and disorganization as processes of metabolism

These questions may best be answered, perhaps, by thinking of urban growth as a resultant of organization and disorganization analogous to the anabolic and katabolic processes of metabolism in the body. In what way are individuals incorporated into the life of a city? By what process does a person become an organic part of his society? The natural process of acquiring culture is by birth. A person is born into a family already adjusted to a social environment – in this case the modern city. The natural rate of increase of population most favorable for assimilation may then be taken as the excess of the birth-rate over the death-rate, but is this the normal rate of city growth? Certainly, modern cities have increased and are increasing in population at a far higher rate. However, the natural rate of growth may be used to measure the disturbances of metabolism caused by any excessive increase, as those which followed the great influx of southern Negroes into northern cities since the war. In a similar way all cities show deviations in composition by age and sex and from a standard population such as that of Sweden, unaffected in recent years by any great emigration or immigration. Here again, marked variations, as any great excess of males over females, or of females over males, or in the proportion of children, or of grown men or women, are symptomatic of abnormalities in social metabolism.

Normally the processes of disorganization and organization may be thought of as in reciprocal relationship to each other, and as co-operating in a moving equilibrium of social order toward an end vaguely or definitely regarded as progressive. So far as disorganization points to reorganization and makes for more efficient adjustment, disorganization must be conceived not as pathological, but as normal. Disorganization as preliminary to reorganization of attitudes and conduct is almost invariably the lot of the newcomer to the city, and the discarding of the habitual, and often of what has been to him the moral, is not infrequently accompanied by sharp mental conflict and sense of personal loss. Oftener, perhaps, the change gives sooner or later a feeling of emancipation and an urge toward new goals.

In the expansion of the city a process of distribution takes place which sifts and sorts and relocates individuals and groups by residence and occupation. The resulting differentiation of the cosmopolitan American city into areas is typically all from one pattern, with only interesting minor modifications. Within the central business district or on an adjoining street is the 'main stem' of 'hobohemia', the teeming Rialto of the homeless migratory man of the Middle West. In the zone of deterioration encircling the central business section are always to be found the so-called 'slums' and 'bad lands', with their submerged regions of poverty, degradation, and disease, and their underworlds of crime and vice. Within a deteriorating area are rooming-house districts, the purgatory of 'lost souls'. Near by is the Latin Quarter, where creative and rebellious spirits resort. The slums are also crowded to overflowing with immigrant colonies – the Ghetto, Little Sicily, Greek-town, Chinatown – fascinatingly combining old world heritages and American adaptations. Wedging out from here is the Black Belt, with its free and disorderly life. The area of deterioration, while essentially one of decay, of stationary or declining population, is also one of regeneration, as witness the mission, the settlement, the artists' colony, radical centers – all obsessed with the vision of a new and better world.

The next zone is also inhabited predominatingly by factory and shop workers, but skilled and thrifty. This is an area of second immigrant settlement, generally of the second generation. It is the region of escape from the slum, the *Deutschland* of the aspiring Ghetto family. For *Deutschland* (literally 'Germany') is the name given, half in envy, half in derision, to that region beyond the Ghetto where successful neighbors appear to be imitating German Jewish standards of living. But the inhabitant of this area in turn looks to the 'Promised Land' beyond, to its residential hotels, its apartment-house region, its 'satellite loops', and its 'bright light' areas.

This differentiation into natural economic and cultural groupings gives form and character to the city. For segregation offers the group, and thereby the individuals who compose the group, a place and a rôle in the total organization of city life. Segregation limits development in certain directions, but releases it in others. These areas tend to accentuate certain traits, to attract and develop their kind of individuals, and so to become further differentiated.

The division of labor in the city likewise illustrates disorganization, reorganization, and increasing differentiation. The immigrant from rural communities in Europe and America seldom brings with him economic skill of any great value in our industrial, commercial, or professional life. Yet interesting occupational selection has taken place by nationality, explainable more by racial temperament or circumstance than by old-world economic background, as Irish policemen, Greek ice-cream parlors, Chinese laundries, Negro porters, Belgian janitors, etc.

The facts that in Chicago one million (996,589) individuals gainfully employed reported 509 occupations, and that over 1,000 men and women in *Who's Who* gave 116 different vocations, give some notion of how in the city the minute differentiation of occupation 'analyzes and sifts the population, separating and classifying the diverse elements'.[2] These figures also afford some intimation of the complexity and complication of the modern industrial mechanism and the intricate segregation and isolation of divergent economic groups. Interrelated with this economic division of labor is a corresponding division into social classes and into cultural and recreational groups. From this multiplicity of groups, with their different patterns of life, the person finds his congenial social world and – what is not feasible in the narrow confines of a village – may move and live in widely separated, and perchance conflicting, worlds. Personal disorganization may be but the failure to harmonize the canons of conduct of two divergent groups.

If the phenomena of expansion and metabolism indicate that a moderate degree of disorganization may and does facilitate social organization, they indicate as well that rapid urban expansion is accompanied by excessive increases in disease, crime, disorder, vice, insanity, and suicide, rough indexes of social disorganization. But what are the indexes of the causes, rather than of the effects, of the disordered social metabolism of the city? The excess of the actual over the natural increase of population has already been suggested as a criterion. The significance of this increase consists in the immigration into a metropolitan city like New York and Chicago of tens of thousands of persons annually. Their invasion of the city has the effect of a tidal wave inundating first the immigrant colonies, the ports of first entry, dislodging thousands of inhabitants who overflow into the next zone, and so on and on until the momentum of the wave has spent its force on the last urban zone. The whole effect is to speed up expansion, to speed up industry, to speed up the 'junking' process in the area of deterioration (II). These internal movements of the population become the more significant for study. What movement is going on in the city, and how may this movement be measured? It is easier, of course, to classify movement within the city than to measure it. There is the movement from residence to residence, change of occupation, labor turnover, movement to and from work, movement for recreation and adventure. This leads to the question: What is the significant aspect of movement for the study of the changes in city life? The answer to this question leads directly to the important distinction between movement and mobility.

Mobility as the pulse of the community

Movement, per se, is not an evidence of change or of growth. In fact, movement may be a fixed and unchanging order of motion, designed to control a constant situation, as in routine movement. Movement that is significant for growth implies a change of movement in response to a new stimulus or situation. Change of movement of this type is called *mobility*. Movement of the nature of routine finds its typical expression in work. Change of movement, or mobility, is characteristically expressed in adventure. The great city, with its 'bright lights', its emporiums of novelties and bargains, its palaces of amusement, its underworld of vice and crime, its risks of life and property from accident, robbery, and homicide, has become the region of the most intense degree of adventure and danger, excitement and thrill.

Mobility, it is evident, involves change, new experience, stimulation. Stimulation induces a response of the person to those objects in his environment which afford expression for his wishes. For the person, as for the physical organism, stimulation is essential to growth. Response to stimulation is wholesome so long as it is a correlated *integral* reaction of the entire personality. When the reaction is *segmental*, that is, detached from, and uncontrolled by, the organization of personality, it tends to become disorganizing or pathological. That is why stimulation for the sake of stimulation, as in the restless pursuit of pleasure, partakes of the nature of vice.

The mobility of city life, with its increase in the number and intensity of stimulations, tends inevitably to confuse and to demoralize the person. For an essential element in the mores and in personal morality is consistency, consistency of the type that is natural in the social control of the primary group. Where mobility is the greatest, and where in consequence primary controls break down completely, as in the zone of deterioration in the modern city, there develop areas of demoralization, of promiscuity, and of vice.

In our studies of the city it is found that areas of mobility are also the regions in which are found juvenile delinquency, boys' gangs, crime, poverty, wife desertion, divorce, abandoned infants, vice.

These concrete situations show why mobility is perhaps the best index of the state of metabolism of the city. Mobility may be thought of in more than a fanciful sense, as the 'pulse of the community'. Like the pulse of the human body, it is a process which reflects and is indicative of all the changes that are taking place in the community, and which is susceptible of analysis into elements which may be stated numerically.

The elements entering into mobility may be classified under two main heads: (1) the state of mutability of the person, and (2) the number and kind of contacts or stimulations in his environment. The mutability of city populations varies with sex and age composition, the degree of detachment of the person from the family and from other groups. All these factors may be expressed numerically. The new stimulations to which a population responds can be measured in terms of change of movement or of increasing contacts.

Land values, since they reflect movement, afford one of the most sensitive indexes of mobility. The highest land values in Chicago are at the point of greatest mobility in the city, at the corner of State and Madison streets, in the Loop.[3] Our investigations so far seem to indicate that variations in land values, especially where correlated with differences in rents, offer perhaps the best single measure of mobility, and so of all the changes taking place in the expansion and growth of the city.

In general outline, I have attempted to present the point of view and methods of investigation which the department of sociology is employing in its studies in the growth of the city, namely, to describe urban expansion in terms of extension, succession, and concentration; to determine how expansion disturbs metabolism when disorganization is in excess of organization; and, finally, to define mobility and to propose it as a measure both of expansion and metabolism, susceptible to precise quantitative formulation, so that it may be regarded almost literally as the pulse of the community. The project in which I am directly engaged is an attempt to apply these methods of investigation to a cross-section of the city – to put this area, as it were, under the microscope, and so to study in more detail and with greater control and precision the processes which have been described here in the large. For this purpose the West Side Jewish community has been selected. This community includes the so-called 'Ghetto', or area of first settlement, and Lawndale, the so-called 'Deutschland', or area of second settlement. This area has certain obvious advantages for this study, from the standpoint of expansion, metabolism, and mobility. It exemplifies the tendency to expansion radially from the business center of the city. It is now relatively a homogeneous cultural group. Lawndale is itself an area of flux, with the tide of migrants still flowing in from the Ghetto and a constant egress to more desirable regions of the residential zone. In this area, too, it is also possible to study how the expected outcome of this high rate of mobility in social and personal disorganization is counteracted in large measure by the efficient communal organization of the Jewish community.

Notes

1 'British conurbations in 1921', *Sociological Review*, XIV (April, 1922), 111–12.
2 Weber, Adua Ferrin. 1899: *The growth of cities in the nineteenth century: a study in statistics*. New York, p. 442.
3 From 1912 to 1923, land values per front foot increased in Bridgeport from $600 to $1,250; in Division-Ashland-Milwaukee district, from $2,000 to $4,500; in 'Back of the Yards', from $1,000 to $3,000; in Englewood, from $2,500 to $8,000; in Wilson Avenue, from $1,000 to $6,000; but decreased in the Loop from $20,000 to $16,500.

2 David Harvey and Lata Chatterjee, 'Absolute Rent and the Structuring of Space by Governmental and Financial Institutions'

Reprinted from: *Antipode* **6**(1) 22–36 (1973)

In this paper we sketch the answers to two questions: (1) how do the macro and micro features of housing markets relate to each other in theory and in practice? (2) How is absolute rent realized in the housing markets of large metropolitan areas? We can afford the luxury of two questions because the same materials suffice to answer both.

I The macro and micro features of housing markets

There is a considerable body of theory and a mass of empirical information on the macro-economic aspects of housing markets. The same can be said with respect to the micro-economic modelling of housing choices – including locational choices through which residential differentiation of metropolitan areas is thought to be achieved. But there is little theory or information on the relationship between these two aspects of housing markets. What there is usually gets lost in the formalism of aggregation theory in which it is assumed that the relationship between national aggregates and local individual behaviours is a technical problem that has a formal (usually mathematical) solution. In the face of problems of this kind, we believe that it is imperative to investigate human practice, for that practice will likely reveal what formal analysis seems helpless to resolve. Economies do not stop working because of the aggregation problem and there are innumerable procedures in practice which serve to link decisions made at the national level to decisions made at lower levels. Elaborate devices exist to integrate the national and local aspects of economies. These devices are to a large degree embedded in the structure of governmental and financial institutions. In order, therefore, to understand the links between the national and the local, we have to examine in detail the structure of these institutions.

It is important to establish at the outset an appropriate methodological stance for such an investigation. We require a methodology capable of dealing with the relationship between the individual and society viewed as a totality of some sort. The methodology appropriate to this purpose has been generally neglected in the social sciences and much of our inability to deal with the 'aggregation problem' must be traced to that eclecticism and methodological myopia of which Barnbrock[1] speaks, which insists that we have to understand phenomena through conventional filters of cause-and-effect, functionalism and the like. The relations which we need to understand are, however, *contextual relations* or, as Ollman and others have proposed, *internal relations* through which the social totality is structured and transformed by individuals or entities (such as corporations) simultaneously exhibiting both 'learning' behaviours which sustain the social structure as a whole and 'instructing' behaviours through which the social structure is itself transformed.[2] This conception of things is neither widely accepted nor even understood. But from it we gain an immediate object for enquiry – viz., the *processes* of structuring and transformation together with the determinate structures that mediate these processes.[3] We will now adopt such a methodological stance with respect to the housing market.

Typical concerns of housing policy at the national level are:

1 the relationship between construction, economic growth, and new household formation (population growth);
2 the behaviour of the construction industry and the housing sector as a Keynesian regulator through which cyclical swings in the economy at large are ironed out;
3 the relationship between housing provision and the distribution of income (welfare) in society.

Since the 1930s, these concerns have generated aims that have, by and large, been successfully met.[4] Economic growth has been accompanied and to some degree accomplished by rapid suburbanization – a process that has been facilitated by national housing policies. Much of the growth in GNP (both absolute and per capita) since the 1930s is wrapped up in the suburbanization process (taking into account the construction of highways and utilities, housing, the effective demand generated by the automobile, etc.). Cyclical swings have been broadly contained since the 1930s and the construction industry appears to have functioned effectively as one of several major counter-cyclical tools. The evident social discontent of the 1930s has to some extent been successfully defused by a government policy that has created a large wedge of middle-income people who are 'debt-encumbered homeowners' and who are unlikely to rock the boat because they are both debt-encumbered and reasonably well satisfied with their housing. The discontent of the 1960s exhibited by blacks and the poor provoked a similar political response to that of the 1930s in the housing sector – a response that has not been particularly successful in obtaining 'a decent house in a decent living environment' for many of the poor and the black, although social instability of the 1960s appears to have been defused. At the national level, then, policies are designed to maintain an existing structure of society intact in its basic configurations, while facilitating economic growth and capitalist accumulation, eliminating cyclical influences, and defusing social discontent. Housing provides a vital and effective tool for stabilizing and perpetuating the social structure of a market-based, capitalist system.

How are these general programs and policies transmitted to the local level and ultimately to individuals making choices with respect to housing services in different locations? The mechanisms are very complex and we can do no more than sketch-in some of the basic relations between financial and governmental institutions, through which policies are filtered and transmitted to the local level. State Savings and Loans, Federal Savings and Loans, mortgage bankers, savings banks and commercial banks all operate in the housing market. The operations of State and Federal S&Ls are confined to housing and these institutions are designed to 'promote the thrift of the people *locally* to finance their own homes and the homes of their neighbors.'[5] The rules and regulations vary but in general State S&Ls tend to be small scale, community based and depositor controlled, whereas the larger Federal S&Ls tend to be 'professionally' managed.[6] But Federal S&Ls are usually restricted to financial operations within 100 miles of their head office. Import and export of funds for housing from one market to another cannot occur through these institutions unless depositors shift their funds. The Federal S&Ls are, for the most part, under the control and guidance of the Federal Home Loan Bank Board which has tended in the past to regulate the flow of funds into the mortgage market in a counter-cyclical fashion with respect to the economy as a whole. FNMA and GNMA (the government institutions that buy up mortgages from the financial institutions to provide the latter with the liquidity for further investment in housing), on the other hand, operate usually to dampen cycles in housing construction which means that two sets of governmental institutions tend to follow contradictory policies.[7]

Other institutions such as mortgage bankers, savings banks, and commercial banks also vary in the way they operate in the housing market. None of these institutions, however, are confined to housing. Mortgage bankers have complete freedom in the geographic transfer of mortgage funds, whereas the other institutions are, to some degree or other, more restricted by Federal or State regulation. In these institutions it is the competition for funds which dictates the flow into the housing market; housing finance is best regarded as a residual that is left over after basic corporate needs are met. The counter-cyclical flow that this generates is emphasized by State and Federal policies that put interest-rate ceilings on housing mortgages. If there is a heavy demand for credit and the interest rate rises (to, say, 10 per cent) then FHA or State interest ceilings of 8 per cent effectively dry up the flow of funds into housing (although a 'points' system exists to offset this). This drying up may be offset by the release of more funds through FNMA, GNMA or even through the Home Loan Bank Board. Depositors are likewise sensitive to interest rates and are likely to move their funds out of, for example, a savings bank (with, say, a 5 per cent interest rate ceiling) to higher-yielding investments.

How do all of these financial and governmental complexities relate to the individual? In the typical micro-economic models of residential differentiation and housing markets, it is assumed that income is the relevant determinant of housing choice.[8] In fact, it is the ability to obtain credit and a mortgage that is, for most people, the immediate determinant. This ability is income-related, of course, but the ability to obtain a mortgage under suitable terms is also a function of the policies of financial and governmental institutions. Because servicing costs are constant no matter what the price of a house, financial institutions (particularly those geared to profit-making) prefer to finance the more expensive housing. In Baltimore, for example, savings banks rarely finance transactions in the price range below $20,000. Mortgage bankers have not gone below $7,000 very often and have recently decided as a matter of policy to try and stay above the $15,000 mark. Different institutions have distinctive policies with respect to downpayments, credit-worthiness, and the like, while government policies (particularly those of the FHA) also intervene in these respects and play a major role in those sectors of the population with moderate to marginal incomes or slender resources for downpayment. There are also distinctive policies with respect to the nature of the housing stock which financial institutions are willing to finance (above and beyond the general operating rule, that appears to have held good for the last decade or so, that 'new means better and safer' investment). There is also no question that different institutions, both financial and government, exhibit strong 'neighborhood biases' (both pro and con) over and beyond that inherent in the community basis found in the State and some Federal S&Ls.

We will now describe some of the ways in which all of these factors coalesce in the Baltimore housing market. Consider first of all the behaviour of the different institutions with respect to house sales in different price categories (Table 2.1). There is clearly a structured relationship that leads the 'commercial' institutions to operate in the higher price ranges while the State S&Ls, which tend not to be very strongly profit oriented, take up the housing in the lower price categories. The health of the housing market in this lower price category appears to be attached entirely to the fate of the State S&Ls and a perpetuation of their community-based non-profit orientation. Should these institutions collapse, or come to have a strong profit orientation (as seems to be happening), then the housing market in the below $15,000 price range will suffer irreparable damage, particularly if the mortgage bankers implement a policy of a $15,000 minimum. There is, clearly, a highly structured relationship between household characteristics (particularly income) and the availability of mortgage funds (in appropriate price categories).[9]

This structured relationship has a geographical manifestation.[10] To demonstrate this we have divided the housing market in Baltimore City into 13

Table 2.1 Distribution of mortage activity in different price categories by type of institution, Baltimore City, 1972*

	under $7,000	*$7,000– $9,999*	*$10,000– $11,999*	*$12,000– $14,999*	*over $15,000*
Private	39	16	13	7	7
State S&Ls	42	33	21	21	20
Federal S&Ls	10	22	30	31	35
Mortgage banks	7	24	29	23	12
Savings banks	—	3	5	15	19
Commercial banks	1	1	2	3	7
Per cent of City's transactions in category	21	19	15	20	24

* Source: 'Homeownership and the Baltimore Mortgage Market', Draft Report of the Home Ownership Development Program, Department of Housing and Community Development, Baltimore City, 1973.

sub-markets which can be further aggregated to eight primary sub-market types (see Figure 2.1). We have tabulated data concerning the financing of housing in each of these sub-markets (see Table 2.2). We have also tabulated some information on house prices and socio-economic composition in the sub-markets. It is plain that the housing market in Baltimore City is highly structured geographically in terms of institutional involvement and FHA insured mortgage activity. The main features in this structuring are:

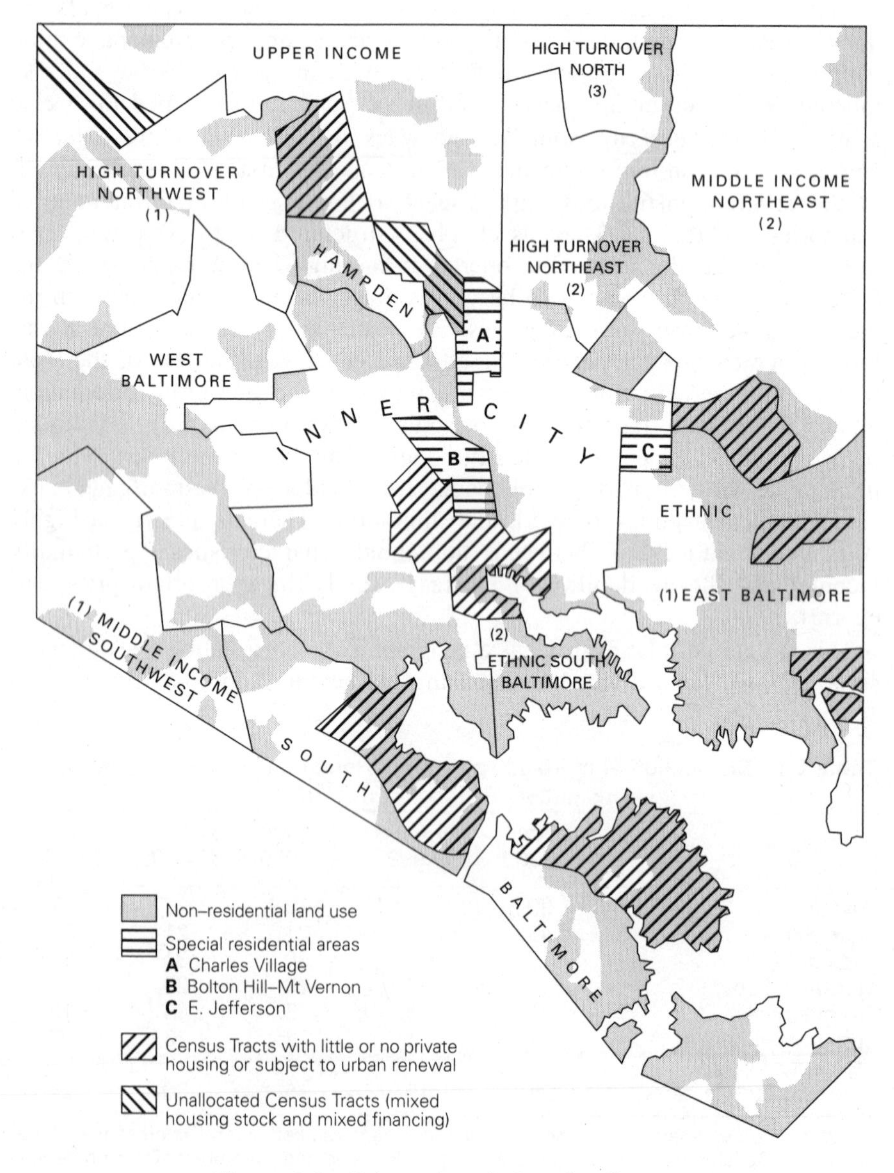

Figure 2.1 Sub-markets in housing finance

Table 2.2(i) Housing sub-markets, Baltimore City, 1970

	Total houses sold	*Sales per 100 properties*	*% Transactions by source of funds:*								*% Sales insured*		*Average sale price ($)***
			Cash	*Private*	*Fed S&L*	*State S&L*	*Mtge Bank*	*Comm Bank*	*Sav Bank*	*Other**	*FHA*	*VA*	
Inner city	1,199	1.86	*65.7*	15.0	3.0	12.0	2.2	0.5	0.2	1.7	2.9	1.1	3,498
1. East	646	2.33	*64.7*	15.0	2.2	14.3	2.2	0.5	0.1	1.2	3.4	1.4	3,437
2. West	553	1.51	*67.0*	15.1	4.0	9.2	2.3	0.4	0.4	2.2	2.3	0.6	3,568
Ethnic	760	3.34	*39.9*	5.5	6.1	*43.2*	2.0	0.8	0.9	2.2	2.6	0.7	6,372
1. E. Baltimore	579	3.40	*39.7*	4.8	5.5	*43.7*	2.4	1.0	1.2	2.2	3.2	0.7	6,769
2. S. Baltimore	181	3.20	*40.3*	7.7	7.7	*41.4*	0.6			2.2	0.6	0.6	5,102
Hampden	99	2.40	*40.4*	8.1	18.2	*26.3*	4.0		3.0		14.1	2.0	7,059
West Baltimore	497	2.32	*30.6*	12.5	12.1	11.7	*22.3*	1.6	3.1	6.0	*25.8*	4.2	8,664
South Baltimore	322	3.16	*28.3*	7.4	*22.7*	13.4	13.4	1.9	4.0	9.0	*22.7*	10.6	8,751
High turnover	2,072	5.28	*19.1*	6.1	13.6	14.9	*32.8*	1.2	5.7	6.2	*38.2*	9.5	9,902
1. Northwest	1,071	5.42	*20.0*	7.2	9.7	13.8	*40.9*	1.1	2.9	4.5	*46.8*	7.4	9,312
2. Northeast	693	5.07	*20.6*	6.4	14.4	16.5	*29.0*	1.4	5.6	5.9	*34.5*	10.2	9,779
3. North	308	5.35	12.7	1.4	*25.3*	18.1	13.3	0.7	15.9	12.7	*31.5*	15.5	12,330
Middle income	1,077	3.15	*20.8*	4.4	*29.8*	17.0	8.6	1.9	8.7	9.0	17.7	11.1	12,760
1. Southwest	212	3.46	17.0	6.6	*29.2*	8.5	15.1	1.0	10.8	11.7	*30.2*	17.0	12,848
2. Northeast	865	3.09	*21.7*	3.8	*30.0*	19.2	7.0	2.0	8.2	8.2	14.7	9.7	12,751
Upper income	361	3.84	19.4	6.9	*23.5*	10.5	8.6	7.2	*21.1*	2.8	11.9	3.6	27,413

* Assumed mortgages and subject to mortgage.

** Ground rent is sometimes included in the sales price and this distorts the averages in certain respects. The relative differentials between the sub-markets are of the right order however.

Source: City Planning Department Tabulations from Lusk Reports.

Table 2.2(ii) Census data

	*Median income**	*% Black occupied d.u.'s***	*% Units owner occupied*	*Mean $ value of own. occ.*	*% Renter occupied*	*Mean monthly rent*
Inner city	6,259	72.2	28.5	6,259	71.5	77.5
1. East	6,201	65.1	29.3	6,380	70.7	75.2
2. West	6,297	76.9	27.9	6,963	72.1	78.9
Ethnic	8,822	1.0	66.0	8,005	34.0	76.8
1. E. Baltimore	8,836	1.2	66.3	8,368	33.7	78.7
2. S. Baltimore	8,785	0.2	64.7	6,504	35.3	69.6
Hampden	8,730	0.3	58.8	7,860	41.2	76.8
West Baltimore	9,566	84.1	50.0	13,842	50.0	103.7
South Baltimore	8,941	0.1	56.9	9,741	43.1	82.0
High Turnover	10,413	34.3	53.5	11,886	46.5	113.8
1. Northwest	9,483	55.4	49.3	11,867	50.7	110.6
2. Northeast	10,753	30.4	58.5	11,533	41.5	111.5
3. North	11,510	1.3	49.0	12,726	51.0	125.1
Middle income	10,639	2.8	62.6	13,221	37.5	104.1
1. Southwest	10,655	4.4	48.8	13,470	51.2	108.1
2. Northeast	10,634	2.3	66.2	13,174	33.8	103.0
Upper income	17,577	1.7	50.8	27,097	49.2	141.4

*Weighted average of median incomes for census tracts in sub-market.
**d.u. = dwelling unit.

Source: 1970 Census.

1 The inner city is dominated by cash and private transactions (with scarcely a vestige of institutional or governmental involvement) accompanied by a low purchase price, low incomes and a high proportion of tenants and blacks;
2 The ethnic areas are dominated by small community and neighborhood State S&Ls which circulate money within the community and in some cases finance migration to, for example, the middle-income sub-market of northeast Baltimore;
3 Black residential areas are serviced mainly by mortgage bankers operating under FHA guarantees (a response to the social discontent of the 1960s). Community-based State S&Ls are absent and without a strong sense of community will be difficult to bring into being. The Federal S&Ls apparently are reluctant to become involved in financing black home ownership.
4 Mortgage bankers using the FHA guarantees (often of the no-downpayment sort) are the predominant source of finance in areas of high turnover and racial change.
5 The white middle class, largely brought into homeownership through the FHA programs of the 1930s, occupies a solid area of northeast and southwest Baltimore as well as much of Baltimore County (which surrounds much of the City). Federal S&Ls here dominate with FHA guarantees supporting the market in traditional fashion.

6 The more affluent groups make greater use of savings banks and commercial banks and rarely make use of FHA guarantees.

This geographical structure forms a 'decision environment' in the context of which individual households make housing choices. These choices are likely, by and large, to conform to the structure and to reinforce it. The structure itself is a product of history. Attempts to change the structure can in fact generate considerable social conflict. For example, middle-income buyers, disillusioned with the suburban dream, have to fight the institutional policies of the lending industry and the FHA if they wish to renovate an inner city neighborhood.[11] Likewise, low-income blacks cannot be turned into 'debt-encumbered homeowners' painlessly. By struggles of this sort the structure can be transformed. In practice, therefore, we find that the geographic structure is continuously being transformed by the ebb and flow of market forces, the operations of speculators and realtors, the changing potential for homeownership, the changing profitability of landlordism, the pressures emanating from community action, the interventions and disruptions brought about by changing governmental and institutional policies, and the like. It is this process of transformation of and within a structure that must be the focus for understanding residential differentiation and, as we will later show, provides the basis for understanding how absolute rent is realized in the housing market. We will examine two facets of this transformation process in Baltimore very briefly.

(1) The FHA in Baltimore City[12]

Since the 1930s the Federal Housing Administration has administered a variety of programs designed to facilitate homeownership by making institutional investment in housing risk-free. Until the 1960s these programs serviced the needs of middle-income buyers and were instrumental in financing the suburbanization process. During the 1960s various programs with 'social' objectives were initiated as the government attempted to create a debt-encumbered, socially stable class of homeowners amongst the poor and the black. A mix of old and new programs were applied for this purpose and these – together with administrative directives to end, for example, the discriminatory 'red-lining' of low-income and black neighborhoods – led to the creation of FHA-insured black and low-income housing sub-markets during the 1960s. The main tool in Baltimore was the 221 (d) (2) program (D2s) which permits the financing of home purchase for moderate- or low-income people of housing below $18,000 with negligible or no downpayment. A map of FHA insured mortgages (Figure 2.2) indicates areas of high concentration within Baltimore City in 1970, while Table 2.2 indicates the extent of FHA involvement in the different sub-markets.

The creation of such a housing sub-market has not been without its problems, although Baltimore has not suffered from the speculative intervention that proved so disastrous in Detroit (where redistributive rent, in the sense that Walker argues for it,[13] was extracted with a vengeance). Foreclosure rates

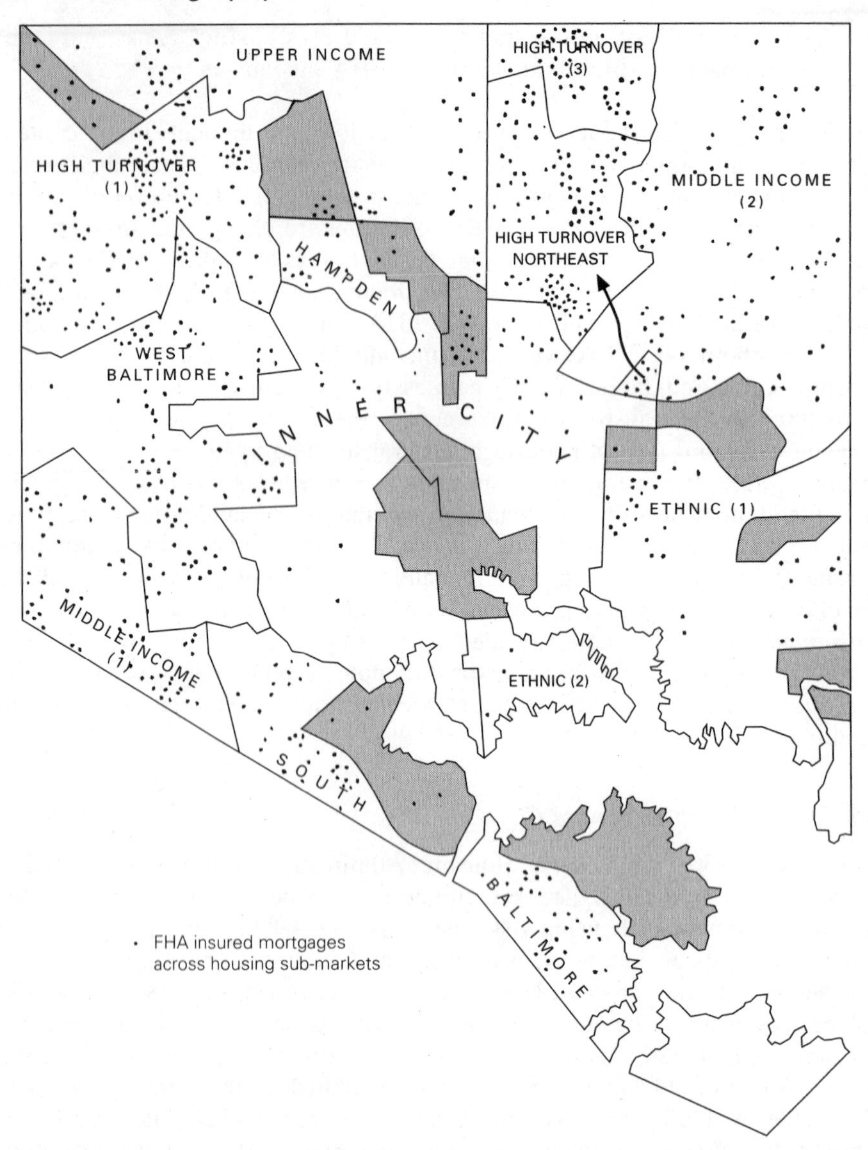

Figure 2.2 FHA insured mortgages across housing sub-markets, Baltimore, 1970

have been low in Baltimore compared to Detroit. The main problems in Baltimore have arisen from supporting the movement of a low-income and predominantly black population into middle-class white neighborhoods, from forming a specific geographical sub-market within which a specific kind of financing and population are confined, and from risk-aversion practices, designed to prevent a high rate of foreclosure, which effectively deny government insurance to inner-city neighborhoods. These problems arise for a variety of reasons. The FHA programs, particularly the D2s, are a 'last resort' for housing finance – they are only made use of when conventional sources of financing are unavailable. The FHA has to 'draw a line' somewhere in administering these programs (a line set up in terms of credit characteristics of the purchaser, the nature of the housing and its physical condition, etc.). This 'drawing of the line somewhere' in fact means a line between mortgage finance and cash transactions which becomes, by design or accident, a 'red-line' on the map of housing market activity. Within this line, in the inner city, mortgages cannot be obtained and immediately outside of it the FHA supports the market. The jump of housing prices across this line is clearly evident (see Table 2.2). A similar discontinuity emerges at the boundary between the D2 market and the middle-income sub-market. Middle-income whites, nervous at the growing effective demand of predominantly low-income blacks, move under the perceived threat of a 'declining neighborhood' while conventional sources of mortgage finance react similarly to this perceived danger and withdraw support from that neighborhood. A vacuum may form into which the coalition of mortgage banker and FHA insurance moves if the speculator does not get there first. This amounts to a self-fulfilling prophecy of neighborhood change. If a vacuum does not form, pressure may build up at the boundary between the middle-income and the D2 sub-markets and social conflict may result. Something akin to blockbusting takes place. For example, on one block of a street in northwest Baltimore, on the edge of the D2 sub-market, there were 19 houses sold in 1970, 17 of which were insured by the FHA (13 with no downpayment) and 14 of which were financed by mortgage bankers. The combination of mortgage banker and the FHA programs appears to act like a blow-torch peeling back the middle-income sub-market as the D2 sub-market expands. Across this boundary there is a market discontinuity of house prices, with prices higher in the middle-income areas compared with the D2 market, and considerable confusion and speculative activity at the interface between them.

The net effect is that FHA programs and policies create a plateau of house prices between the 'disinvestment sink' of the inner city and the stable middle-income areas. On this plateau we find a mixture of low-income and predominantly black homeownership and landlordism. At the edges of the plateau we find social conflict as processes of 'filter-down' and 'blow-out' erode the geographical structure of housing sub-markets.[14]

(2) Landlordism and the Baltimore City housing market[15]

There are many different types of landlord varying from the individual who has one house to the professional who owns a large number, uses professional management techniques, and is very sensitive to profits, losses and the rate of return on capital. We will ignore the complexities and confine attention to professional landlords who own and manage about a quarter of Baltimore City's rental inventory. These professional landlords make their decisions in terms of a structured decision environment and closely gear their operations to the characteristics of sub-markets as they perceive and experience them.

We asked four landlords to distinguish areas in the city which they regarded as 'good,' 'moderately good,' 'moderately bad,' or 'bad' for investment. We aggregated their maps of perceived investment opportunity to form a composite map (Figure 2.3) and analyzed their costs, expenditures, rates of return, etc. across these different perceived investment situations. It was found that there was a rational adjustment of landlord operations to sub-market characteristics and that this adjustment had certain important consequences for the maintenance of the housing stock, rent levels, and the like. We will not discuss the details here, but the most important element in the landlord's decision is the availability or non-availability of landlord finance. In areas

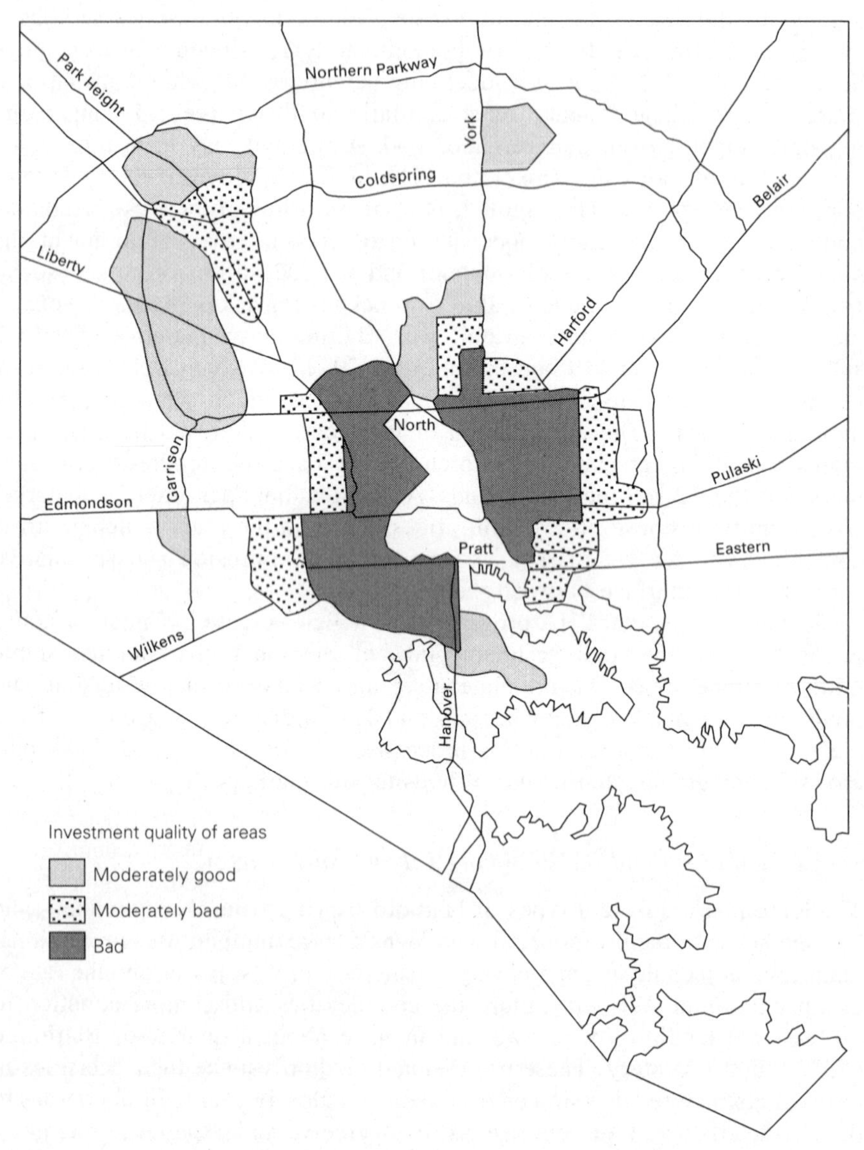

Figure 2.3 Aggregate perceptions of four professional landlords concerning areas of investment opportunity in housing in Baltimore City (1970)

where the landlords can obtain mortgage funds they work on the basis of 'leverage' which works as follows. A $10,000 house is purchased with a $1,000 downpayment and a $9,000 mortgage at, say, a 6 per cent interest rate. The landlord manages the property to yield an overall 8 per cent rate of return on the $10,000 (which is $800), pays off the interest of $540, which leaves $260 to be applied to the landlord's investment of $1,000 which yields an effective return of 26 per cent on the landlord's part of the capital. This calculation is more complicated when worked out over the lifetime of the mortgage, of course, but the principle of leverage operates over each round and the average rate of return is much higher for the landlord (it usually comes to between 15 and 18 per cent) than is indicated by the rate of return calculated against the total value of the house. The tenant is also advantaged by this, for the tenant lives in a $10,000 house for a rent that yields only 8 per cent on that value, while the landlord, interested in capital accumulation in the physical form of the house, will keep it well maintained.

If the landlord cannot obtain mortgage finance, the story becomes rather different. The landlord has to obtain a 'satisfactory' yield on the total value of the house. What is meant by a 'satisfactory' yield is set by convention on the part of professional landlords but it is obviously sensitive to the rate of return possible in the capital market in general.[16] Professional landlords in Baltimore, in fact, look for a 20 per cent rate of return on their capital, regard 15 per cent as 'normal' and will still stay in operation at 11–12 per cent (this is *after* all expenses are met including interest payments *and* an imputed managerial wage to the landlord as manager). Using leverage, landlords can gain a 15 per cent rate of return for themselves by taking, say, an 8 per cent rate of return on the total value of the property. Without leverage they have to take a 15 per cent rate of return directly. Leverage is not possible in the inner city because mortgage funds are not generally available; so tenants here have to pay a much greater rent relative to the total value of the property than would be the case if mortgage finance were available. But the inner city is the worst sub-market in terms of quality of housing stock, etc. Landlords therefore find it difficult to dispose of their properties in this sub-market, except through urban renewal schemes (which landlords frequently actively promote as a means of bailing out of the worst sub-market). If profit levels fall or if there is even an expectation that they will fall below the limit of, say 11–12 per cent (because of oversupply of poor housing, rising management costs, vandalism, and the like), then landlords will seek to get their money out through a disinvestment process (economizing on maintenance, milking properties, etc.). Hence there arises a housing sub-market characterized by landlord disinvestment, housing abandonment, and severe neighborhood decay. Landlords will still, on occasion, purchase in this sub-market, but if they do so, it is at a very low market price (note the prices in Table 2.2) for a good-quality dwelling which they operate to regain their outlay over a short time-horizon.

It is evident that landlords 'structure' their behaviour according to their decision environment. There is a 'rational' (profit maximizing) adjustment of landlord behaviour to sub-market characteristics. This behaviour, in turn, structures outcomes with respect to the renter, the maintenance of the housing stock, reinvestment and disinvestment, neighborhood decay and the like. This structuring activity is not without its implications for the way in which financial institutions and government institutions together formulate, in turn, rationales for investment and intervention.

II The realization of absolute rent

The category of absolute rent has been completely ignored in locational analysis until recently.[17] We will not discuss its theoretical aspects here, but will attempt to explain the actual processes whereby it is realized in the housing markets of a large urban area. It is important to realize, however, that rent essentially represents a *transfer payment* between individuals, interest groups or classes and that it does not represent any increase in value through production.

Absolute rent implies class monopoly power of some sort. By a 'class monopoly' we mean a class of producers (or consumers) who have power over a class of consumers (or producers) in a situation of structured scarcity.[18] We have first to define the basis for such class monopoly power in the housing market. The materials we have assembled in Part I of this paper in fact provide a description of this necessary basis. Through the structuring activity of governmental and financial institutions, urban space is differentiated into specific sub-markets. If absolute rent is to be realized we have to show that there are absolute limits of some sort operating over different segments of the housing market. These absolute limits can be set by the joint attributes of housing, of financiers, of housing suppliers and of consumers. To become a low-income owner of a $5,000 house in Baltimore, for example, means either a cash transaction or access to a community-based State S&L. A low- or moderate-income black will purchase a $10,000 house through a mortgage banker operating under an FHA guarantee. An upper-income person will typically go through a savings bank or a Federal S&L and purchase a house priced above $25,000. In each of these cases, the opportunities are restricted in terms of the structure. But the absolute limits are also set geographically through the structured pattern of housing sub-markets within which specific conditions hold. 'Absolute limits' means in this case the creation of absolute urban spaces within which producers and consumers of housing services face each other as classes in conflict.[19] What transpires within each sub-market depends (1) on the internal conditions within that sub-market and (2) on the interaction between sub-markets. We will consider these separately and make the link to the realization of absolute rent.

Consider first a situation in which a sub-market is completely isolated from all other sub-markets so that consumers and providers of housing are all locked into a specific situation. Suppose, for the sake of simplicity, that we are dealing with an inner city sub-market where low-income tenants cannot possibly find alternative accommodation and from which landlords cannot possibly extract themselves. Suppose no leverage is possible because financial institutions and government intervention play no role. In such a sub-market rent levels will be set by the relative power of the landlord and the tenant. If the landlord dominates, high rates of return will be extracted while tenants will be forced to spend a large proportion of their already low disposable income on rent, leaving little over for food and other needs. The only way in which tenants can adapt to this situation is to 'double up' in dwelling units to save on rent. If we assume a fixed population, this doubling up will create vacancies, rents will fall and eventually an equilibrium will be achieved in which exactly that rent is charged which extracts the maximum rate of return on capital without generating overcrowding through doubling up. With an increasing population and fixed housing supply rents will rise even higher with doubling up and overcrowding. In all of these cases rent has to be conceived of as an absolute rent which accrues to the monopoly power of landlords as a class *vis-à-vis* the collective power and condition of the tenantry. It is set, in short, by a 'class' conflict within a restricted geographical area (within an absolute space).

If we now relax some of the restrictions, the siuation becomes somewhat more complex. Landlords can pull out their capital and place it elsewhere either in another sub-market (where, for example, leverage is possible) or on the capital market in general. If tenant power rises to the point where rates of return fall below, say, 11 or 12 per cent, the landlord will simply take steps to withdraw capital from this particular sub-market. Tenants have a counter-power. They can, through the exercise of social pressure, expand their own sub-market and increase the supply of housing available to them, they can seek alternative accommodation in other sub-markets, or they can, by the exercise of political muscle, put restraints on landlord profits by such means as rent control legislation. But there are limits to both landlord and tenant power. Landlords often experience difficulty in getting their capital out of the inner city housing sub-market. Tenants have virtually no opportunity to obtain accommodation in other sub-markets if they are poor and uncreditworthy by the standards of the FHA programs. Expanding the sub-market can help to reduce rents, and during the 1960s social unrest effectively did just that. But in the process landlord profits began to fall. Professional landlords began to disinvest as rates of return dipped to the 11–12 per cent level and non-professional landlords, lacking efficient management and advantages of economies of scale, got into deep financial trouble. Abandonment became widespread as a result. The housing stock has consequently begun to shrink to levels at which landlords can once more obtain an effective rate of return on their capital invested.

The processes are evidently complex when we take into account the relationships both within and between sub-markets, and as we introduce more actors into the housing drama (financial and governmental institutions, the construction industry, and so on). But the principle remains the same. The geographical and social structure creates conditions within which absolute rents can be realized to varying degrees depending upon the relative power of those supplying housing and those consuming it, as that relative power is mediated by institutional policies (legal, political, financial, governmental, etc.).

It is possible to document these relative powers with respect to sub-markets in particular historical situations. The relative powers shift continuously. Consider, for example, the matter of consumer preference. In the more privileged sub-markets the constraints to consumer choice are social – consumers have a wide range of alternatives, but considerations of prestige and status make it unlikely that individuals will consider all potential choices. If these social constraints are strongly developed, then it is possible for producers of housing to gain considerable absolute rent as high-income consumers vie with each other for prestigious housing in the 'right' neighborhoods. The same considerations permit absolute rent to be extracted from suburban construction. In the inner city, on the other hand, we find a low-income and uncreditworthy population confined to tenant occupancy in a situation where landlords cannot use leverage and must extract a high rate of return out of bad housing. Absolute rent is here gained by the landlord from a population that has no other choice (unless, in the Baltimore case, it migrates back to the rural South). This population is trapped in the structure of absolute space. On the next rung up the ladder, it is possible to become either a low-income homeowner through the FHA D2 program or a low-income but 'reliable' tenant in a situation where landlords can use leverage to provide housing for a lower rent than would otherwise be the case. In this sub-market, absolute rent accrues either to the intermediary, who frequently buys cheap from the fleeing white and sells dear to the incoming black, or to the landlord – and in both cases the financial institutions take their interest.

In all of these situations – and we could specify the situation in detail with respect to each sub-market if we so wished – the rates of return and the potential for obtaining absolute rent are structured by the opposing forces within each sub-market and the interactions between sub-markets. Class conflict within a sub-market is tempered by class differentials and class conflicts between sub-markets. And it is in such a structured situation that absolute rent is realized in the housing market in general.

III A concluding comment

We accept the view that rent is a transfer payment out of a social rate of return on capital. We also accept the view that rent accrues to the inherent monopoly power of private property.[20] To understand how rent is in practice realized is therefore to understand all of those situations and conditions which permit transfer payments to occur and which affect the actual amounts transferred. The categories of differential, absolute and monopoly rent, to which Walker adds redistributive rent, provide generalized descriptions of the theoretical circumstances which contribute to the processes of transfer. The explanatory power of these categories depends, however, on our ability to specify in practice how actual transfer payments arise under the conditions theoretically specified in each category. In this paper we have shown that the category of absolute rent is meaningful in the contemporary urban scene. We have shown that the conditions for its realization are automatically generated by the way in which institutional arrangements are structured to integrate national and local aspects of housing market behaviour.

The absolute spaces created by institutional arrangements form a geographical framework within which absolute rent can be realized. Within this geographical framework we find different interest groups facing each other as classes. The relative power of these classes with respect to each other, together with the possibilities open for substitution in other sub-markets, provides the social setting within which the realization of absolute rent becomes possible. The classes we identify here are not, of course, the classes relevant to understanding the production process. Rent is not, after all, inherent in production but arises only because the legal institution of private property is a necessary feature in the capitalist mode of production and because it proves difficult or impossible to restrict that legal right to production solely. The classes we are here concerned with are, when set against the broader class structure of society, perhaps best interpreted as sub-classes in conflict with each other over the transfer payment that rent represents. The rich, for example, may be forced to yield a relatively high transfer payment to other members of their own general social class (the company director may yield up absolute rent to the class monopoly power of the developer). A low-income tenant may likewise gain absolute rent by sub-letting. The general pattern of transfer payments is, however, fairly obvious – poorer groups yield a net transfer payment to richer groups because the former have little power or possibility for substitution, while the latter have considerable class monopoly power and a greater range of choice.

We are not claiming that rent in an urban situation has to be totally understood in absolute terms. Differential, monopoly, absolute and redistributive rent all contribute to the formation of actual rent. But we believe we have shown, quite conclusively, that absolute rent can contribute substantially to actual rent in large urban areas. We have also shown that this contribution is made possible by the way in which social, institutional and geographical structures are created for the purpose of integrating local and national aspects of economies. We believe that we have also shown that a great deal can be learned if we are prepared to adopt a methodology appropriate for understanding society as a totality fashioned through a structured set of internal relations.

Notes

1 'A prologomenon to a methodological debate on location theory – the case of von Thünen', *Antipode*, 6, No. 1.
2 See Harvey, D., *Social Justice and the City* (Johns Hopkins University Press, Baltimore; 1973) chapter 7; Ollman, B., *Alienation: Marx's Conception of Man in Capitalist Society* (Cambridge University Press, London; 1971) and 'Marxism and political science: prologomenon to a debate on Marx's method' (Department of Politics, New York University; 1972).
3 The concept of structure which we adopt is set out in detail by Godelier, M., *Rationality and Irrationality in Economics* (NLB, London; 1972). See also Piaget, J., *Structuralism* (Harper, New York; 1970).

4 The views set forth in this paragraph can be documented in detail from the Douglas Commission Report, *Building the American City* (Government Printing Office, Washington, D.C.; 1968). More detailed analyses of cyclical swings are collected together in Page, A.N. and Seyfried, W.R., *Urban Analysis* (Scott Foresman, Glenview, Illinois; 1970).
5 See *Homeownership and the Baltimore Mortgage Market* (Draft Report of the Homeownership Development Program, Department of Housing and Community Development, Baltimore City; 1973).
6 The management characteristics of Federal S&Ls are documented in Rose, S., 'The S&Ls break out of their shell', *Fortune*, 86, No. 3, pp. 152–70.
7 The details can be documented from the Douglas Commission Report (*op. cit.*).
8 For a general critical evaluation of these models see Harvey, D. (*op.cit.*) chapter 5.
9 See *Homeownership and the Baltimore Mortgage Market* (*op. cit.*).
10 This material is summarized from Harvey, D., Chatterjee, L. and Klugman, L., *Effects of FHA Policies on the Housing Market in Baltimore City* (Draft Report to the Urban Observatory, 222 E. Saratoga St., Baltimore, Md.; 1973).
11 A typical struggle of this sort is documented by the Citizens Planning and Housing Association, *FHA: an Unsatisfactory Status Quo* (Baltimore, Md.; 1973).
12 This material is summarized from Harvey, D., Chatterjee, L. and Klugman, L. (*op. cit.*).
13 Walker, R.A., 'Urban ground rent – building a new conceptual framework', *Antipode* 6, No. 1.
14 The concepts of filter-down and blow-out are discussed in detail in Harvey, D. (*op. cit.*).
15 This material is derived from Chatterjee, L., *Real Estate Investment and Deterioration of Housing in Baltimore* (Doctoral Dissertation, Department of Geography and Environmental Engineering, The Johns Hopkins University; 1973).
16 Exactly how a 'satisfactory' rate of return on capital is defined in general is, of course, a subject of dispute. We follow in general the view put forward by Rhaduri, A., 'Recent controversies in capital theory: a Marxian view', *Economic Journal*, 79, 1969, pp. 532–9.
17 See Harvey, D. (*op. cit.*) and Walker, R.A. (*op. cit.*) for recent attempts to explicate the rental concept in a contemporary context.
18 The use of the word 'class' here may generate some confusion. We use the word flexibly to mean any group of individuals who find themselves collectively in opposition to any other collectivity of individuals with respect to the transfer payment that rent represents. The 'class distinctions' which we here make use of are not, therefore, of the sort that stem from the organization of production and the division of labor (for a modern examination of these see Poulantzas, N., 'Marxism and social classes', *New Left Review*, 78, 1973, pp. 27–55). We do not believe our use of the concept of class in this special sense is inconsistent with Marx's relational definition and usage of the concept. Certainly, the concept of class monopoly power is explicitly formulated in Marx, K., *Capital*, Volume 3 (International Publishers Edition, New York; 1967) pp. 194–5, while it is clear from Marx's various analyses of the rental concept that distinctive classes (such as that of the rentier) may emerge outside of the process of production to reflect the various transfers that can occur in the circulation of the surplus.
19 The concept of absolute space is further elaborated on in Harvey, D. (*op. cit.*).
20 This, of course, was Marx's main point – see Marx, K. (*op. cit.*) and *The Poverty of Philosophy* (International Publishers Edition, New York; 1963) pp. 154–66.

3 David Ley, 'Social Geography and the Taken-for-Granted World'

Excerpts from: *Transactions of the Institute of British Geographers* **2**(4), 498–512 (1977)

Humanisons la géographie humaine. (Max Sorre)

The most important thing, therefore, that we can know about a man is what he takes for granted, and the most elemental and important facts about a society are those that are seldom debated and generally regarded as settled. (Louis Wirth)

The present impasse

The recent growth of interest in social geography once again raises the question of the field's latent ambiguity. Despite a proliferation of empirical studies, there is neither a well-developed body of theory nor explicit discussion of philosophical underpinnings. More conspicuous is the complete equivocation concerning the relative roles of spatial form and social process. Review articles over the past decade have increasingly inclined to the view that while the map may be the first step it should not be the last word.[1] Yet the precise avenues for process studies have not been explicitly discussed and even current research seems preoccupied with the 'frail structure'[2] of spatial fact rather than social process. Pahl's inclusive definition of social geography remains more a declaration of faith than of actuality: '. . . the processes and patterns involved in an understanding of socially defined populations in their spatial setting';[3] more appropriate is Buttimer's less specific statement, '. . . a multi-faceted perspective on the spatial organization of mankind'.[4]

This paper argues that the lack of firm direction in contemporary social geography is based upon a fundamental distinction between spatial form and social process. These concepts find themselves on opposite sides of a philosophical divide separating fact from value, object from subject, and natural science from social science. It is only by establishing a firm and appropriate philosophical underpinning that social geography will pass successfully to a concern with the social processes antecedent to a spatial fact and beyond that to the development of limited generalizations and ultimately theory. The history of geography has several instructive examples of the false blooming of a social geography of man, failures, it will be argued, due to the discipline's traditional preoccupation with the objective, and neglect of the lesson of everyday experience where every object is an object *for* a subject. These examples from the past provide useful lessons now as the epistemological impasse is once again engaged, and contribute toward a possible solution within the philosophy of phenomenology, where object and subject reassume the unity they share in our naïve realms of experience, those realms, we will argue, which form the subject matter and central concern for a social geography of man.[5]

Lessons from the past

The school of human geography founded by Paul Vidal de la Blache has often been suggested as a forerunner to the present development of a social geography concerned with the relationships and processes underlying a landscape fact. His dual concepts of *genre de vie* and *milieu* were the building blocks for exploring the reciprocal relationship between social group and environment, and his possibilist stance acknowledged that landscapes were the outcomes of choice-making social groups. Each region, each place, was to be considered holistically as an intimate amalgam of environment and decision-making men, as an object with a subject. Men were rooted, they 'saturated themselves with the environment';[6] yet the dialectic was continuously emphasized, as in Vidal's famous description of a landscape as a medal cast in the likeness of a people. Within this relationship there were few independent phenomena; each feature pointed beyond itself to other parts of the whole. In undertaking such research, Vidal was sceptical of the appropriateness of the methodology of the physical sciences and of being able to develop general laws of human behaviour. Here he was close to the position of incremental middle-order generalization held by Weber, whose interpretative sociology has acted as an important bridge in the introduction of phenomenological perspectives to the science of man.

The passing of Vidal's system reveals an instructive methodological turn. It had been under criticism, often bitter, from the new field of social morphology, led by Durkheim and other sociologists.[7] This conflict had philosophical undertones. The Durkheimian method adopted much of the positivist model of the natural sciences, insisting upon a definition of each science which incorporated a distinctive set of phenomena and a body of laws. This ran at odds with the Vidalian perspective which emphasized more a point of view of place, and the uncovering of relationships which did not force deterministic statements, while pointing to likelihood and probability.[8]

But the more severe blow came from a gradual shift of emphasis within the French school itself. Concern passed from interpretative statements of place to the more formal categorization of landscape facts in the work of Jean Brunhes and others.[9] These landscape facts were categorized, removed from their context, so that they expressed only the objective meaning of functionalism. Vidal's balanced humanism was replaced by an increasing materialist orientation, culminating in Demangeon's quantitative coding of rural settlement,[10] and in North America represented in aspects of Carl Sauer's landscape school.[11] The material facts had sprung loose from their everyday human world; artifact replaced attitude, and land-use morphology triumphed over regional personality. The transition from a science of man in place to a science of phenomena prepared the way for a scientism which ultimately abstracted place to a geometry of space, and reduced man to a pallid, entrepreneurial figure. Unlike Weber's interpretative sociology,

Vidal's interpretative human geography did not engage epistemological and philosophical questions. Consequently it found no firm base to counter the criticisms of idiosyncrasy and an atheoretical content and method of a later generation attuned to the fiats of 'positive' science. These criticisms can, however, be met in geography as they have been in sociology from the foundation of a philosophy embracing both subject and object such as that offered by phenomenology.

Turning from the rural French school to the urban Chicago school there are some instructive developmental parallels. The methodological similarities between Park's urban sociology and Vidal's human geography are striking, and negate too easy a dismissal of the Vidalian system in an urban setting. Park was certainly familiar with French human geography which he quoted in his earlier writing. Like Vidal, Park saw the man–environment relationship as reciprocal. Though the prompting influence of environment was not to be disregarded,[12] a land-use unit only became a neighbourhood when it was 'inevitably stained with the peculiar sentiments of its population',[13] a phrase reminiscent of Vidal's humanistic definition of landscape. The natural area was the equivalent of the region, and the concept of social world parallel to *genre de vie*, a localized set of people, attitudes and behaviour in a given setting. Park's view was as holistic as the French school, emphasizing the role of regional monographs of natural urban areas, a synthetic, inductive approach in which generalization could only be incremental. The regional monograph was to be concerned with the everyday, experienced world of the city dweller, the moral order, 'man in his habitat and under the conditions in which he actually lives',[14] incorporating his attitudes, subjective experiences, and conception of self relative to his milieu.[15] The role of man the creator and of human decision-making were as prominent in the Chicago urban studies as in rural France, for what was the city other than 'a product of nature, and particularly of human nature'?[16] The humanist motif is clear in these statements, and it is important to note the influence of both Dewey and Mead upon Park and the Chicago school, North American scholars whose thought is closely associated with the phenomenological stream.

In this context the debate of the 1940s when the Chicago school was accused of narrowminded economic determinism and the neglect of social and cultural variables appears incongruous.[17] But in the same way that the French school had passed into a period of materialistic functionalism, so Chicago sociology had become preoccupied with human ecology, a science of spatial relations heavily impregnated with biological metaphor in which social groups and land uses were allocated spatially according to economic gradients in the city.[18] This change of emphasis permitted a methodological about-turn from Park's earlier statements concerning the place of theory:

> Reduce all social relations to relations of space and it would be possible to apply to human relations the fundamental logic of the physical sciences.[19]

In this climate the geographical response to the Chicago school is illuminating. While the regional monographs were ignored, human ecology emphasizing material phenomena and economic relations was quickly claimed from within the geographical bailiwick, an emphasis which has continued into current writing. Barrows, writing in 1923, claimed that 'upon economic geography for the most part the other divisions of the subject must be based' while, as much of the material of social geography is 'intangible . . . this body of relationships appears to form a potential field for geography rather than an assured one'.[20]

Similar currents diverted both the early French and early Chicago schools from common initial methodological goals. In each instance, a humanistic perspective with a holism incorporating subject and object was compromised for a materialist treatment drawing upon a physical science tradition and encouraging deterministic thinking.[21] Social relations were suppressed in the interest of spatial facts, and social geography became preoccupied with man's material works and the irresistible objectivity of the map. It was not until 1960 that Firey's injunction concerning the role of social and cultural variables in urban land use was broadly advertised in the geographic literature.[22] Before an intellectual climate underscoring economic materialism, social geography has substituted objective description of spatial facts for a *human* geography which was no less concerned with the *meaning* of those facts in the life-world of social groups.

The lesson of behavioural geography

At first glance it would seem as if the balance has been righted in recent years. Vidal's humanism was revived in Max Sorre's development of the concept of social space, with its subjective as well as objective components.[23] Both seminal and empirical research in perception, often emphasizing holistic microscale settings, promised to reassert the role of human values and creativity in interaction with the environment.[24] But most of all, the emergence of behavioural geography witnessed the explicit commitment to delve beneath the distribution maps and spatial facts to an examination of social and cognitive process in their everyday context: '. . . behavioural approaches emphasize the decision processes that were responsible for locations rather than concentrating on the topological relation between locations themselves'.[25]

At the heart of behavioural geography, however, lies a methodological ambiguity which represents yet another example of the subject–object dichotomy, the dualism of cognitive process and spatial form. On the one hand is the recognition that behavioural geography is less concerned with the geometry of spatial relations than it is with the motives and social processes operating to prompt group and individual behaviour. From the other perspective is the concern with the science of spatial geometry, macro-scale patterns, in which cognitive variables absorb only individual exceptions, 'a "residual" domain of events which cannot be handled by the normative or stochastic location theories'.[26]

This locational perspective has its familiar methodological entourage which is lucidly described in *Explanation in Geography*.[27] Its underpinning is positivism, its model the model of natural science, with a concern for a set of phenomena particular to each discipline, precisely stated concepts, high order measurement, and the formulation of theory and ultimately general laws. In geography it is associated with the properties of aggregates, macro-spatial structure, and implicit economic determinism, with a pale spectre of man responsively following economic gradients. In this manner, the locational school is not unrelated to human ecology, albeit in more sophisticated costume and separate from the cruder biological analogies. Mackenzie's definition of human ecology captures much of the flavour of the spatial school: 'Human ecology deals with the spatial aspects of the symbiotic relations of human beings and human institutions.'[28] The interconnections are more fully underlined in Park's statement that human ecology seeks to emphasize not so much geography as space, for now geography has itself claimed the mantle of the geometric science of space.[29]

As has occurred in other social sciences, it is this methodological set which has been extended to the behavioural approach in geography.[30] Yet positivism's uneasiness in dealing with the cognitive is apparent, and leads to the conclusion that 'we may get further more quickly in developing economic and stochastic theory than we will in developing the cognitive-behavioural theory'.[31] The marked similarity of this conclusion with that of Barrows a half century earlier is an important commentary on a common epistemological foundation which is not well suited to the examination of social and cognitive processes. To the positivist, the subjective has been seen as metaphysical, and therefore unknowable, irrational or private, and beyond the range of theory. Alternatively, the mental world has been reduced by the kind of psychologism challenged by Husserl which, in its imitation of natural science, destroys those situational aspects which are integral to the meaning of experience. 'In the realm of the mental we cannot understand the whole from the parts.'[32]

Behavioural geography has tended to follow the model of psychologism and it is for this reason that Olsson has described the field as still-born. In the passage from spatial fact to social process an important threshold is crossed as the objective world of facts and material phenomena is joined by the subjective world of ideas and values, and the logical language for discussing the former does not necessarily cover the latter: '. . . in the oblique realm of intentions, hopes and fears, two times two is not always equal to four'.[33] The deterministic (or stochastic) relations implicit in the positivist perspective suggest a precision that is foreign to the world of human action which is more properly characterized by relations which are fuzzy, ambiguous and evolving, where 'rational' behaviour is clouded by a myriad of subjective influences. These influences are rarely revealed by the types of group variables contained in the census or mass surveys; these variables are convenient for scientific analysis but rarely salient for human decision-making. To understand social process one must encounter the situation of the decision-maker, which includes incomplete and inconsistent information, values and partisan attitudes, short-term motives and long-range beliefs.

The phenomenological traditions

The preceding discussion has isolated two methodological positions, the one concerned with holism, a man–environment dialectic, and the incorporation of social and cognitive variables, and the other committed to material phenomena, explicit reductionism and implicit determinism, and analysis which separates fact from value. In each instance there has been a tendency for human geography to pursue the latter rather than the former route, so that even initially man-centred fields such as Vidalian geography, Chicago sociology, and behavioural geography have evolved toward the inappropriate format of physical science.[34]

The debate between these two positions should be drawn into its much broader epistemological context. The emergence of the Vidalian school as a corrective to the implicit materialism and determinism of Ratzel was a local example of a widespread protest against the scientism of the late nineteenth century. Dilthey's emphasis on historical consciousness, Weber's methodology of *Verstehen*, Mannheim's sociology of knowledge, were all reactions against a too severe positivism which in disassociating subject from object, had removed the human context from behaviour, and even from life. In North America a second and largely independent scholarship represented by Dewey, Cooley and Mead similarly reacted against the strictures of stimulus–response behaviourism and founded what was later to be known as symbolic interactionism with its concern upon the essential meaningfulness of human interaction. Though there is often a difference in emphasis among these writers, their position is often parallel with, if not derivative of, their contemporary, Edmund Husserl, founder of phenomenology:

> In the second half of the nineteenth century the world-view of modern man was determined by the positive sciences. That meant a turning away from the questions which are decisive for a genuine humanity. . . . Mere factual knowledge makes for factual men. . . . In our desperate need, this science has nothing to say to us.[35]

The second generation of phenomenologists, including Merleau-Ponty and Alfred Schutz, have maintained its fundamental critique of positivism, while also synthesizing the phenomenological traditions; Schutz, for example, has successfully built a bridge between Weber and Husserl, and thus established a firm philosophical underpinning for Weber's interpretative sociology.[36]

If phenomenology begins in radical protest, its assertion of 'back to the things themselves' is a constructive rebuilding of the relations between subject and object. Action is regarded as intended, as meaningless when divorced from its subject. Thus phenomenology is a holistic philosophy: 'The world is all in us, and I am all outside myself.'[37] Merleau-Ponty and Schutz have also been prominent in asserting the primacy of human relations, of encounter with the world; before them, Mannheim had emphasized the relational nature of knowledge, its 'existential relativity' to a historical era. Likewise Schutz's concept of multiple realities points to the relational nature of both objects and concepts; every object, including knowledge, is an object for a subject.[38] It follows then that in all phenomenological traditions the question of meaning is a central concern, for meaning and perception speak of existence, of a subject in encounter with an object.

This summary account of some of the themes of phenomenology serves to locate the discussion of social geography within its philosophical context. The Vidalian commitment to holism, and reciprocity between man and environment, subject and object overlapped with this philosophical stream in its disagreement with Durkheim's positivism. Sympathetic understanding as a mode of enquiry, the creative role of man, the use of artifacts as products, and the interest in the genius or personality of place which is the spatial equivalent of Mannheim's *Weltanschauung*, the global outlook of an historical era, were all features of Vidalian geography shared with phenomenological enquiry. These remarks apply equally to the early Chicago school, and here the philosophical influence is more direct, for Park and his colleagues were in close contact with the work of Dewey and Mead.

But what of the present day? Does the phenomenological stream continue to offer an underpinning to a reinvigorated social geography concerned with group action which culminates in a landscape effect? The shortcomings of the positivist approach in behavioural geography have been suggested. What is the phenomenological alternative?

Social science as social behaviour

Since its beginnings with Husserl, the first task of phenomenology has been the unmasking of the founding asssumptions of alternative philosophies; foundational criticism is often regarded as one of the ways to phenomenology.[39] In this manner it can be shown that at the root of an empirical science, there are necessary taken-for-granted assumptions, the same subjective naïvety as occurs within our own private life-worlds. The revelation of the subjective at the root of a science which empirically rejects the role of subjectivity, prepares the way for a constructive synthesis in which subject and object are re-united as they are in naïve experience. A phenomenological examination of social science thus begins with an analysis of presuppositions, with the exposure of assumptions which are unselfconsciously taken for granted.

In science as in life, the subjective basis for the objective is most visible in circumstances where an individual or group encounters the unknown, and into the informational vacuum projects its own definition of reality.[40] An instructive case was the geographical misinterpretation endemic to many scientific expeditionary parties of the eighteenth and nineteenth centuries. An example was the search for the Southern Continent, an intellectual myth which was perpetuated for centuries as Pacific explorers resolved ambiguous cues in the physical environment before the clarity of their own unambiguous mental image. Thus low clouds on the horizon were *interpreted* as the edge of a continuous land surface, the presence of sea birds and floating vegetation were *regarded* as indicators of the proximity of a landmass, and the island archipelagoes were *perceived* as its northernmost promontories and offshore islands. Exploratory behaviour became an amalgam of fact and value, object and subject. To contemporaries the image of the Southern Continent was neither idiosyncratic nor whimsical; it was a socially constructed reality, endorsed and maintained by intellectual and scientific fiat.

These processes do not disappear in recent intellectual exploration. Consider for example the unchanging social science image of black America. The attempt by intellectual explorers to order their own conceptual uncertainty has characteristically been from a distance with the use of secondary sources. At whatever period they have written, social scientists have claimed a contemporary new unity in the black movement.

At each period social scientists imposed organization upon the black community, claiming that at the time of *their* writing new mobilizing forces were at work. Nobody recognized that what was happening was not a new awakening of black America in their day, but rather a cyclic reinforcement and perpetuation of a tenacious image of social science. It was the socially endorsed image which was highly organized and possessed a clear teleology, not black America. It is difficult to account for the disarray revealed by more perceptive participant observation studies in black inner-city neighbourhoods if real forces of unity have been active for at least 50 years. Social scientists, in their own search for order in chaos, imposed a conceptual structure which was then unwittingly projected on to the black community. Scientific behaviour then followed the pilot light of its own misplaced image.

It is from this perspective that the black sociologist Billingsley comments that 'American social scientists are much more American than social and much more social than scientific'.[41] Subjective in-group values have pervaded objective analysis; indeed we learn as much about the cognitive categories of the researcher as we do of the 'facts' he is examining. The *a priori* 'outside view' of behaviour acts as a mirror reflecting the social values of the observer. Behind the objective is the subjective, the mind which selects it, names it, classifies it, interprets it, or, alternatively, dismisses it. The check against such bias is, of course, to bring back scientific theory against the test of common experience. But when the everyday world is perceived only from afar by an objective observer, and its fabric is defined in secondary sources by easily derived variables of convenience, testing is often far from complete, and the 'sighting' of a Southern Continent too easily occurs.

In social science neither questions of technique nor of concept are endorsed in a social vacuum. As we now become a stranger to them, is this not exactly what we see in parts of the methodological fervour in geography during the 1960s? The diffusion of an innovative technique was as much a matter of in-group endorsement as it was of pragmatic application to the everyday world.[42] Or consider spatial theory with its isolation of transportation costs and behaviour characterized by distance minimization. Why these? Is overcoming distance *the* central pre-occupation for man in his selection of home and workplace, or as he shops for clothes, or is it a socially underscored variable in the collective mind of a group of mutually attendant social scientists?[43] Perhaps there is a deeper irony than we have yet recognized to the oft-quoted maxim of geographers who follow the natural science model of explanation: 'by our theories you shall know us'.[44]

The intersubjective life-world

Even in the most attentively objective of behaviours – the scientific – it is impossible to disregard the powerful influence of the subjective. Geographical 'thinking as usual', the taken-for-granted building blocks of the scientific method, reveal the full interplay of fact and value within a milieu of intersubjective social conventions. As social geography follows its agenda and dips beneath spatial facts and the unambiguous objectivity of the map, it encounters the same group-centred world of events, relations and places infused with meaning and often ambiguity. Husserl, in his later writing, characterized this realm as the *life-world*. More recent philosophers like Schutz and Merleau-Ponty have urged that this reality encompassing mundane experience is not irrational and impossible to study; but neither, they argue, should it be investigated with an inappropriate methodology which violates the integral unity of the things themselves. The phenomenological method provides a logic for understanding the life-world. Its basic question becomes 'what does this social world mean for the observed actor within this world and what did he mean by his acting within it?'[45] Actions are intentional and purposive, they have meaning, but access to this meaning requires knowledge of the motives and perception of the actor, his definition of the situation.[46]

Meanings are rarely fully private, but are invariably shared and reinforced in peer group action. Unlike the lonely wasteland of the economic rationalist and of some existential notions of man, phenomenological man is unavowedly social. His life-world is an intersubjective one of shared meanings, of *fellow men* with whom he engages in face-to-face *we-relationships*. These relationships are entered into by choice and show, in pure form, the familiar pattern of selective interaction between like-minded individuals, birds of a feather flocking together.[47] The social group is not of course autonomous in its decision-making, but is impinged upon to varying degrees by society at large. For some men, the macro social structure does not permit a wide range of action. As one graffiti artist of the American inner city put it:

> There isn't much choice of what to do. . . . I did it because there was nothing else. I wasn't goin' to get involved with no gangs or shoot no dope, so I started writin' on buses. I just started with a magic marker an' worked up.[48]

Each individual has a history and geography which imposes constraints within his life-world; so begins the dialectic between creativity and determinism, charisma and institution, a dialectic which for the geographer becomes that between man and place.

A second, and often more binding, set of constraints upon action in everyday life are forces internal to the life-world of the individual and group. In the process of group consolidation its collective view of the world becomes more telling on the individual, as he becomes successively more 'included' within it. So too his action becomes increasingly identified with group norms. At the extreme, a common reality is enacted by repeated interaction and shared tasks, a reality which becomes socially defined and may appear quite eccentric to the outsider who does not share its taken-for-granted norms.[49] The phenomenological model of man is one of a life-world with a group-centred reality.

Unlike economic man, social man has highly biased information sources, an information map characterized by a core area of familiar and credible subjective knowledge, which is succeeded by realms where knowledge is increasingly objective and typified, and passing finally into realms of ignorance.[50] This information map is closely related to behaviour, for the stock of knowledge is essentially pragmatic in origin and is acquired in achieving mastery over recurrent situations in the life-world; standard procedures, or *recipes*, are devised to deal with the more repetitive or routine matters. Not only is knowledge within the life-world incomplete, but also it can be contradictory; consistency and certainty of information are by no means to be assumed even for a single individual, for as social roles are changed so both the relevance and the interpretation of knowledge may shift.

Schutz's model of the intersubjective life-world can usefully be extended beyond informal groups to organizations. While recent analysis has stressed the manner in which organizations impersonally stand over man, yet they too retain a less visible internal environment, a taken-for-granted life-world which is the seed-bed for action and decision-making: 'the objectivity of the institutional world, however massive it may appear to the individual, is a humanly produced, constructed objectivity'.[51] The economic model of the firm is as incomplete as the lonely figure of rational man; as soon as analysis passes from organizational facts to the meaning of organizational decisions, the analyst enters the same fuzzy, intersubjective world.[52]

Decisions in the organizational world are also coloured by both an external and an internal environment. Organizations establish a tradition of goals, priorities and strategies and all of these reveal a corporate set of values, a climate which influences decision-making. Corporate action displays the social basis of this climate as past allies and familiar networks are activated in routine problem-solving.[53] A common world view is constructed through intersubjective we-relationships; isolation, however, permits the survival of alternative values even between different departments of the same institution.[54]

Perception is a central concern in understanding organizational behaviour: 'it is perhaps through the study of perception and information flows that progress can be made in understanding the process whereby organizations respond to their changing environment'.[55] Information fields are group-centred for organizations as for informal groups, biased in favour of socially and spatially proximate sources, and once again knowledge in the life-world need be neither complete nor consistent; 'in fact . . . consistency or completeness, at times, create problems in finding feasible solutions'.[56] Decisions are usually made following simple, standardized and pragmatic rules, following closely Schutz's model of the recipe. The social basis of decision-making is reflected by widespread mimicry in problem-solving, whereby organizations imitate the strategies of their 'fellow men', those which are proximate to them in social space.[57] Finally, we note how newcomers to an institution are socialized to its construction of reality, its view of the world. To some extent this is formalized in training and orientation, but probably of greater significance is the incremental experience of daily contact, as newcomers are attuned to the taken-for-granted world of the organization, and its definition of the situation becomes their own.[58]

Thus the social model of man derived from Schutz's theory of social action within the life-world is equally useful for charting the naïve realms of experience which constitute both the presuppositions and the source of scientific, informal and institutional behaviour. This is not to assert a new determinism where the social milieu necessarily controls individual behaviour, but rather to identify a set of baseline characteristics within the life-world from which there will invariably be local departures. The theory of social action provides a pertinent framework and underpinning for a social geography which examines the social and cognitive contexts antecedent to a spatial fact.

The meaning of place

One of Mannheim's achievements was to reveal the relational nature of history. This was not to present a relativism where everybody and nobody was right, but rather to stress that every historic truth was also a truth for a subject, that the existential relativity of historic facts should be remembered in their interpretation: 'history is a creative medium of meanings and not merely the passive medium'.[59]

There is a meaning to space as well as to time. Each place should equally be seen phenomenologically, in its relational context, as an object for a subject. To speak of a place is not to speak of an object alone, but of an image and an intent, of a landscape much in the Vidalian sense of the word.[60] Thus place always has meaning, it is always 'for' its subject, and this meaning carries back not only to the intent of the subject, but also forward as a separate variable prompting the behaviour of a new generation of fellow men and contemporaries.

But if a place is meaningless without a subject, so too a person removed from his own place is a man of uncertain identity.[61] This simple dialectic permits an understanding of the human dimensions of, for example, long-range migration. The distant metropolis is never perceived in the perfect material terms that the gravity model with its economic determinism would have us believe. The metropolis has a meaning, it is, in Park's words, a state of mind, and it is always this meaning for the subject that precedes action; creative decision-making is not pre-empted by a mechanistic gravity field. In the same manner, the newly arrived urban immigrant finds himself initially disassociated from his new environment; the place is not a place *for him*.[62]

In unravelling the personality of place, the interpretative geography of the French school, with (like Max Weber's) its strong empirical and intuitive base, was primarily concerned with the relationships between man and his physical environment. Vidal's successor, Max Sorre, recognized that in an urbanizing age the critical forces in defining *genre de vie* were increasingly becoming social rather than physical. Merleau-Ponty has commented that history is other people; in an urban age, geography too has become other people. Place is increasingly being defined in terms of relations between men; the province of British Columbia has an added meaning to a Toronto investment company since the electoral defeat of its socialist-leaning government, as does Mexico City to a potential New York vacationer named Goldstein since Mexico's alignment against Israel at the United Nations.

In contemporary urbanism a place may commonly have a multiple reality; its meaning may change with the intent of its subject, and a plurality of subjects may simultaneously hold a different meaning for the same place. Usually, however, a dominant meaning holds sway, and the landscape can then act in the phenomenological sense of a product, as an indicator of the subjective intentions which moulded it. In the same way, mundane and taken-for-granted features in the environment can point beyond themselves to local societal values. Wall graffiti, for example, present a mirror on intangible attitudes which govern everyday relations between groups and between groups and space in the American inner city. They also of course have an immediate audience and act as an indirect form of communication between social groups. Many landscape settings present such cues of appropriate behaviour within their bounds, through micro-design features and the creation of a 'mood' atmosphere.[63] The meaning of a place systematically attracts groups with similar interests and lifestyles; places are selected and retained by reflective decision-makers on the basis of their perceived image and stock of knowledge. The result is that the city becomes a mosaic of social worlds each supporting a group of similar intent, who in their habitual interaction reinforce the character both of their group and of their place.

This personality of place ranges in scale from the nation state to the neighbourhood church; any habitually interacting group of people convey a character to the place they occupy which is immediately apparent to an outsider, though unquestioned and taken for granted by habitués. Thus a newcomer to a place is under strong pressures to adopt the local world view, and empirically there is a tendency for such attitudinal convergence to occur.[64]

Within the life-world there is a spatial counterpart to the social designation of fellow men and contemporaries. At the simplest level in our experience space is partitioned into near and far portions, those that are private and public. Near space is well known, predictable and protective, anchored, though not limited, to the home; it 'concentrates being within limits that protect'.[65] The home is the core of the taken-for-granted world. Bureaucracies speak of housing programmes, of geometry and dollars; home is a word of experience, a word of the people.

There are circumstances in which private space extends beyond the home to a community. This extension is rarely determined by objective land use or socio-economic characteristics but through an intersubjective consensus derived from a shared perception of solidarity and identity.[66] A distinctive status on the map is not enough to define community; consider the comment of a frustrated resident in a spatially well-identified, ethnically homogenous black inner-city neighbourhood:

> The community? Huh, what is the community? There ain't no community. People are just watching out for themselves. . . . Who cares about the community? Give people a meal and a can of beer and they're satisfied. They don't need no community.

To have a meaning and thereby reality in experience, community must be an amalgam of fact and value; it must be a common object for a plurality of subjects.

The separation of subject from object is a characteristic of the anomie of the post-industrial metropolis. Dehumanized urban settings comprise little more than space, little more than geometry; they excite no commitment, they are not 'for' a collectivity. They are the landscape realization of a mode of analysis which has emphasized functional materialism; such analysis translated into a planning paradigm creates a landscape of objective meaning only, a form without a subject.[67] Both the capitalist and the socialist city display manifestations of a materialist philosophy which has forgotten man. In response, many contemporary grassroots movements are attempting to reinstate meaning to the land: the protection of historic buildings, the preservation of neighbourhoods, opposition to demolition for motorways and public works, asserting the sanctity of nature and open space; all are protests against the objectification of the land, its reduction to a fact which has been torn free of its human context. For some groups such as nationalist minorities in peacetime, and whole nations in time of war, the land is a surrogate for identity; 'if we lose the land, we lose ourselves'.[68] The relationship between identity and landscape is profound, though largely unexamined, and lends credence to the phrase of French phenomenologists that a man *is* his place. The deterministic implications of this watchword fade when we recall its dialectical dimension, that the meaning of a place is itself derived initially in relation, from the intent of a human group.

Conclusion

In conclusion the omissions of this paper should be acknowledged. There has been no discussion of a number of important methodological issues, including the nature of generalization in a phenomenologically based social geography, *Verstehen* as an explanatory method, the role of structural as against situational forces in social action, and the controversy of existence and essence in phenomenology. These are matters which require far more space than is possible in this paper.[69] It should be added also that phenomenological analysis is not prescriptive; it encounters reality as it *is*, not as it *should* be.

The objective of this paper has been more modest. It has argued that the present equivocal status of social geography is not a new problem but arose at least as early as Barrows's presidential address in 1922. It stems from the discipline's pre-occupation with material phenomena, economic forces, and the physical science model in explanation. Successively this tradition has diverted the Vidalian school, the Chicago school of urban sociology, and behavioural geography, as it has either disregarded the subjective or else has forced essentially subjective problems into an inappropriate *a priori* hypothetico-deductive cast. This is not a promising avenue for a progression beyond spatial fact to social process, a progression which leads from questions of form to questions of meaning and intent.

The first step in a reformulation is a radical description of the things themselves which recognizes the pervasive presence of the subjective as well as the objective in all areas of behaviour: the informal, the scientific, the institutional. The second is to adopt a philosophical underpinning which embraces both object and subject, fact and value. Phenomenology restores to these troubling dualisms the unity they carry in the everyday world; indeed it is exactly this taken-for-granted realm of experience which is its constant reference point. Third is the recognition that the life-world is not solitary but a place of fellow believers; intersubjectivity is the basis for a social model of man. Fourthly, place should be viewed in relation, as an amalgam of fact and value, comprising both the objectivity of the map and the subjectivity of experience. For the social geographer a materialist preoccupation with spatial facts is not enough; in Sorre's words: 'A geographer is not a collector of shells which are no longer the home of a living being.'[70]

As social geography delves beneath a spatial fact to a social group, these prescriptions will draw the researcher back to the ground of behaviour, to the world as it is naïvely known, 'back to the "forgotten man" of the social sciences, to the actor in the so1ial world whose doing and feeling lies at the bottom of the whole system'.[71]

Notes

1 For example, Watson, J. W. (1957) 'The sociological aspects of geography', in Taylor, G. (ed.) *Geography in the twentieth century* (London) 463–99; Pahl, R. (1965) 'Trends in social geography', in Chorley, R. and Haggett, P. (eds) *Frontiers in geographical teaching* (London) 81–100; Buttimer, A. (1968) 'Social geography', in Sills, D. (ed.) *Int. Encyc. Soc. Sci.* Vol. 6 (New York) 134–45; Claval, P. (1973) 'Problèmes théoriques en géographie sociale', *Can. Geogr.* 17, 103–12.

2 Jones, E. (1975) *Readings in social geography* (London) ref. on p. 2.

3 Pahl (1965) p. 81 (note 1).

4 Buttimer (1968) p. 138 (note 1).

5 The small and as yet exploratory literature introducing phenomenology to geography includes Relph, E. (1970) 'An inquiry into the relations between phenomenology and geography', *Can. Geogr.* 14, 193–201; Mercer, D. and Powell, J. (1972) *Phenomenology and other non-positivist approaches in geography* (Melbourne); Buttimer, A. (1974) 'Values in geography', *Ass. Am. Geogr. Commn Coll. Geogr. Resourc. Pap.* 24.

6 Vidal de la Blache, P. (1926) *Principles of human geography* (New York) p. 163. Compare, in interpretative sociology, Max Weber's dialectic between individual creativity and social institutionalization: Weber, M. (1968) *On charisma and*

institution building: selected papers, edited and introduced by Eisenstadt, S. N. (Chicago).

7 Berdoulay, V. (1975) 'Human geography versus social morphology', presented at Can. Ass. Geogr. meeting, Vancouver; Buttimer, A. (1971) 'Society and milieu in the French geographic tradition', *Ass. Am. Geogr. Monogr. Ser.* 6.

8 A North American echo of this debate appeared in a little-known paper: Hayes, E. (1908) 'Sociology and psychology; sociology and geography', *Am. J. Sociol.* 14, 371–407.

9 Buttimer (1971) p. 63 (note 7).

10 Buttimer (1971) p. 103.

11 For the source of this characterization, see: Brookfield, H. (1964) 'Questions on the human frontiers of geography', *Econ. Geogr.* 40, 283–303.

12 Indeed isolated extracts from Park, as from Vidal, could give the misleading impression of a deterministic inclination. For example, '. . . while temperament is inherited, character and habit are formed under the influence of environment'. Park, R. (1967) 'The city as a social laboratory', in Turner, R. (ed.) *On social control and collective behavior* (Chicago) 3–18.

13 Park, R. (1916) 'The city: suggestions for the investigation of human behavior in the urban environment', *Am. J. Sociol.* 20, 597–612.

14 Park (1967) p. 5 (note 12); for an outstanding example of an urban regional study see: Zorbaugh, H. (1929) *The gold coast and the slum* (Chicago).

15 The fullest studies of personality and milieu were carried out by other members of the Chicago school, notably Thomas and Znaniecki. See the useful review volumes: Thomas, W. I. (1966) *On social organization and social personality*, edited and introduced by Janowitz, M. (Chicago); Znaniecki, F. (1969) *On humanistic sociology: selected papers*, edited and introduced by Bierstedt, R. (Chicago).

16 Park (1916) (note 13).

17 For example, Firey, W. (1945) 'Sentiment and symbolism as ecological variables', *Am. sociol. Rev.* 10, 140–8; Jonassen, C. (1949) 'Cultural variables in the ecology of an ethnic group', *Am. sociol. Rev.* 14, 32–41.

18 Park, R. (1936) 'Human ecology', *Am. J. Sociol.* 42, 1–15. In effect Park retained the concept of the moral order even at the height of his social Darwinism, though this qualification was often overlooked by both outside reviewers and his students.

19 Ibid.; but again Park went on to moderate this strong claim, with the statement that the vagaries of human idiosyncrasy would make this development unlikely.

20 Barrows, H. (1923) 'Geography as human ecology', *Ann. Ass. Am. Geogr.* 13, 1–14, see p. 13 and pp. 7–8.

21 At this period, geographic determinism continued to be physical rather than economic as in human ecology. Compare Visher's 'Democracy is interfered with by exceptionally fertile soil . . .' in his remarkably deterministic introduction of social geography to a broader audience in 1932: Visher, S. (1932) 'Social geography', *Soc. Forces* 10, 351–4.

22 Jones, E. (1960) *A social geography of Belfast* (London).

23 Sorre, M. (1957) *Rencontres de la géographie et de la sociologie* (Paris); Buttimer, A. (1969) 'Social space in interdisciplinary perspective', *Geogrl Rev.* 59, 417–26.

24 For example, Kirk, W. (1963) 'Problems of geography', *Geography* 48, 357–71; Lowenthal, D. (1961) 'Geography, experience, and imagination', *Ann. Ass. Am. Geogr.* 51, 241–60.

25 Golledge, R., Brown, L. and Williamson, F. (1972) 'Behavioural approaches in geography: an overview', *Aust. Geogr.* 12, 59–79.

26 Harvey, D. (1969) 'Conceptual and measurement problems in the cognitive-behavioral approach to location theory', in Cox, K. and Golledge, R. (eds) 'Behavioral problems in geography', *N. West. Univ. Stud. Geogr.* 17, 35–67, see p. 39.

27 Harvey, D. (1969) *Explanation in geography* (London); see also Alonso, W. (1964) *Location and land use* (Cambridge, Mass.). In the frank introduction to this influential work in locational analysis, Alonso sets out the economic model of man upon which his theoretical structure is based; as Olsson has pointed out, the logic of Alonso's model has been retained in Harvey's recent socialist reformulations.

28 Mackenzie, R. (1931) 'Human ecology', in *Encycl. Soc. Sci.* Vol. 5, p. 314.

29 Park, R. (1967) 'The urban community as a spatial pattern and a moral order', reprinted in Turner (1967) op. cit. 55–68, see p. 56 (note 12); the fullest statement of the primacy of geography as spatial science appears in Bunge, W. (1966) *Theoretical geography* (Lund). The convergence of the two literatures is made explicit in Duncan, O. D., Cuzzort, R. P. and Duncan, B. (1961) *Statistical geography* (Glencoe, Ill.); Timms, D. (1965) 'Quantitative techniques in urban social geography', in Chorley and Haggett, op. cit. 239–65 (note 1).

30 Golledge, R. and Amadeo, D. (1968) 'On laws in geography', *Ann. Ass. Am. Geogr.* 58, 760–74; compare the critique of behaviouralism in other social sciences in Natanson, M. (ed.) (1973) *Phenomenology and the social sciences* (Evanston, Ill.)

31 Harvey (1969) p. 63 (note 26).

32 Mannheim, K. (1952) *Essays on the sociology of knowledge* (London) 82.

33 Olsson, G. (1974) 'The dialectics of spatial analysis', *Antipode* 6 (3) 50–62.

34 Compare here Zelinsky's recognition and protest against this trend in geography, and also his inability to offer an alternative: Zelinsky, W. (1975) 'The demigod's dilemma', *Ann. Ass. Am. Geogr.* 65, 123–43.

35 Quoted by Buytendijk, F. (1967) 'Husserl's phenomenology and its significance for contemporary psychology', in Lawrence, N. and O'Connor, D. (eds) *Readings in existential phenomenology* (Englewood Cliffs, N.J.) 352–64, see p. 364.

36 For this aspect of Schutz's work, a recent review appears in Williame, R. (1973) *Les fondements phénoménologiques de la sociologie compréhensive: Alfred Schutz et Max Weber* (The Hague).

37 Quoted by Spiegelberg, H. (1969) *The phenomenological movement*, Vol. 2 (The Hague) 551; also Samuels, M. (1976) 'Human geography and existential space', Dep. Geogr., Univ. British Columbia.

38 Schutz, A. (1945) 'On multiple realities', *Philosophy phenom. Res.* 5, 533–76.

39 For example, Zaner, R. (1970) *The way of phenomenology* (New York).

40 For a further empirical discussion see Ley, D. (1974) 'The black inner city as frontier outpost: images and behaviour of a Philadelphia neighborhood', *Ass. Am. Geogr. Monogr. Ser.* 7, esp. Chap. 9.

41 Billingsley, A. (1970) 'Black families and white social science', *J. Soc. Issues* 26, 127–42; for a generic discussion of such 'false universes' see Schutz, A. (1954) 'Concept and theory formation in the social sciences', *J. Philos.* 51, 257–74.

42 Note, for example, that the spatial and social diffusion of the new geography both in Britain and the United States during the 1960s show parallels with the diffusion of other phenomena whose adoption is heavily dependent upon social consensus: Whitehand, J. (1970) 'Innovation diffusion in an academic discipline: the case of the "new geography"', *Area 2* (3) 19–30; Lavalle, P., McConnell, H. and Brown,

R. (1967) 'Certain aspects of the expansion of quantitative methodology in American geography', *Ann. Ass. Am. Geog.* 57, 423–36.

43 For an example of the potentially distorting effects from overemphasis of the distance variable see Smith, C. (1976) 'Distance and the location of community mental health facilities: a divergent viewpoint', *Econ. Geogr.* 52, 181–91. Zelinsky's recent critique of scientism in geography (note 34) continually underscores the social nature of the geographical endeavour. The literature of symbolic interactionism offers a more formal development of the institutionalization of social norms: Shibutani, T. (1961) *Society and personality* (Englewood Ciffs, N.J.).

44 Harvey (1969) p. 486 (note 27).

45 Schutz, A. (1960) 'The social world and the theory of social action', *Soc. Res.* 27, 205–21, see pp. 207–8: This provides the most succinct overview of a significant part of Schutz's thinking; another useful review is Gurwitsch, A. (1962) 'The commonsense world as social reality', *Soc. Res.* 29, 50–72; for fuller statements, see Schutz, A. (1967) *The phenomenology of the social world* (Evanston, Ill.); Schutz, A. (1970) *On phenomenology and social relations*, Wagner, H. (ed.) (Chicago); Schutz, A. and Luckmann, T. (1973) *The structures of the life world* (Evanston, Ill.)

46 This well-known concept was developed by W. I. Thomas, one of Park's contemporaries at Chicago, and points again to the appropriateness of the situational philosophies of phenomenology and symbolic interactionism as underpinnings for the early empirical schools of Vidal and Park. See also Mercer, D. (1972) 'Behavioural geography and the sociology of social action', *Area* 4, 48–51.

47 This is a central tenet of symbolic interactionism; see, for example, Bulmer, H. (1969) *Symbolic interactionism* (Englewood Cliffs, N.J.); for empirical validation, see Laumann, E. (1973) *Bonds of pluralism: the form and substance of urban social networks* (New York).

48 Ley, D. and Cybriwsky, R. (1974) 'Urban graffiti as territorial markers', *Ann. Ass. Am. Geogr.* 64, 491–505.

49 For an excellent discussion, see Berger, P. and Luckmann, T. (1966) *The social construction of reality* (New York); for an empirical example, see Ley, D. (1975) 'The street gang in its milieu', in Rose, H. and Gappert, G. (eds) *The social economy of cities* (Beverly Hills, Calif.) 247–73.

50 Schutz and Luckmann (1973) esp. pp. 178–9 (note 45).

51 Berger and Luckmann (1966) p. 57 (note 49).

52 Cyert, R. and March, J. (1963) *A behavioral theory of the firm* (Englewood Cliffs, N.J.); Hamilton, F. (ed.) (1974) *Spatial perspectives on industrial organization and decision-making* (London). Despite the subjective nature of their data, these authors nevertheless largely follow the standard procedure in behavioural studies of working within a positivist framework; compare note 30.

53 See, for example, Hartmann, C. (1974) *Yerba Buena: land grab and community resistance in San Francisco* (San Francisco).

54 Weiner and Deak, for example, have shown the very different world views of the Connecticut state highways departments, and the state planning commissioners as they jointly plan the state's transportation network: Weiner, P. and Deak, E. (1972) *Environmental factors in transportation planning* (Lexington, Mass.).

55 Dicken, P. (1971) 'Some aspects of the decision-making behavior of business organizations', *Econ. Geogr.* 47, 426–37.

56 Cyert and March (1963) p. 78 (note 52).

57 For example, Mercer, J. (1974) 'City manager communities in the Montreal metropolitan community', *Can. Geogr.* 18, 352–66.

58 This would help explain the familiar pattern of co-optation of organizational watchdogs set up in the public interest. Repeated interaction with the organization being regulated leads to a convergence of attitudes, and ultimately to a sharing of its view of the world.
59 Mannheim (1952) p. 187 (note 32).
60 See, for example, Tuan, Yi-fu (1971) 'Geography, phenomenology, and the study of human nature', *Can. Geogr.* 15, 181–92.
61 See Schutz's analysis of the existential dilemma of the stranger: Schutz, A. (1944) 'The stranger', *Am. J. Sociol.* 49, 499–507.
62 The classic study of adjustment problems remains one of the early publications of the Chicago school: Thomas, W. and Znaniecki, F. (1927) *The Polish peasant in Europe and America*, 2nd edn (New York).
63 For example, Cressey, P. (1932) *The taxi-dance hall* (Chicago); Hugill, P. (1975) 'Social conduct on the golden mile', *Ann. Ass. Am. Geogr.* 65, 214–28.
64 This contextual effect has often been observed in the geographic literature, for example, Robson, B. (1969) *Urban analysis* (Cambridge); Schutz has examined it from the subjective perspective, Schutz (1944) op. cit. (note 61).
65 Bachelard, G. (1969) *The poetics of space* (Boston) p. xxxiii; also Tuan, Yi-fu (1975) 'Place: an experiential perspective', *Geogrl Rev.* 65, 151–65.
66 Clark, D. (1973) 'The concept of community: a re-examination', *Sociol. Rev.* 21, 397–416.
67 Sommer, R. (1974) *Tight spaces: hard architecture and how to humanize it* (Englewood Cliffs, N.J.).
68 The slogan on a banner at a recent rally for native land rights in Canada. Place may also lead to a stigmatized identity: for example, the wrong side of the tracks, the clues of certain regional dialects, and, as the television series reminds us, in Victorian households, below stairs as opposed to upstairs.
69 These questions are taken up in detail in two volumes currently in preparation: Buttimer, A. *Rhythm, routine, and symbol*; Ley, D. and Samuels, M. (eds) *Humanistic orientations in geography.*
70 Sorre (1957) p. 199 (note 23).
71 Schutz (1960) p. 207 (note 45).

4 K. H. Halfacree, 'Locality and Social Representation: Space, Discourse and Alternative Definitions of the Rural'

Excerpts from: *Journal of Rural Studies* **9**(1), 23–37 (1993)

Introduction

In his introductory editorial for this journal in 1985, Cloke spent some time examining the definition of the rural and concluded that further analysis was necessary before such an apparently straightforward issue – one that 'trip[s] very easily off the tongue' (p. 4) – could be satisfactorily resolved. A similar point has been made by Falk and Pinhey (1978) and, much earlier, by Bealer *et al.* (1965). Indeed, Gilbert (1982) observed that the definition of the rural has been in dispute for at least 70 years. This paper is meant as a contribution to this ongoing debate and as a companion piece to Hoggart's (1990) response to Cloke's challenge.

The paper is divided into two principal sections. In the first, I outline three broad approaches to defining the rural and conclude that only the last of these, discussing the problem of the rural as a locality, is adequate. The other two alternatives, namely descriptive definitions and definitions based on socio-cultural characteristics, rest upon a false conceptualization of space. However, there is yet another way of defining the rural, which can be regarded as existing in parallel to the locality-based definition, shifting attention away from a concentration on tangible space in favour of the non-tangible space of 'social representations'. This alternative focus provides the subject matter for the second section of the paper, where I introduce the theory of social representations and its relevance to the rural definition debate. I conclude by suggesting, somewhat provocatively, that the social representation of the rural may be assuming increasing significance *vis-à-vis* the rural as locality as we move into an increasingly post-modern era.

The rural as space: description, spatial determinism and rural localities

Descriptive definitions

Just as there are 'urban areas', 'residential areas', 'suburban areas' and a host of other types of area, so too can we define 'rural areas' according to their socio-spatial characteristics. This way of defining the rural concentrates upon that which is observable and measurable and, hence, leads to descriptive definitions. Such empiricism accepts that the rural exists and concerns itself with the correct selection of parameters with which to define it (Pratt, 1989).

The most widely known example of this type of definition is Paul Cloke's 'index of rurality for England and Wales' (Cloke, 1977, 1978; Cloke and Edwards, 1986). This developed a classification scheme devised by the Department of the Environment (DOE, 1971) and is based upon 16 census variables covering employment, population, migration, housing conditions, land use and remoteness. Degrees of rurality were obtained from the scores calculated using a principal components analysis of the variables. These indices were divided into quartiles in order to simplify analysis.

Table 4.1 lists a number of other descriptive definitions of the rural, each focusing on a different aspect of the socio-spatial environment which the author felt was the 'essence' of rurality. Typical uses of these definitions included justifying a study as being 'rural' and the production of national maps of rurality. With notable exceptions, the definitions were not utilized within a more theoretical framework for distinguishing the rural from the urban but concentrated upon their own coherence.

In spite of the technical consideration which went into the production of these varied definitions, there are a number of criticisms which can be levelled

Table 4.1 Descriptive definitions of the rural

Statistical	Cloke, 1977, 1978; Cloke and Edwards, 1986 Openshaw, 1985 Harper, 1987a; Donnelly and Harper, 1987
Administrative	Stevens, 1946 Lassey, 1977 OPCS in Openshaw, 1985
Built-up area	Cherry, 1976 Denham, 1984 OPCS, 1986 Craig, 1987
Functional regions	Coombes *et al.*, 1982; CURDS, 1983–84; Champion and Coombes, 1984; Champion *et al.*, 1984, 1987
Agricultural	Vince, 1952 Robertson, 1961 Wibberley, 1972 Newby, 1977, 1978, 1979, 1980, 1987; Newby *et al.*, 1978; Newby and Buttel, 1980 Friedland, 1982 Gilbert, 1982
Population size/density	Clout, 1972 Lewis, 1979 Urry, 1984 Fothergill *et al.*, 1985

Sources: Halfacree, 1992, pp. 63–6; Cloke, 1985, p. 4.

at them (see also Halfacree, 1992, pp. 66–73; Clark, 1984; Openshaw, 1985; Openshaw *et al.*, 1980). These include the historical relativism of the classifications; the sensitivity of the classifications to the definitions of the variables, the choice of variables, the data quality and the statistical technique; the question of the scale at which the rural is defined and the importance of context; and the neglect of qualitative data. By way of a summary, it can be argued that each of the definitions given in Table 4.1 relate to a specialized use and are therefore not a general measure of rurality. Hence, statistical definitions are geared towards socio-economic studies; administrative definitions towards political studies; built-up area definitions towards land-use studies; functional definitions towards economic studies; agricultural definitions towards land-use and social relations studies; and population density definitions towards service provision studies. More generally, the definitions are all geared towards various planning and academic purposes (Craig, 1988; see also Rodwin and Hollister, 1984).

This last observation is not really a criticism of the descriptive definitions themselves but it cautions us against using any one of them as *the* means of defining the rural, even if the 'technical problems' could be overcome satisfactorily. The 'definitions' are better seen as research tools for the articulation of specific aspects of the rural than as ways of defining the rural. Their methods involve trying to fit a definition to *what we already intuitively consider to be rural*, in the absence of any other justification as to why they should be regarded as representing the rural. In other words, they are trying to put the cart before the horse, the rural having been already 'defined' by those doing the classifying. As Shields expresses it, in the context of discussing similar attempts to 'define' the 'Far North' of Canada:

> the appeal to popular perception is indicative of a tautological circle in all of these studies: starting out from commonsensical intuition, statistics are gathered and then interpreted in the light of commonsense. Thus ennobled by the clothes of empiricism, commonsense is represented as scientific conclusions (1991, p. 168).

Descriptive methods only describe the rural, they do not define it themselves. Attention must thus be given to *what* it is that they are trying to articulate, a point to which we shall return after considering the socio-cultural way of defining the rural.

Socio-cultural definitions

A second approach to defining the rural concentrates on highlighting the extent to which people's socio-cultural characteristics vary with the type of environment in which they live. In short, socio-cultural definitions of the rural assume that population density affects behaviour and attitudes (Hoggart and Buller, 1987). This assumed correlation between social and spatial attributes has been a strong force in the organization and development of rural studies but has changed in character and sophistication over the years.

In 1938, Louis Wirth set forth his proposal that 'urbanism' represented a distinct 'way of life'. 'Urbanism' was characterized as being dynamic, unstable, mobile in stratification and impersonal, with contacts being determined by one's precise situation at the time (work, home, leisure). Counterpoised to this is ruralism, characterized by stability, integration and rigid stratification, with individuals coming into contact with the same people in a variety of situations. Wirth's dichotomy was one of many proposed throughout the early part of this century in an attempt to distinguish between urban and non-urban societies, in order to map the rapid pace of social change. The principal dichotomies are listed in Table 4.2. These theoretical ideas influenced some of the early studies of 'rural' communities, such as Rees (1950) and Williams (1956), and have been reviewed by Bell and Newby (1971, 1974) and Symes (1981). However, after pioneering studies by Williams (1963) and Littlejohn (1963), the emphasis on integrated rural stability was increasingly challenged (Symes, 1981).

Researchers soon realized that there was no simple dichotomy between 'rural' and 'urban' areas. Instead, they recognized a variety of communities conforming to various levels of urbanism and ruralism (as defined by Wirth and others). Therefore, the notion of a rural–urban continuum was devised (Redfield, 1941), which positioned settlements along a spectrum ranging from very remote rural areas, through transitional areas, to the modern city.

Table 4.2 Some urban/rural dichotomies

Author	*Urban category*	*Non-urban category*
Becker	Secular	Sacred
Durkheim	Organic solidarity	Mechanical solidarity
Maine	Contract	Status
Redfield	Urban	Folk
Spencer	Industrial	Military
Tonnies	*Gesellschaft*	*Gemeinschaft*
Weber	Rational	Traditional

Source: Reissman (1964), in Phillips and Williams (1984), p. 10.

Frankenberg (1966) and Duncan and Reiss (1976) provide a series of studies supposedly situated at various points along this continuum. These studies, in particular, illustrate the continuum's usage as a classificatory device as well as a measure of development.

Pahl (1966, 1970) pioneered the discrediting of the rural–urban continuum principle within British studies. He pointed out that previous work had neglected 'urban' aspects of rural society. Instead, he proposed that we should concentrate on the people living in rural areas rather than on rural areas themselves. Hence, he outlined the characteristics of various 'social groups in the countryside', with the balance between these groups in any specific place being the key. None the less, whilst there are considerable benefits to be gained from concentrating on the 'groups' or, more specifically, the classes which comprise rural populations, such a focus manages to avoid providing any definition of the rural itself, unless it is statistically associated with a certain mix of groups.

Consolidating various criticisms to be made of the rural–urban continuum, Newby (1986) points out how the idea that population density, population size and the general social and physical environment gave rise to a distinct form of society was rapidly demolished by the weight of contradictory evidence. The sociological characteristics of a place could not simply be 'read off' from its relative location on a continuum. For example, Young and Wilmott (1957) found supposedly 'rural' societies in the East End of London (see also Gans, 1962), whilst Pahl (1965) and Connell (1978) found 'urban' societies in Hertfordshire and Surrey, respectively. Nevertheless, the continuum idea survived well into the 1970s (for example, Carlson *et al.*, 1981), aided by the institutional framework in which rural studies was embedded, contributing to what some consider to be a crisis in rural social science (Newby, 1980; Newby and Buttel, 1980; Bradley and Lowe, 1984). Even today, the idea of the urban penetrating the rural community lingers (Hoggart and Buller, 1987); Hoggart (1988, 1990) sees Cloke's index as an expression of continuum ideas.

Whilst neither simple rural–urban dichotomies nor more complex continua work well empirically, these approaches can also be criticized as being theoretically flawed. Studies reflected the penetration of geographical (environmental) determinism into social science, whereby human behaviour and character is determined by the physical environment in which it exists. The specific development of such determinism in rural studies was aided by the combination of structural–functional anthropology and the romanticist defence of supposedly rural culture and traditions (Newby, 1986). Briefly, this isolated the rural as symbolizing a stable, harmonious community, everything 'positive' that urban life seemed to lack. This was a myth, since the determinists had a misconception of space (and time). Instead, we need an appreciation of rural space which neither prioritizes its empirical structure nor relies upon a false dichotomy between space and society. Defining the rural comes after this.

Conceptualizing space

The principal theoretical criticism of both the descriptive and socio-cultural definitions of the rural is that they demonstrate an erroneous conceptualization of the relationship between space and society. There has recently been a very active debate concerning this relationship (for example, see Gregory and Urry, 1985) and, rather than start from a basic philosophical discussion of space (see Harvey, 1973; Smith, 1984), I shall advocate and seek to justify a position which claims that whilst space has no inherent causal powers (i.e. it is not absolute), whereby a spatial formation can give rise to social practices, neither can it be reduced to the sum of relationships (distances) between objects (i.e. it is not relative). Instead, space and spatial relations are both expressions of underlying structures – space is produced (Smith, 1984) – and a means of creating further spaces – space is a resource (Smith, 1981). This conceptualization of space aims to overcome both spatial determinism, inherent in socio-cultural definitions' association between the rural and a distinctive 'way of life', and spatial indifference, inherent in descriptive definitions' ontological shallowness (Bhaskar, 1975, 1979; Keat and Urry, 1981; Sayer, 1984, 1985; Gregory, 1985).

Spatial (or geographical or environmental) determinism holds an absolute conception of space, whereby space itself possesses causal powers. The rural–urban continuum is an example of such thinking, whereby properties inherent in the rural environment were thought to produce a distinctive rural character. Whilst this 'environment' includes social aspects, therefore going beyond the physical confines of 'environmentalism' (Gold, 1980), the environment is still treated as a 'given' factor, rather than something which is produced (ibid.; Smith, 1984).

In rejecting all that had gone before, many social scientists, particularly in the 1970s, relegated space to the position of being a mere 'reflection' of society, an unimportant detail (Massey, 1985). An example of this 'spatial indifference' comes in a paper by Short (1974) who, following the entreaties of Castells (1977), investigated the city as a spatial product of the economic, political and ideological levels of society (cf. Gregory, 1978a, pp. 118–120). In brief, this alternative argues that environments do not produce societies as these are ultimately defined by the dominant mode of production (Gilbert, 1982). Such a framework could also be used to investigate the rural, but is open to at least three criticisms.

First, although this perspective emphasizes how space is produced (Smith, 1984), it neglects consideration of the extent to which *all* space is produced. The idea that all space is reducible, even ultimately, to the capitalist mode of production appears unduly deterministic and restrictive of human agency (see Sayer, 1989). This criticism applies even if we can still distinguish between socially produced space (spatiality), physical space and mental space (Soja, 1985; cf. Smith, 1984).

Secondly, and similarly, emphasis on 'society' as a whole leads to a neglect of the diversity which exists at a host of levels. Even disaggregating society into a number of levels, which together might be shown to give rise to a distinctive rural society, is inadequate, notwithstanding the need for some generalization. Gould has commented on how generalization destroys so much of what is of interest, asking 'how can we ever believe in the search for laws of human behaviour again?' (1982, p. 78). Reductionism to the lowest common denominator is wide but not very deep.

The third and most serious criticism of spatial indifference is that this conceptualization fails to appreciate the dynamism of space. This comes about through its treatment of society and space as a dualism rather than as a duality (see Giddens, 1984). Whilst it acknowledges that space is produced, it leaves the matter there: space is the *residue* of structures. (This implication also applies when it is accepted that there is no identity between society and space, and even if adequate flexibility of spatial form is allowed.) From this perspective, although looking at space might be useful in providing clues to the underlying structures, study of space itself has little merit.

It has become increasingly clear that indifference to space led to it being unduly marginalized (Massey, 1985). For example, Castells was concerned that:

> It has been a custom in the recent literature . . . to use the formula according to which space is expression of society. Although such a perspective is a healthy reaction against the technological determinism and the short-sighted emphasis too frequently prominent in the space-related disciplines, it is clearly an insufficient formulation of the problem . . . Space is not a 'reflection of society', it *is* society . . . (1983, p. 4, emphasis in original).

Two closely inter-related issues are involved here. First, space is produced, reproduced and transformed by society (Smith, 1984; Harvey, 1985) but this does not make it an 'environment' of social residues, since society only has material existence in a spatio-temporal form; space is society. *Space* represents the meshing together of structures but it also *delineates the structures themselves* (Giddens, 1984). Only in total abstraction are structures aspatial (Sayer, 1985); at the concrete level they take on spatial form, as determined by the spatial arrangement of the contingent forces which activate and sustain them. Giddens (1984) draws attention to the significance of this 'stretching' of social systems, or 'time–space distanciation', for the organization of society and power. In this way, Soja is correct in saying that 'social life is both space-forming and space-contingent' (1985, p. 98), with the social world consisting of 'interdependent, mutually modifying, four-dimensional, space–time entities' (Urry, 1985, p. 28).

Secondly, a dynamic theory recognizes that the space produced is not simply dumped aside as unimportant but is used to (re)produce space, to (re)produce structures and ultimately, to (re)produce society. As Harvey recognized:

> Geographical differentiations . . . frequently appear to be what they are not: mere historical residuals rather than actively reconstituted features within the capitalist mode of production (1982, p. 416).

More fundamentally, space is essential for sustaining structures since, to remain current, 'general structures do not float above particular contexts but are always (re)produced within them' (Sayer, 1989, p. 255). Hence, space can be regarded as a *means of production* (see Lefebvre, 1976), most obviously in the transportation industry but also more generally in relation to the quantitative and qualitative form of the land involved in the production process (Smith, 1984, pp. 86–86). The spatial properties of matter are a crucial part of its use value, where:

> The use-value of an object is not inherent in the thing itself but is determined in practice by the object's utility, yet at the same time the use-value has no existence separate from the object. Concerning space, two things follow from this. First, space and spatial properties have no meaning separate from concrete pieces of matter, and second, geographical space is nothing independent of its use (Smith, 1981, p. 115).

In other words, when we consider (rural) space, we must not only consider the structures producing that space but also the way in which that space is subsequently used to produce other space and, fundamentally, to reproduce the original causal structures themselves. Of course, there is still room for the existence of 'unused' relic space, the legacy of past structures, but this space is to be treated as any other object with no use value.

The rural locality

> There is no rural and there is no rural economy. It is merely our analytic distinction, our rhetorical device. Unfortunately we tend to be victims of our own terminological duplicity. We tend to ignore the import of what happens in the total economy and society as it affects the rural sector. We tend to think of the rural sector as a separate entity which can be developed while the non-rural sector is held constant. Our thinking is ensnared by our own words (Copp, 1972, p. 519, quoted in Newby, 1980).

If we are to develop a conception of rural space within social science based on the principles outlined above then we have to satisfy two criteria (Hoggart, 1990, p. 248; see also Cloke, 1980, p. 197, and 1985, p. 6):

1 There are significant structures operating which are unambiguously associated with the local level.
2 Looking at these structures enables us to distinguish clearly between an urban and a rural environment.

In short:

> the designation rural would have to identify locations with distinctive causal forces. If these specifications are met, then genuinely local, distinctively rural causal forces can be said to exist (Hoggart, 1990, p. 248).

There is considerable dispute as regards the extent to which the two criteria given above are fulfilled. For example, whilst the current industrial restructuring of space appears to exhibit a rural–urban dimension (for example, see Gould and Keeble, 1984; Fothergill *et al.*, 1985), this appears to be more a reflection of properties contingently found in what we regard as rural areas than an inherent reflection of 'the rural'. Movement of industry out of the cities in recent years reflects, *inter alia*, the search for lower labour costs. If labour costs were to fall in the cities then we might expect much of this outflow to be stemmed and reversed. In other words, the 'ruralization of industry' (Fothergill *et al.*, 1985) appears to be a very precarious phenomenon. The crucial issue is the movement to *low-wage* areas; whether these are 'rural' or 'urban' is largely contingent (see Massey, 1984). Similarly, Moseley (1980a, b) highlighted the overlap between rural and urban deprivation, whilst Friedland (1982) pointed out the need to see the urban-to-rural population turnaround as being urban-based (see also the examples of 'overlap' given in Hoggart, 1990).

In general, the structuring and restructuring of space and society does not distinguish necessarily between rural and urban areas. The key feature is that it is an *uneven* development, and can take place at a variety of spatial scales. Indeed, when Marx discussed the tendency of space to be annihilated by time, he possibly did not realize the subsequent significance of this process. The separation of town and country, which was a sedimented division of labour, has for many purposes been obliterated (Smith, 1984). The implications from the 'restructuring thesis' (for example, Massey, 1984) is that abstract spatial criteria now have decreased significance compared to factors such as the relative quality and cost of human labour power in different places (Marsden and Murdoch, 1989).

In most circumstances, such as employment, leisure and shopping, the created space of the rural is absorbed within a created space which transcends any rural–urban divide (for example, Coombes *et al.*, 1982; CURDS, 1983–84; Smith, 1984, p. 136). This is because the spatial 'moments' (dimensions) of the processes with which we are concerned themselves transcend the rural–urban distinction. Indeed, many transcend the distinction between cities and exist within the created absolute space of the nation state, whilst others are still more transcendent. Like a distorted Russian doll, there are a myriad of sructural levels in society, attached to which are distinct geographical spaces

(Urry, 1985) or 'fields of effectivity' (Gregory, 1985). For Thrift, the urban-to-rural shift in population merely reflects the infilling of urban space, since 'the South of England is now one vast created and manicured urban–suburban space' (1987, p. 77).

Giddens (1984 and elsewhere) says much the same, although he expresses it somewhat more relativistically when he claims that there are dominant 'locales', defined as settings for interaction (created absolute spaces), which are made heterogeneous through 'regionalization', either temporally, spatially or temporal–spatially. Whilst in a 'class-divided society' the dominant locale organization centred around the 'symbiosis of city and countryside', by the time we get to a 'class society' (capitalism) the dominant locale organization is the 'created environment', for which most authors use the somewhat confusing euphemism 'urban'.

Rural *localities* (see Urry, 1986; cf. Newby, 1986; Jonas, 1988; Duncan and Savage, 1989), if they are to be recognized and studied as categories in their own right, must therefore be carefully defined according to that which makes them *rural*. None of the previously discussed definitions has adequately achieved this, although the rigorousness introduced by the society–space debate should enable us to further this goal more adequately in the future.

What features are likely to assume prime importance in the creation of such a definition (see also Pahl, 1970; Moseley, 1980b; Cloke and Park, 1984)? First, there is the link between the rural and the *agricultural* (Newby, 1978). From an economic viewpoint, this is to associate rural space with primary production (Gilbert, 1982) or with 'the competitive sector' (Hoggart and Buller, 1987). However, many urban localities could be similarly classified (ibid.). The social distinctiveness of agriculture necessary for such a definition has also been questioned (Bradley, 1981; Barlow, 1986; Hoggart and Buller, 1987).

Secondly, there is the connection between the rural and issues of 'collective consumption', the key being the effect of *low population densities* (Urry, 1984). This is clearly very socially and historically specific and its overall importance is debatable, especially given the decline in the importance of the friction of distance (Hoggart and Buller, 1987; Marsden and Murdoch, 1989).

Thirdly, following Bradley (1981), Thrift (1987) and Marsden and Murdoch (1989), we may wish to stress rural space's more general role in *consumption*, whether collective (for example, national parks) or privatized (for example, residence). As such, this requires us to more fully examine competing social representations of rurality (see Mormont, 1987), as introduced in the next section (see also Halfacree, 1992, pp. 110–203).

Not being convinced by any of these arguments, Hoggart (1988, 1990; Hoggart and Buller, 1987) wants to 'do away with rural', as he sees it as a confusing 'chaotic conception' lacking explanatory power. Instead, we are better off looking at extra-rural structures and processes, notwithstanding that these are moulded by local circumstances. The key issue for him is that these local processes are not generalizable into the terms 'rural' or 'urban'. 'Rural' places are not identical, with any similarities that they do possess not being significantly different from many features of 'urban' places:

> the undifferentiated use of 'rural' in a research context is detrimental to the advancement of social theory . . . The broad category 'rural' is obfuscatory, whether the aim is description or theoretical evaluation, since intra-rural differences can be enormous and rural–urban similarities can be sharp (Hoggart, 1990, p. 245).

In short, 'causal processes do not stop at one side of the urban–rural divide' (Hoggart, 1988, p. 36; see also Zelinsky, 1977). Nevertheless, if Hoggart's thesis becomes widely accepted then this does *not* have the necessary corollary that we have to assign 'the rural' to the dustbin of research history. This is because there is yet another means of defining the rural, which directs our attention to the realm of discourse and the theory of social representation.

The rural as social representation: 'invisible only to the clever'?

There is an alternative way of defining rurality which, initially, does not require us to abstract causal structures operating at the rural scale. This alternative comes about because 'the rural' and its synonyms are *words and concepts understood and used by people in everyday talk.* Copp's claim that 'Our thinking is ensnared by our own words' (1972, p. 519) is thus extremely astute and requires us to investigate the status of the rural in discourse since 'rural is just a symbolic shorthand (as are all concepts) by which we mean to encapsulate something' (Falk and Pinhey, 1978, p. 553). First, however, it is necessary to introduce the theory of 'social representations', which I believe provides a necessary theoretical background for such an investigation but has been largely neglected in the current theoretical debates within the 'rural' literature (but see Harper, 1987b; Falk and Pinhey, 1978).

The theory of social representations

The theory of social representations, as developed principally in the work of Moscovici (1976, 1981, 1982, 1984; see also Potter and Wetherell, 1987), attempts to outline how people understand, explain and articulate the complexity of stimuli and experiences emanating from the social and physical environment in which they are immersed (cf. Schutz, 1967). In contrast to orthodox social psychology, social representation theory rejects the idea that everyday behaviour involves a 'scientific' approach to objects, people and events, where understanding is merely information processing (Moscovici, 1984). This 'academic fallacy' (Thrift, 1986; see also Shotter, 1984) assumes that people act logically, systematically and rationally, as if guided by an 'inner, on-board computer' (Thrift, 1986, p. 87), abstracting from the everyday realities of the social world. Instead, social representation theory argues that renewed importance needs to be attached to the 'paramount realities of everyday life' (Schutz, 1970), an important component of which is what

Giddens (1984) has termed 'practical consciousness'. This is in line with Falk's and Pinhey's (1978) call for rural studies to pay greater attention to Schutz's constitutive phenomenology, with its concern with the socially and situationally specific 'actor's view of the world' (see also Harper, 1987b).

Following the 'symbolic interaction' tradition of Mead (1934), Moscovici (1984) proposes that people use social representations to deal with the complexity of the social world. These are defined as organizational mental constructs which guide us towards what is 'visible' and must be responded to, relate appearance and reality, and even define reality itself. The world is organized, understood and mediated through these basic cognitive units. Social representations consist of both concrete images and abstract concepts, organized around 'figurative nuclei' which are 'a complex of images that visibly reproduce . . . a complex of ideas' (ibid., p. 38). Therefore, whilst they are partly a description of the physical material world, social representations are irreducible to it. They are both iconic and symbolic.[1]

We use social representations in two main ways. First, they enable us to conventionalize the objects, persons and events encountered. Secondly, they help us to prescribe and organize our subsequent behaviour and responses (see also Brewer, 1988, p. 155, on 'descriptive vocabularies'). In Shields' (1991) terms,[2] they are both referential and anticipatory. Prescription/anticipation makes social representations not merely neutral and reactive but also creative and transformative, through their usage by people trying to go about their everyday lives:

> Rather than motivations, aspirations, cognitive principles and the other factors that are usually put forward, it is our representations which, in the last resort, determine our reactions, and their significance is, thus, that of an actual cause (Moscovici, 1984, p. 65).

This usage, which takes place in changing world contexts, makes social representations dynamic and ever-adapting to new circumstances.

Moscovici's social representations, although concentrating on signification and meaning, display similarities with the 'virtual' structures proposed by Giddens (1984) – in their lack of a subject, in the necessity for their constant reproduction, and in their status as 'rules' (interpretive schemes and norms) and 'resources'. Representations are not tied to specific 'individuals', be they people, objects or places. This is in contrast to studies in the spatial cognition literature, such as Lee (1968) and Lynch (1960), where spatial images tend to be related to specific city areas and, as a result, lose some of their powerful flexibility (Canter, 1977).

It is important to stress the *social* aspect of representations (Gregory, 1978b). They are consensual means of making the unfamiliar familiar, but this consensus is group-specific. Only those who share a representation will use it the same way, having a similar understanding and evaluation of the aspects of the world being covered (Potter and Wetherell, 1987). Moreover, social representations are also inherently social through their linkage to communication processes and because they require an agreed code for this communication (ibid.) Hence, issues of knowledgeability and the political economy of the structural acquisition of this knowledge (Thrift, 1985) are crucial to social representations.

Potter and Wetherell (1987; see also Potter and Litton, 1985) outline the difficulties of linking specific social representations with specific groups and of defining a satisfactory degree of consensus within such a group. They also question the overly cognitive status of the representations, as this underplays their context-dependency. In sum, they argue that the role of *language* has been underplayed and they introduce a rival concept, namely the 'interpretive repertoire'. This is 'a lexicon or register of terms and metaphors drawn upon to characterize and evaluate actions and events' (Potter and Wetherell, 1987, p. 138). The repertoire exists within discourse alone, rather than referring back to an image (see Gilbert and Mulkay, 1984).

Unlike social representations, interpretive repertoires are not intrinsically linked to social groups, these being constructed in the course of the accounts themselves. In addition, there is no search for consensus, as people frequently switch between repertoires in any account and use different aspects of the repertoire in different circumstances. Finally, there is no cognitive reductionism in repertoires, concern being almost solely with language use and function.

It would appear that both parties in this debate tend to overstate their respective cases. This makes them appear unduly incompatible. Criticisms of Moscovici are well-founded if the theory of social representations attempts to extend itself directly into the realm of discourse. However, in a similar manner, problems emerge when interpretive repertoires are extended back from the realm of discourse into the structural realm. The latter is attempted negatively through considering nothing apart from 'talk'. As Bowers (1988) points out, the whole tone of Potter's and Wetherell's (1987) book is relativist and over-idealist, resulting in an unwarranted rejection of realist ontology and truth claims. Excessive power is attributed to language as a creative force, neglecting its role as a communicator of information *about* the world, whether 'correctly' perceived or not. The groups which language creates usually have some material referent, whether people or things.

A development of this idealism is that Potter and Wetherell prioritize the 'hubbub' of language at the expense of the more 'structured' but 'quieter' mental realm of representation. The latter is less transient and context-specific than language, and provides the foundation for talk; interpretive repertoires thus have a referent. This is not to claim that discourse is determined by mental constructs but to assert that language is not as free-floating and self-referential as Potter seems to claim. There is more to the process of discourse than the shallowness of language. Moreover, as Moscovici points out, 'all that is image or concept does not entirely pass into language' (1985, p. 92) but nevertheless can influence the course of social interaction. Articulation of ideas within language should not be equated with articulation of ideas (Hewstone, 1985).

Both Moscovici and Potter neglect the socio-political processes involved in the production, reproduction and mediation of social representations (Semin,

1985; Bowers, 1988). They appear over-voluntaristic, although probably this mostly reflects the academic division of labour. As Moscovici makes clear, social representations are independent stimuli for social psychologists but require explanation within other disciplines (1984, p. 61). Nevertheless, this constraint does not justify the lack of attention given to the process of choice between competing social representations or interpretive repertoires in specific circumstances (Bowers, 1988). This issue, including the link between social representations and groups, requires that more attention is given to the 'political economy of social representations' than has been the case to date.

I would like to suggest that Moscovici's original theory of social representations needs to restrict itself to the realm of mental constructs – the disembodied structures – informing discourse and action. Similarly, Potter's interpretive repertoires should be confined to the realm of discourse, where social representations are drawn upon in a specific, context-dependent and partial manner, in order to pursue discursive actions. They may also be drawn upon non-discursively, as they inform physical actions as well. In short, there is an ontological difference between social representations and interpretive repertoires. The latter are the modalities of social representations within a specific instance of discourse; they link structures and interaction (see Giddens, 1984, p. 29). Both merit study, especially the content of the representations and the specificities of their usage.

Classifying discourses

Our attempts at defining the rural can be termed 'academic discourses' (Sayer, 1989) because they are the constructs of academics attempting to understand, explain and manipulate the social world (cf. Schutz, 1970). They are mostly carefully formulated 'theories', possessing a rigour not apparent in everyday 'accounts' (Thrift, 1986) or 'lay narratives' (Sayer, 1989).

Academic discourses are pivoted between talking about their objects (for example, the rural) and talking about lay discourses on these objects (see also Falk and Pinhey, 1978). Neglect of how lay discourses underpin academic discourses leads to a certain self-referentiality and, hence, to a circularity when attempts at definition are involved. For example, as illustrated earlier, in trying to establish an academic definition of the rural, many of the descriptive definitions referred back to our 'commonsense', intuitive idea of the rural: they tried to express the rural rather than define it.

Failure to recognize the non-objectivity underpinning academic discourses has a number of other negative implications. Most seriously, it overlooks the way in which different interests are promoted through the objectification of these discourses (Rodwin and Hollister, 1984). For example, Appleyard (1979) comments on how a 'professional and scientific' view of the environment suppresses meaning and how its deliberate de-symbolization legitimates dealing with it technically and apolitically. These issues have been most fully developed by Habermas (for example, Habermas, 1972) in his investigation of the cognitive interests implied by different types of science and forms of knowledge. Eliot Hurst (1980, 1985) has taken up this issue in his deconstruction of 'the church of geography'.

Hoggart's (Hoggart, 1988, 1990; Hoggart and Buller 1987) criticism of attempts at defining the rural can thus be seen as a call for a more direct engagement between academic discourse and its object (the rural). However, to proceed in this manner alone would be to neglect the lay discourses of rurality, which would be effectively marginalized as partial/inaccurate versions of the rural-as-locality object. Moreover, as Hoggart questions whether the latter has any significant existence, the lay discourses can be regarded as myths which need to be discarded.

There is a growing realization of the necessity of studying people's accounts of the form, content and context of their daily lives as the relationship between structure and agency is reassessed after the excesses of structuralism (for example, see Thrift, 1983; Giddens, 1984, 1986; Shotter, 1984). As Giddens puts it: 'the technical concepts of social science are, and must be, parasitical upon lay concepts' (1987, p. 70). Rural social scientists have at least been as guilty as anyone else in neglecting the ordinary person's view of the world, leading to a 'theoretical–empirical myopia' (Falk and Pinhey, 1978). The social world must be regarded as an ongoing accomplishment, not a taken-for-granted facticity (ibid). A retreat from theory is not involved here:

> Lay-knowledge or practical consciousness are only atheoretical in the sense that their conceptualisations and claims are relatively unexamined. Since lay-knowledge is both part of our object, and a rival account of it, our response to it must not be to dismiss it, but rather to examine it. There are no philosophically defensible grounds for using the term 'theory' as many (often otherwise sophisticated) social scientists do – where the only distinguishable feature of 'theoretical' terms is their unfamiliar or esoteric character in relation to lay vocabularies . . . [The] double hermeneutic of social science requires us to understand and negotiate with 'actors' accounts', not to dismiss them in order to maintain the purity of what we like to call theory (Sayer, 1989, p. 267).

Similar sentiments have been expressed by Geertz:

> Some of the most crucial properties of the world are . . . invisible only to the clever . . . [The] world is a various place . . . and much is to be gained . . . by confronting that grand actuality rather than wishing it away in a haze of faceless generalities and false comforts (1983, pp. 89, 234).

The theory of social representations allows us to reconcile Hoggart's doubts regarding the existence of the rural locality with this call for greater attention to be paid to lay discourses. This is because *'the rural' should not be seen as a singular object of discourse*: Hoggart's object of discourse is the rural as locality, whereas the object of discourse implicated in the lay discourses is the social representation of rurality. Whilst these two objects of discourse should not be regarded as mutually exclusive – hence the descriptive and socio-cultural definitions' rootedness both in a commonsense understanding of the rural and in the rural as locality – they are none the less different. Lay discourses are thus not to be regarded as being rooted in a probable myth but should be seen as interpretive repertoires derived from a disembodied but none the less real social representation of the rural.

Given the disembodied and virtual character of the social representation, to attempt to interrogate them directly would be to neglect the interpretive repertoires that grounded the representation by giving it a subject through discourse, and in the process serve to articulate, reproduce and sometimes transform the underlying representation itself. Hence, the significance of their contextual usage would be missed (Shotter, 1985). Sayer came to a similar conclusion, albeit as regards a somewhat different subject matter:

> to be good causal analysts we need to explain things in relation to their 'causal contexts', abstracting from unrelated phenomena operating in different causal contexts. A purely causal analysis of the industrial base of Los Angeles would instantly make us forget its ironies by resolving them into separate causal groups. But what we would overlook is precisely the fact that the society in question does split into these contexts of determination rather than other ones; we would miss the significance of the fact that the division of areas of action is made *this* way, in practice (1989, p. 273, emphasis in original).

Whilst our social representations of 'the rural' may be fetishized and misplaced, distorted, idealized and generalized, they nevertheless produce very 'real' effects. Social representations provide resources for both discursive and non-discursive actions. The rural social representation[3] – a 'symbolic shorthand' (Falk and Pinhey, 1978) – both guides and constrains action (Shields, 1991, p. 30). It must be seen as causative, 'channelling' causation, although not causal (ibid., p. 57; see also Sayer, 1985).

Social representations of the rural: problems and prospects

Constructing a representational definition of the rural requires a strongly hermeneutical and interpretive methodology (Falk and Pinhey, 1978; Harper, 1987b; Shields, 1991; see also Pile, 1991) because of the need to deal directly with lay discourses. This task, difficult in itself, is compounded by a number of factors. First, the context-dependence of the interpretive repertoires employed (see also Gilbert and Mulkay, 1984; Potter and Wetherell, 1987) means that the social representation from which they derive may contain diverse and even contradictory elements. The latter is possible not only because of such components not being expressed together through individual interpretive repertoires but also because of the location of much of the 'knowledge' contained in the interpretive repertoires at the level of 'practical consciousness' (Thrift, 1985; see also Giddens, 1984, 1986; Shotter, 1984, 1985).

Social representations of space provide 'a means to express ideas' (Shields, 1991, p. 46) through interpretive repertoires but the latter provide:

> *an intellectual shorthand* whereby spatial metaphors and place images can convey a complex set of associations *without the speaker having to think deeply and to specify exactly* which associations or images he or she intends (ibid., my emphases).

Indeed, the construction and usage of social representations of space 'is not a primarily conceptual or logical process' but is 'internalised through the "embodied" memory of habit, gesture, and spatial practice' (ibid., p. 264). As such, we can see immediately the challenge faced by researchers in bringing to the surface and exposing the usage of social representations of the rural through discourse.

Another difficulty facing researchers stems from the abstractness and disembodied character of social representations, which means they may not be at all clearly delineated. Such a lack of empirical clarity will make it harder to reconstruct these representations, especially given their probable contradictory character, as outlined above. However, some help can be gained by seeing what Shields rather grandly terms a 'modern geomancy' in the interrelation between representations, where:

> places or regions mean something only in relation to other places as a constellation of meanings, that is, the North makes sense only with reference to other regions: the 'urban jungle', the southern agricultural fringe, or the commodified consumer landscape of Toronto's suburban strip developments. The images are oriented towards each other in a mutually supporting dialogical exchange (Bakhtin, 1984, pp. 10–12) (1991, p. 199).

The key task may not be to define the rural social representation in the abstract – indeed, this is not the implication of the contextual model of society – but the apparently more relativist task of comparing and contrasting them, through their interpretive repertoires, with other spatial social representations. Hence, we must not be too eager to discard the idea of the social representation of the rural just because our research discovers that people do not hold a clear, well-defined and well-structured 'image' of 'the rural'.

A final compounding factor in our attempt to expose social representations of the rural requires us to consider the distinction between the sign and both its significant (meaning) and its referent (Barthes, 1967). In his attempt to promote a representational definition of the rural, Pratt (1989) saw the sign ('rurality') as being increasingly detached from its signification – in my terminology, social representations and interpretive repertoires of the rural are becoming increasingly diverse. In addition, the sign and its significant are also becoming divorced from their referent (the rural locality), leading to cross-cutting discourse about rurality (ibid.). This promotes the everyday significance of the rural as a social representation *vis-à-vis* the rural as locality (see also Shields, 1991, p. 50). More controversially, it can also be seen as symptomatic of a turn towards socio-cultural post-modernism, where the symbolic assumes precedence over the material (see Baudrillard, 1983, 1988). Finally, the severing of the social representation of the rural from any material referent also makes the former a site of social struggle within discourse, as promoters of competing representations strive for hegemony.

The dynamism of these representations – indeed the very emergence of 'the rural' as a representation – further undermines the validity of using a purely 'material' definition of the rural. This is because Mormont, too, recognized that the social representation of the rural was becoming increasingly divorced from the rural space which it nominally represented. Hence, he asks:

> Where earlier we talked of the rural world in order to define a separate and distinct universe . . . are we not now moving towards another representation which considers things rural not as belonging to another social world, but as a different mode of social relationship? (1987, p. 17).

The physical use of rural space, which has resulted in its 'reduction . . . to a mosaic of countrysides' (ibid., p. 18) primarily dictated to from within the overall urban space, is to be distinguished from the representation of rural space. For Mormont, the latter has an ideological component as it informs a social project which is not confined to the rural space of physical use. Rurality is materially despatialized, as the 'sense of' the rural is separated from its 'existential realness' (Falk and Pinhey, 1978).

Finally, the precise structure of our social representation of rurality is not easy to indicate. It will be an amalgam of personal experiences and 'traditional' handed-down beliefs propagated through literature, the media, the state, family, friends and institutions (Gould, 1977; Pocock and Hudson, 1978; for example, see Blishen, 1984). This 'knowledge' will be both spatially and socially variable (Thrift, 1985). As most of our daily life is carried on at the empirical level (Tuan, 1989), the world of surface appearances may be expected to predominate within representations of rurality. The definitional wheel may thus appear to have turned full circle, taking us back to regarding the rural as a descriptive category,[4] although this time with a sounder conceptualization of space and an explicit engagement with the 'commonsense' underlying these descriptions. Moreover, the 'new' descriptive category can no longer be regarded as a singular objective category but one which is specific to class, gender, ethnicity, etc.

Conclusion

For Falk and Pinhey (1978), the tendency to discuss the rural in a taken-for-granted manner reifies the concept and neglects any questioning of its 'reality'. Whilst one may wish to be wary of distinguishing between the 'real' and the 'unreal' rural, there is a growing realization in the literature that the quest for any single, all-embracing definition of the rural is neither desirable nor feasible. Increasingly, there is a call for the definition to be used to be tailored to the task at hand (for example, Matthews, 1988; Hoggart, 1990). One response to this would be to propose a compound definition of the rural which would attempt to capture the multiple meanings of the word (for example, Willits and Bealer, 1967; Cloke and Park, 1984) but this would be to downplay the situated specificity of these multiple meanings which is precisely that which must be of interest to researchers (Falk and Pinhey, 1978).

This paper has proposed an alternative response, which should enable us to impose some order on the diversity of the rural without losing the significance of this diversity. I have argued that the problem in the literature stems from a failure to distinguish between the rural as a distinctive type of locality and the rural as a social representation – the rural as space and the rural as representing space – confounded by other difficulties, such as the inadequate conceptualization of space.

In an era described by some as post-modern, where symbols appear increasingly to be 'freed' from their referential moorings, it is increasingly important to acknowledge explicitly this difference between space and its social representation. Indeed, we may have to recognize that whilst the referent – the rural locality – may be withering away in respect to its causal significance and distinctiveness (assuming such a salience ever really existed), its nominally associated social representation may well be flourishing and evolving. With the utilization of social representations of space, 'material space' is 'recoded' (Shields, 1991, p. 264). Space becomes imbued with the characteristics of these representations, not only at an imaginative level but also physically, through the use of these representations in action.

As regards the rural, if the 'material space' of the rural is no longer the rural locality but the 'material space' created through usage of the rural social representation, then the much-derided 'chocolate box' countryside may not be(come) such a myth, after all:

> Abstration today is no longer that of the map, the double, the mirror or the concept. Simulation is no longer that of a territory, a referential being or a substance. It is the generation by models of a real without origin or reality: a hyperreal. The territory no longer precedes the map, nor survives it. Henceforth, it is the map that precedes the territory – *precession of simulacra* – it is the map that engenders the territory . . . (Baudrillard, 1988, p. 166).

Notes

1 Pocock and Hudson (1978) make a similar point by distinguishing between the designative and appraisive aspects of images.

2 Shields' book on 'places on the margin' develops what he terms a theory of social spatialisation, which is very close to the version of the theory of social representations developed here, except that it is restricted to spatial phenomena and contains immediate material as well as mental components:

> I use the term *social spatialisation* to designate the ongoing social construction of the spatial at the level of the social imaginary (collective mythologies, presuppositions) as well as interventions in the landscape (for example, the built environment) (Shields, 1991, p. 31).

3 Shields (1991) would term the rural social representation a 'place-myth' but this perpetuates the idea that such structures are in some way 'unreal' because they lack *direct* material grounding.

4 Rurality as a 'way of life' can also be salvaged in a similar manner. The basis for such a way of life rests on people's usage of social representations to create a specific kind of world; it has nothing to do with any inherent causal powers of the rural environment (see Falk and Pinhey, 1978). As with the 'new' descriptive category, this 'way of life' must be seen as being specific to class, gender, ethnicity, etc.

Selected references

Appleyard, D. 1979: The environment as a social symbol. *Ekistics* 278, 272–82.

Bakhtin, M. M. 1984: *Rabelais and his World.* Bloomington, Indiana University Press.

Barlow, J. 1986: Landowners, property ownership and the rural locality. *International Journal of Urban and Regional Research* 10, 309–29.

Barthes, R. 1967: *Elements of Semiology*. London, Jonathan Cape.

Baudrillard, J. 1983: *Simulations*. New York, Semiotext(e).

Baudrillard, J. 1988: *Selected Writings*. In Poster, M. (ed.). Stanford, Stanford University Press.

Bealer, R., Willits, F. and Kuvlesky, W. 1965: The meaning of 'rurality' in American society. *Rural Sociology* 30, 255–66.

Bell, C. and Newby, H. 1971: *Community Studies*. London, George Allen and Unwin.

Bell, C. and Newby, H. (eds) 1974: *The Sociology of Community*. London, Frank Cass.

Bhaskar, R. 1975: *A Realist Theory of Science*. Leeds, Leeds Books.

Bhaskar, R. 1979: *The Possibility of Naturalism: a Philosophical Critique of the Contemporary Human Sciences*. Brighton, Harvester.

Blishen, E. 1984: Town, bad: country, good. In Mabey, R. (ed.), *Second Nature*. London, Jonathan Cape, 15–24.

Bowers, J. 1988: Review of 'Discourse and Social Psychology'. *British Journal of Social Psychology* 27, 185–92.

Bradley, T. 1981: Capitalism and countryside: rural sociology as political economy. *International Journal of Urban and Regional Research* 5, 581–7.

Bradley, T. and Lowe, P. (eds) 1984: *Locality and Rurality*. Norwich, Geo Books.

Brewer, J. 1988: Micro-sociology and the 'duality of structure'. In Fielding, N. (ed.), *Actions and Structure*. London, Sage, 144–66.

Canter, D. 1977: *The Psychology of Place*. London, Architectural Press.

Carlson, J. E., Lassey, M. L. and Lassey, W. R. 1981: *Rural Society and Environment in America*. New York, McGraw-Hill.

Castells, M. 1977: *The Urban Question*. London, Edward Arnold.

Castells, M. 1983: Crisis, planning, and the quality of life: managing the new historical relationships between space and society. *Society and Space* 1, 3–21.

Champion, A. G. and Coombes, M. 1984: A nation on the move. *Geographical Magazine* 56, 245–8.

Champion, A. G., Coombes, M. and Openshaw, S. 1984: New regions for a new Britain. *Geographical Magazine* 56, 187–90.

Champion, A. G., Green, A., Owen, D., Ellin, D. and Coombes, M. 1987: *Changing Places*. London, Edward Arnold.

Cherry, G. E. (ed.) 1976: *Rural Planning Problems*. London, Hill.

Clark, G. 1984: The meaning of agricultural regions. *Scottish Geographical Magazine* 100, 34–44.

Cloke, P. J. 1977: An index of rurality for England and Wales. *Regional Studies* 11, 31–46.

Cloke, P. J. 1978: Changing patterns of urbanisation in rural areas of England and Wales, 1961–1971. *Regional Studies* 12, 603–17.
Cloke, P. J. 1980: New emphases for applied rural geography. *Progress in Human Geography* 4, 181–217.
Cloke, P. J. 1985: Whither rural studies? *Journal of Rural Studies* 1, 1–9.
Cloke, P. J. and Edwards, G. 1986: Rurality in England and Wales 1981: a replication of the 1971 index. *Regional Studies* 20, 289–306.
Cloke, P. J. and Park, C. C. 1984: *Rural Resource Management: a Geographical Perspective*. London, Croom Helm.
Clout, H. D. 1972: *Rural Geography: an Introductory Survey*. Oxford, Pergamon Press.
Connell, J. 1978: *The End of Tradition. Country Life in Central Surrey*. London, Routledge and Kegan Paul.
Coombes, M., Dixon, J., Goddard, J., Openshaw, S. and Taylor, P. 1982: Functional regions for the population census of Great Britain. In Herbert, D. T. and Johnston, R. J. (eds), *Geography and the Urban Environment*. London, Wiley, 63–112.
Copp, J. 1972: Rural sociology and rural development. *Rural Sociology* 37, 515–33.
Craig, J. 1987: An urban–rural categorisation for wards and local authorities. *Population Trends* 47, 6–11.
Craig, J. 1988: Local authority urban–rural indicators compared. *Population Trends* 51, 30–8.
CURDS (Centre for Urban and Regional Development Studies) 1983–84: Functional regions. *CURDS Factsheets* 1–21.
Denham, C. 1984: Urban Britain. *Population Trends* 36, 10–18.
DOE (Department of the Environment) 1971: The nature of rural areas of England and Wales. DOE Internal Working Paper.
Donnelly, P. and Harper, S. 1987: British rural settlements in the hinterland of conurbations: a classification. *Geografiska Annaler B* 69, 55–63.
Duncan, O. and Reiss, A. 1976: *Social Characteristics of Rural and Urban Communities 1950*. New York, Russell and Russell.
Duncan, S. S. and Savage, M. 1989: Space, scale and locality. *Antipode* 21, 179–206.
Eliot Hurst, M. 1980: Geography, social science and society: towards a de-definition. *Australian Geographical Studies* 18, 3–21.
Eliot Hurst, M. 1985: Geography has neither existence nor future. In Johnston, R. (ed.), *The Future of Geography*. London, Methuen, 59–91.
Falk, W. and Pinhey, T. 1978: Making sense of the concept rural and doing rural sociology: an interpretive perspective. *Rural Sociology* 43, 547–58.
Fothergill, S., Gudgin, G., Kitson, M. and Monk, S. 1985: Rural industrialization: trends and causes. In Healey, M. and Ilbery, B. (eds), *The Industrialization of the Countryside*. Norwich, Geo Books, 147–59.
Frankenberg, R. 1966: *Communities in Britain*. Harmondsworth, Penguin.
Friedland, W. 1982: The end of rural society and the future of rural sociology. *Rural Sociology* 47, 589–608.
Gans, H. 1962: *The Urban Villagers*. Glencoe, Illinois, Free Press.
Geertz, C. 1983: *Local Knowledge*. New York, Basic Books.
Giddens, A. 1984: *The Constitution of Society*. Cambridge, Polity Press.
Giddens, A. 1986: Action, subjectivity, and the constitution of meaning. *Social Research* 53, 529–45.
Giddens, A. 1987: *Social Theory and Modern Sociology*. Cambridge, Polity Press.
Gilbert, G. and Mulkay, M. 1984: *Opening Pandora's Box: a Sociological Analysis of Scientists' Discourses*. Cambridge, Cambridge University Press.

Gilbert, J. 1982: Rural theory: the grounding of rural sociology. *Rural Sociology* 47, 609–33.

Gold, J. R. 1980: *An Introduction to Behavioural Geography*. Oxford, Oxford University Press.

Gould, A. and Keeble, D. 1984: New firms and rural industrialization in East Anglia. *Regional Studies* 18, 189–201.

Gould, P. 1977: Changing mental maps: childhood to adulthood. *Ekistics* 255, 111–19.

Gould, P. 1982: Is it necessary to choose? Some technical, hermeneutic and emancipatory thoughts on enquiry. In Gould, P. and Olsson, G. (eds), *A Search for Common Ground*. London, Pion, 71–104.

Gregory, D. 1978a: *Ideology, Science and Human Geography*. London, Hutchinson.

Gregory, D. 1978b: The discourse of the past: phenomenology, structuralism and human geography. *Journal of Historical Geography* 4, 161–73.

Gregory, D. 1985: Suspended animation: the stasis of diffusion theory. In Gregory, D. and Urry, J. (eds), *Social Relations and Spatial Structures*. London, Macmillan, 296–336.

Gregory, D. and Urry, J. (eds) 1985: *Social Relations and Spatial Structures*. London, Macmillan.

Habermas, J. 1972: *Knowledge and Human Interests*. London, Heinemann.

Halfacree, K. H. 1992: The importance of spatial representations in residential migration to rural England in the 1980s. A quest for 'sophisticated simplicity' in a postmodern world? Ph.D. thesis, Department of Geography, Lancaster University, UK.

Harper, S. 1987a: The rural–urban interface in England: a framework for analysis. *Transactions of the Institute of British Geographers NS* 12, 284–302.

Harper, S. 1987b: A humanistic approach to the study of rural populations. *Journal of Rural Studies* 3, 309–19.

Harvey, D. 1973: *Social Justice and the City*. London, Edward Arnold.

Harvey, D. 1982: *The Limits to Capital*. Oxford, Blackwell.

Harvey, D. 1985: *Consciousness and the Urban Experience*. Oxford, Blackwell.

Hewstone, M. 1985: On common-sense and social representations: a reply to Potter and Litton. *British Journal of Social Psychology* 24, 95–7.

Hoggart, K. 1988: Not a definition of rural. *Area* 20, 35–40.

Hoggart, K. 1990: Let's do away with rural. *Journal of Rural Studies* 6, 245–57.

Hoggart, K. and Buller, H. 1987: *Rural Development. A Geographical Perspective*. London, Croom Helm.

Jonas, A. 1988: A new regional geography of localities? *Area* 20, 101–10.

Keat, R. and Urry, J. 1981: *Social Theory as Science*. London, Routledge and Kegan Paul.

Lassey, W. R. 1977: *Planning in Rural Environments*. New York, McGraw-Hill.

Lee, T. 1968: Urban neighbourhood as a socio-spatial schema. *Human Relations* 21, 241–68.

Lefebvre, H. 1976: *The Survival of Capitalism*. London, Allison and Busby.

Lewis, G. J. 1979: *Rural Communities: a Social Geography*. Newton Abbot, David and Charles.

Littlejohn, J. 1963: *Westrigg: the Sociology of a Cheviot Parish*. London, Routledge and Kegan Paul.

Lynch, K. 1960: *The Image of the City*. Cambridge, Massachusetts, MIT Press.

Marsden, T. and Murdoch, J. 1989: Restructuring rurality: key areas for development in assessing rural change. Paper presented at the Rural Economy and Society Study Group Conference, University of Bristol.

Massey, D. 1984: *Spatial Divisions of Labour*. London, Macmillan.

Massey, D. 1985: New directions in space. In Gregory, D. and Urry, J. (eds), *Social Relations and Spatial Structures*. London, Macmillan, 9–19.

Matthews, A. M. 1988: Variations in the conceptualization and measurement of rurality: conflicting findings on the elderly widowed. *Journal of Rural Studies* 4, 141–50.

Mead, G. H. 1934: *Mind, Self and Society*. Chicago, University of Chicago Press.

Mormont, M. 1987: Rural nature and urban natures. *Sociologia Ruralis* 27, 3–20.

Moscovici, S. 1976: *La Psychoanalyse: son Image et son Public*. Paris, Presses Universitaires de France.

Moscovici, S. 1981: On social representation. In Forgas, J. (ed.), *Social Cognition: Perspectives on Everyday Understanding*. London, Academic Press, 181–209.

Moscovici, S. 1982: The coming era of representations. In Codol, J.-P. and Lyons, J.-P. (eds), *Cognitive Analysis of Social Behaviour*. The Hague, Nijhoff.

Moscovici, S. 1984: The phenomenon of social representations. In Farr, R. and Moscovici, S. (eds), *Social Representations*. Cambridge, Cambridge University Press, 3–69.

Moscovici, S. 1985: Comment on Potter and Litton. *British Journal of Social Psychology* 24, 91–3.

Moseley, M. 1980a: Is rural deprivation really rural? *The Planner* 66, 97.

Moseley, M. 1980b: *Rural Development and its Relevance to the Inner City Debate*. The Inner City in Context 9. London, Social Science Research Council.

Newby, H. 1977: *The Deferential Worker*. London, Allen Lane.

Newby, H. 1978: The rural sociology of advanced capitalist societies. In Newby, H. (ed.), *International Perspectives in Rural Sociology*. Chichester, Wiley, 3–30.

Newby, H. 1979: *Green and Pleasant Land*. London, Wildwood House.

Newby, H. 1980: Rural sociology – a trend report. *Current Sociology* 28, 1–141.

Newby, H. 1986: Locality and rurality: the restructuring of rural social relations. *Regional Studies* 20, 209–15.

Newby, H. 1987: *Country Life*. London, Weidenfeld and Nicolson.

Newby, H., Bell, C., Rose, D. and Saunders, P. 1978: *Property, Paternalism and Power*. London, Hutchinson.

Newby, H. and Buttel, F. 1980: Towards a critical rural sociology. In Buttel, F. and Newby, H. (eds), *The Rural Sociology of Advanced Societies*. London, Croom Helm, 1–35.

OPCS (Office of Population Censuses and Surveys) 1986: Urban/rural ward categorisation. England and Wales. OPCS Census 1981, User Guide 232.

Openshaw, S. 1985: Rural area classification using census data. *Geographia Polonica* 51, 285–99.

Openshaw, S., Cullingford, D. and Gillard, A. 1980: A critique of the national classifications of OPCS/PRAG. *Town Planning Review* 51, 421–39.

Pahl, R. E. 1965: Urbs in rure. *London School of Economics Geographical Paper* 2.

Pahl, R. E. 1966: The rural–urban continuum. *Sociologia Ruralis* 6, 299–327.

Pahl, R. E. 1970: *Whose City?* London, Longman.

Phillips, D. and Williams, A. 1984: *Rural Britain. A Social Geography*. Oxford, Blackwell.

Pile, S. 1991: Practising interpretive geography. *Transactions of the Institute of British Geographers NS* 16, 458–69.

Pocock, D. and Hudson, R. 1978: *Images of the Urban Environment*. London, Macmillan.

Potter, J. and Litton, I. 1985: Some problems underlying the theory of social representations. *British Journal of Social Psychology* 24, 81–90.

Potter, J. and Wetherell, M. 1987: *Discourse and Social Psychology*. London, Sage.

Pratt, A. 1989: Rurality: loose talk or social struggle? Paper presented at the Rural Economy and Society Study Group Conference, University of Bristol.

Redfield, R. 1941: *The Folk Culture of Yucatan*. Chicago, University of Chicago Press.

Rees, A. 1950: *Life in a Welsh Countryside*. Cardiff, University of Wales Press.

Reissman, L. 1964: *The Urban Process*. New York, Free Press.

Robertson, I. 1961: The occupational structure and distribution of rural population in England and Wales. *Scottish Geographical Magazine* 77, 165–79.

Rodwin, L. and Hollister, R. 1984: *Cities of the Mind*. London, Plenum Press.

Sayer, A. 1984: *Method in Social Science: a Realist Approach*. London, Hutchinson.

Sayer, A. 1985: The difference that space makes. In Gregory, D. and Urry, J. (eds), *Social Relations and Spatial Structures*. London, Macmillan, 49–66.

Sayer, A. 1989: The 'new' regional geography and the problems of narrative. *Society and Space* 7, 253–76.

Schutz, A. 1967: *The Phenomenology of the Social World*. Evanston, North Western University Press.

Schutz, A. 1970: *On Phenomenology and Social Relations*. Chicago, University of Chicago Press.

Semin, G. 1985: The 'phenomenon of social representations': a comment on Potter and Litton. *British Journal of Social Psychology* 24, 93–4.

Shields, R. 1991: *Places on the Margin*. London, Routledge.

Short, J. 1974: Social systems and spatial patterns. *Antipode* 6, 77–83.

Shotter, J. 1984: *Social Accountability and Selfhood*. Oxford, Blackwell.

Shotter, J. 1985: Accounting for place and space. *Society and Space* 3, 447–60.

Smith, N. 1981: Degeneracy in theory and practice: spatial interactionism and radical eclecticism. *Progress in Human Geography* 5, 111–18.

Smith, N. 1984: *Uneven Development*. Oxford, Blackwell.

Soja, E. 1985: The spatiality of social life: towards a transformative retheorisation. In Gregory, D. and Urry, J. (eds), *Social Relations and Spatial Structures*. London, Macmillan, 90–127.

Stevens, A. 1946: The distribution of the rural population of Great Britain. *Transactions of the Institute of British Geographers* 11, 21–53.

Symes, D. 1981: Rural community studies in Great Britain. In Durand-Drouhin, J.-L., Szwengrub, L.-M. and Mihailescu, I. (eds), *Rural Community Studies in Europe*. Oxford, Pergamon Press, 17–67.

Thrift, N. 1983: On the determination of social action in space and time. *Society and Space* 1, 23–57.

Thrift, N. 1985: Flies and germs: a geography of knowledge. In Gregory, D. and Urry, J. (eds), *Social Relations and Spatial Structures*. London, Macmillan, 366–403.

Thrift, N. 1986: Little games and big stories: accounting for the practice of personality and politics in the 1945 general election. In Hoggart, K. and Kofman, E. (eds), *Politics, Geography and Social Stratification*. London, Croom Helm, 86–143.

Thrift, N. 1987: Manufacturing rural geography? *Journal of Rural Studies* 3, 77–81.

Tuan, Y.-F. 1989: Surface phenomena and aesthetic experience. *Annals of the Association of American Geographers* 79, 233–41.

Urry, J. 1984: Capitalist restructuring, recomposition and the regions. In Bradley, T. and Lowe, P. (eds), *Locality and Rurality*. Norwich, Geo Books, 45–64.

Urry, J. 1985: Social relations, space and time. In Gregory, D. and Urry, J. (eds), *Social Relations and Spatial Structures*. London, Macmillan, 20–48.

Urry, J. 1986: Locality research: the case of Lancaster. *Regional Studies* 20, 233–42.

Vince, S. 1952: Reflections on the study and distribution of rural population in England and Wales, 1921–31. *Transactions of the Institute of British Geographers* 18, 53–76.

Wibberley, G. P. 1972: Conflicts in the countryside. *Town and Country Planning* 40, 25.

Williams, W. 1956: *The Sociology of an English Village: Gosforth*. London, Routledge and Kegan Paul.

Williams, W. 1963: *A West Country Village. Ashworthy*. London, Routledge and Kegan Paul.

Willits, F. K. and Bealer, R. C. 1967: Evaluation of a composite definition of 'rurality'. *Rural Sociology* 32, 164–77.

Young, M. and Wilmott, P. 1957: *Family and Kinship in East London*. London, Routledge and Kegan Paul.

Zelinsky, W. 1977: Coping with the migration turn-around: the theoretical challenge. *International Regional Science Review* 2, 175–8.

SECTION TWO

GENDER, CLASS, RACE AND SPACE

Editor's introduction

The importance of class, race and ethnicity have long been appreciated by social geographers (Hamnett *et al.*, 1989). In the early 1980s some social geographers realized that hitherto they had neglected the crucial dimension of gender in social geography. Previously, social geography had been largely gender-blind, and therefore dominated by an implicit, if not explicit, male-orientated view of the world. As McDowell (1989) put it: 'geographers have never had much to say about women'. The term 'man' was frequently, and to a lesser extent still is, used to include women, but she pointed out that this in itself implied that gender differences are not significant, and that it is unimportant for geographers to distinguish between men's and women's beliefs, behaviour and activities in space.

Fortunately, this state of affairs has begun to change quite rapidly since the early 1980s and the publication, in 1984, of *Geography and Gender*, by the Women and Geography study group of the Institute of British Geographers. McDowell and Massey (1984) discussed regional differences in the gender division of labour, and McDowell (1983) published the first of a major series of papers on the subject of gender (McDowell, 1991, 1992). I cannot touch on all the various aspects of gender and social geography here; for example, the work of Pred and Palm (1974, 1978; Tivers, 1985) on the activity spaces of women and the time–space constraints imposed by children and problems of accessibility and child-care, or Rose (1984) on the gender dimensions of gentrification. I am restricting the discussion to the gender division of urban space, an issue of major importance for social geographers. McDowell (1983) points out amongst other things that, historically, the separation of home from work, the development of suburbia and the patriarchal ideology that a woman's place is in the home, led to the development of suburban areas as gendered spaces, sites for social reproduction rather than production, and a haven for the male worker. This led to segregation of women in the private space of the home away from the male-dominated public spaces of the city. This thesis has been questioned by Gleeson (1995) at least in the context of the organization of charity relief in nineteenth-century Melbourne, but it does not affect the overall force of the argument (Garmanikow, 1978).

Peake (Chapter 5, this volume) states that the city has been seen as embodying patriarchal principles in four main ways. First, feminists have argued that the city has been shaped, whether by accident or design, to confine women to traditional roles within the family, the site from which women free domestic services and child-care services. Secondly, women's responsibility for domestic labour circumscribes their mobility and they face a wide range of social and spatial inequalities. Thirdly, these inequalities are embedded in the design and organization of urban space, not only in terms of the sexual division of labour in the family and waged employment, but in the form of urban planning and facility provision. Finally, there are a number of social and psychological problems that affect women in specific urban spaces. But Peake suggests that, despite the progress made by feminist geographers, our current conceptions of women's place in the city 'are at worst outmoded, barely advancing beyond the advanced industrial urban spatial divides of "city centre versus suburbs", and at best they are confused' (p. 105). She suggests that this is a result of geographers' seeming reluctance to address the heterosexist and white cultural constructions that pervade our discourses. After discussing this in some detail, she empirically explores these critiques in relation to research with low-income women, both Afro-American and Anglo-American, in two downtown neighbourhoods of Grand Rapids, MI. She concludes that 'whereas women have always adapted to the constraints and limitations of patriarchal social relations, the spatial outcomes of strategies vary within and between "race", sexuality, class and family household status, in a complex manner' and she argues that while building up a social geography of the city from such a microscale juxtaposition of social groups is a difficult task, 'it is only at this local scale that the degree and character of modes of behaviour can be exemplified and meanings of different resources, of the family, of social networks, the state, and the formal and informal economy, can be understood' (p. 116).

I have included Pratt and Hanson's (1988) paper 'Gender, class and space' (Chapter 6, this volume) as it kills two or possibly three birds with one stone. First, it gives something of the flavour of quantitative urban social geography of the 1970s. Second, and much more importantly, it convincingly shows that previous approaches to urban residential differentiation were deficient in neglecting the gendered nature of occupational class. Third, they use the methods of quantitative research to demonstrate the limitations of much previous quantitative work on the social geography of the city.

Pratt and Hanson point out that whereas traditional approaches to urban social geography assumed that residential areas were homogeneous in terms of class, the very different positions held by men and women in the workforce suggest that the geography of social class is different for men and women. They argue that if it is, then the social geography of the city may be far more complex than the standard models of residential structure suggest. Thus, 'census tracts that may be homogeneous with respect to male social class will not be homogeneous when both men's and women's social classes are considered. This . . . holds major implications for the validity of the social reproduction thesis' (p. 123). They note that in the numerous studies of urban residential differentiation, most of which have employed multivariate statistical analysis of a variety of different indicators, the subjective choice of indicators by the researchers has rarely been made explicit. Thus, although the choice of indicators may be as wide-ranging and as representative as possible, 'many factorial ecologies describe how the social class of male but not female workers varies across residential areas in the city. In the most extreme cases, this is because information on occupation type and income is restricted to male workers' (p. 124). But, even if variables on women's labour force participation are included, the researchers' expectations often lead to the results being ignored or wrongly classified. The general argument they make is that women's employment characteristics were not usually considered in the definition of the socio-economic status or social rank dimension. Female labour force participation was generally one of the variables used to define the 'familism' or 'urbanism' dimension, and the question of women's occupational status was ignored.

They go on to point out that it is ironic that in most American metropolitan areas today, female labour-force participation rates are higher in the suburbs – traditionally the bastions of 'familism' – than they are in the central cities – traditionally a bastion of 'urbanism', and they argue that the fact that women now comprise a large proportion of the metropolitan labour force 'renders the dichotomous view of women's labour force participation relatively useless as an indicator of any dimension of contemporary social life' (p. 126).

More generally, they argue that a view of the city in which neighbourhoods are structured into homogeneous class enclaves 'rests on the suppression of . . . gender' (p. 127), and this leaves factorial ecologies on shaky ground given that the 'typical American family' of a male breadwinner, a home-making mother and two children now constitutes only 7 per cent of American households. Similar changes have taken place in most other Western countries. They point to the role of a number of factors which have eroded the prominence of this type of household; the increase in female labour force participation rates, the rising divorce rate, the increase in single-person households, and the increase in the number of female-headed households and single parent families. Also, there has been an increase in the number of dual-earner households. In 1950, only 27 per cent of wives in married couple households worked, whereas this had risen to 68 per cent by 1983. Pratt and Hanson point out that while these trends are well known, and the concern with the relationship between gender and class is not a new one, these changes 'have not . . . been systematically related to space', and their paper examines the significance of trends for 'a reconceptualization of urban residential structure' (p. 128) using evidence from Worcester, Mass. They suggest that these demographic changes 'have . . . probably obliterated the utility of traditional models of urban residential structure', and they suggest that contemporary patterns will probably be quite different from those which are predicted by traditional models and research carried out in the 1960s and 1970s. They argue that it is now unlikely that 'women's labor-force participation is still a valid discriminator of social areas' as child-rearing and employment are no longer mutually exclusive activities (p. 128). The remainder of their paper consists of detailed empirical analysis of several propositions. To those brought up mainly on a diet of postmodern textual analysis, representation and discourse theory, this may prove tough going, but the paper is important, in part, precisely because of its insistence on rigorous empirical analysis to evaluate the validity of certain theoretical propositions. In general, their expectations about changes in urban residential structures were supported by the analysis. 'Residential areas are no longer clearly distinguishable on the basis of whether or not female residents hold paid employment' (p. 139).

Their empirical findings led them to question 'the adequacy of current claims about the functionality of neighborhoods for the social reproduction of classes', arguing that theories of social reproduction which rely on outmoded characterizations of urban residential structure: 'simplify and distort the processes through which political consciousness and social identity are defined'. Most radically, they suggest that the occupational segregation of men and women, along with the increased participation of women in the paid labour force, 'effectively break down any occupation-based residential segregation that may have existed in North American cities in the past' (p. 141). If this is correct, much of the traditional basis of urban social geography may begin to crumble. This is one area where the ritual chant, that more research is needed, is more than a formulaic incantation.

Pratt and Hanson point to some of the major changes in household structure which have taken place in western countries in recent decades. But, as Hilary Winchester (Chapter 7, this volume) has pointed out, with the exception of studies of the elderly, geographers have tended to neglect the study of different household types, falling uneasily as it does between social and population geography. 'Social geographers, although concerned with welfare issues and with the spatial and socio-economic characteristics of various sub-groups, have rarely paid attention to particular household types . . . More commonly, social geographers have studied population sub-groups defined by race, class, or occupational stratification' (p. 145). But, as Winchester points out, the structure of households and families has changed very rapidly in the last 20 years in many industrialized developed societies, and one of the most significant changes has been in the proportion of smaller households.

One of the most important of these changes has been the growing number of one-parent families which constitute such a problem for many right-wing politicians who frequently appear to take the view that one-parent families largely consist of teenage mothers who have got themselves pregnant and want to rely on the state for financial help. In reality, although most single-parent families are female headed, the majority consist of divorced, separated or abandoned women rather than the prototypical teenage mother, and many of them are in financial difficulties. As Winchester points out, at the theoretical level,

> the marginalization of one-parent families exemplifies . . . the feminization of poverty, related to women's dual role in contemporary western society [and] it has immediate material significance for the well-being of a large number of women and children (p. 144).

Her paper examines the feminization of poverty, and the marginal role which single-parent families have in western society. She suggests that polarization is occurring between what Holcomb (1986) has described as the 'new urban woman', skilled, professional and financially secure, on the one hand, and on the other, the dramatic increase in the number of women below the poverty line, both lone female parents and lone elderly women. The paper focuses on the numbers and changing location of one-parent families in both Britain and Australia, and she uses the national trends as a context for detailed studies of two medium-sized cities: Plymouth and Wollongong.

Marginalization

The marginalization of one-parent families exemplifies a more general trend to the marginalization of certain groups in society (Winchester and White, 1988) and to the labelling of these groups as 'outsiders', e.g. new age travellers and gypsies (Sibley, 1988, 1994, 1995). One of the most common and pervasive forms of marginalization is that caused by poverty. Peet (1970) pointed out that: 'In American society the state of being without certain key possessions is viewed as a form of deviance from what is considered "normal" or "acceptable"', but with relatively rare exceptions, geographers have not examined the geography of income or poverty (Rawstron and Coates, 1966). As Coates *et al.* (1977) pointed out: 'geographers have been remarkably slow to investigate regional variations in income levels, even at a fundamental descriptive level' (p. 32). It is thus all the more welcome that Philo (1995) has recently edited a collection on the social geography of poverty in Britain.

But poverty, although extremely important, is only one form of marginality and deprivation, and Winchester and White have examined the identity and location of marginalized groups in the inner city, focusing on case-studies of the impoverished elderly, the lesbian community, and down-and-out groups in Paris in the 1980s. All of these groups can be considered to be outsiders in one way or another (Sibley, 1981) and, as such, they are often subjected to pressures for what Sibley (1988) has termed the 'purification of space'. That is to say, they may be victims of a form of 'spatial cleansing' designed to rid the streets of the socially unwanted. Davis (1990) has analysed this tendency in Los Angeles where the homeless have been subjected to pressure on their urban space. Homelessness has been a growing form of marginality in some western countries in the 1980s, particularly in countries such as the USA and Britain where policies of de-institutionalization and the growth of unemployment and benefit cuts have led to an increasing number of people living on the streets (Hoch, 1991; Rowe and Wolch, 1990; Dear and Wolch, 1987; Dear and Gleeson, 1991). Philo (1992) has recently argued that rural geography has also tended to neglect marginalized groups and communities such as women, children, the elderly, homeless, etc. who do not fall into what Murdoch and Pratt (1993) term 'the category of white, middle-class, middle-aged heterosexual men of sound mind and body', and that issues of diversity and plurality and 'otherness' should be put at the forefront of analysis, although it can be argued that Ray Pahl (1965, 1970a) was alive to these issues in the 1960s.

It is important not to give the impression that marginality or vulnerability are only associated with affluent western countries. Nothing could be further from the truth. Women, children, the elderly and the poor are often the most vulnerable and marginal groups in many societies, and this is particularly true in some third world countries, where poverty and hunger are widespread. The concept of vulnerability, and its relation to hunger and famine, is discussed by Watts and Bohle (1993), and vulnerability in relation to natural disasters and global environment change is discussed by Blaikie *et al.* (1994) and Dow (1992) respectively.

One of the most vulnerable groups of all is refugees (Zetter, 1988), as they have generally had to abandon their jobs, their land, their homes and their wealth, and may be without a source of income. Black (Chapter 8, this volume) looks at the meaning of the concept of vulnerability as it applies to refugees, and the livelihoods of refugee households and individuals. He argues in the context of Greece that vulnerability, when defined in terms of economic stress and dependence, is a powerful issue in refugees' lives, negatively affecting their livelihoods. He states that there are two main ways of defining vulnerability. The commonsense view identifies '"vulnerable" groups such as children and the elderly, or those who are infirm or mentally ill, and who are unable to sustain a livelihood, or respond to threats to their well-being'. In addition, it is possible to see vulnerability as the result of circumstances external to the individuals concerned. He points to Dow's (1992) distinction, in the context of global environmental change, between the coping abilities of individuals and their level of exposure to risk. (A similar distinction is made by Watts and Bohle (1993) in the case of vulnerability to famine in the Third World.)

> Thus vulnerability can be seen to relate not only to the age, health, physical limitations and experience of an individual, but also to issues such as class, gender, ethnicity and market dependency that describe that person's place in the economy or society (Black, p. 168).

Black adds that external factors influencing vulnerability also relate to the ways in which particular groups are perceived and treated by the wider society. To some extent, the dependency and vulnerability of certain groups such as the elderly and the handicapped is socially defined and produced. In the case of refugees, Black notes that there is often an added dimension of political vulnerability 'in the sense that they require protection, normally from the host state, against forcible repatriation to their country of origin'. It might also be noted that they sometimes need protection from the host state itself, particularly where, as in the case of Vietnamese boat people, they are often unwelcome in the host state and face forcible repatriation by the host state.

It is important not to look at social groups in isolation from the environment of which they are a part. This is not to argue for a return to the perspectives adopted by Vidal de la Blache and the early human geographers who sought to locate the ways of life of social groups within the physical context of their region, but to point out that human beings are affected by their environments and have an impact upon them. Black (1994) looks at forced migration and environmental change and the impact of refugees on host environments. He points out that there is considerable interest in the possible consequences of a new category of 'environmental refugee', 'resulting variously from regional environmental conflict, natural disasters or global climatic change' (p. 261). But he notes that the question of potential environmental impacts of refugee flows has received less academic attention. He states that:

> With over 17 million officially recognised refugees in the world today, and as many as 40 million displaced people world wide, the issue of forced displacement has implications for a number of aspects of development throughout the Third World. Given the frequent concentration of such displaced people within fragile or marginal environments, the potential contribution to environmental degradation can be serious (pp. 261–2).

Race, class, migration and space

'Race', ethnicity and segregation have been long-standing areas of research in social geography since the 1920s with Park and Burgess's work on Chicago. Successive waves of European migration, from Russia, Poland, Germany, Scandinavia, Ireland and Italy, rapidly changed the composition of the major cities. More or less simultaneously, though it increased dramatically during the 1950s and 1960s, a continuous flow of Black Americans migrated from the rural, agricultural Southern states to the northern industrial cities such as New York, Chicago and Detroit. Most of the new migrants ended up living in cheap housing in the inner city, close to the ports and the railway stations and the centres of employment. As Burgess (1919) commented of Chicago:

> In the zone of deterioration encircling the central business district are always to be found the so-called 'slums' . . . The slums are crowded to overflowing with immigrant colonies – the Ghetto, Little Sicily, Greektown, Chinatown. The next zone is an area of second immigrant settlement, generally of the second generation. It is the region of escape from the slum.

In the postwar period, many of the original ethnic groups moved out, and the inner areas of most major American cities witnessed a growing concentration of Afro-Americans and, in the case of New York and Los Angeles, Latinos, many of whom were trapped by a combination of low income, poverty and discrimination. This was, and remains, a source of continuing concern in the USA (Rose, 1970, 1972; Fainstein, 1993).

Similarly, Britain and most northern European countries experienced a rapid increase in labour migration from abroad in the 1950s and 1960s, either from former colonies in the case of France (Condon and Ogden, 1991), Britain (Peach, 1968) and the Netherlands, or 'guestworkers' from southern Europe in the case of Germany (O'Loughlin, 1980). The geopolitics of international migration in Europe is explored by Black (1996), Collison (1996), Knights (1996). As in the USA, immigrants were initially concentrated in areas of poor housing, often in the inner city, although this has begun to change (Bonvalet *et al.*, 1995; Peach, 1996), leading to many studies of the extent and the causes of such segregation (Lee, 1973; Peach, 1975; S. Smith, 1988). One of the most important areas of work has been on the position of immigrants in the housing market (Peach and Shah, 1980; Peach and Byron, 1993; Sarre *et al.*, 1989; Western, 1993).

One of the important areas of debate regarding immigrant housing and residential segregation is the extent to which this is the product of discrimination or the class and income characteristics of immigrants and their children. Lee (1973) examined this question in the context of ethnic residential segregation in London, and Peach and Byron (1993) have examined the intersection of 'race', class and gender regarding the representation of Caribbean tenants in council housing. They concluded that although socio-economic position explained a great deal of the over-representation of male household heads in council housing, it explained relatively little of the tenure of female household heads, which is interpreted as the result of a much more complex interplay of class, gender and racism. They noted that female-headed households are the norm in the Caribbean, and pointed out that while only 11 per cent of whites in the council sector in Britain are single parents, the figure for Caribbean heads of household is 44 per cent, the bulk of whom are lone single mothers with dependant children. They comprise an extremely vulnerable group in terms of access to housing and Peach and Byron conclude that it is not surprising that they are concentrated, first, in the council sector, second, in flats rather than houses, and third, in high-rise buildings. They suggest that residualization of single-parent Caribbean families in high-rise flats is a real possibility and note that:

> The impact of the high proportion of female-headed households on housing outcomes seems to have been systematically under-reported in the literature. In particular, the impact on housing outcomes of being a single parent seems not to have been treated with sufficient weight . . . What has been generally seized upon in the literature as racial discrimination, now appears to be more strongly gender and race related. The literature has been gender blind (p. 421).

Ethnic and racial segregation are important in all Western countries but they reached their zenith in South Africa under apartheid, and I have chosen a paper by Hindson, Byerley and Morris (Chapter 9, this volume) on the making of an apartheid city. Using the example of Durban, they argue that the apartheid city was created in response to the urban crises of the 1940s. In order to protect the interests of its white constituents, the National Party government drew on policies of racial segregation and spatial management to restructure city form. Racial groups were channelled into clearly defined spatial zones within the city, or, in the case of illegal entrants, removed from the cities to rural areas or to resettlement camps. They argue that this policy accelerated class mobility by whites and coloured and Indian residents, while blocking it for Africans, and they show that the concentration of resources in the white suburbs and the pushing out of black residents to the urban periphery produced a specific racial and class urban structure. This became increasingly unworkable from the late 1970s due to its economic inefficiency and political unacceptability. Accelerating urbanization from the 1980s and African struggles to improve the quality of urban life have led to increasing challenges to the apartheid city and there are strong pressures to restructure the city to achieve a more equitable distribution of resources and life chances.

5 L. Peake,

'"Race" and Sexuality: Challenging the Patriarchal Structuring of Urban Social Space'

Reprinted in full from: *Society and Space* **11**, 415–32 (1993)

Introduction

The patriarchal structuring of urban social space has been cited as the central theoretical problem of feminist urban research (see Garmanikow, 1978). By challenging accepted, but largely uncritical, notions of this problem I aim to contribute to the debate on how relations of 'race', class, gender, and sexuality are played out across that space. The challenge is posited from two directions: from a critical engagement with feminist theoretical work, in particular that of African-American, Third World, lesbian, and postmodern feminists, and from an interpretation of research work conducted in the late 1980s in a mid-size US city. More broadly, this study is framed by some of the dominant projects of urban geography, namely, social processes of place construction and attempts to increase understanding of the social geography of the city (see Cox, 1991; Hanson and Pratt, 1988; Mackenzie, 1987; Pahl, 1984; Pinch and Storey, 1991). Studies of the latter have exhibited a growing interest in urban-based marginalized groups (Knopp, 1987; Winchester and White, 1988; Wolch, 1990) as well as an emerging concern with what is arguably the most important issue facing women in US cities, that of the feminization of poverty (see Holcomb, 1986). Both these issues are central concerns of this study.

The paper commences with an outline of the ways in which patriarchy has been analyzed in relation to the city followed by a critique of this material, highlighting the epistemological shifts which have both caused this critique to arise and which necessitate addressing the question of how 'we', as academics, as feminists, as women, approach our work. This is followed by an empirical exploration of these critiques in relation to research conducted with low-income women, both African-American and Anglo-American, in two downtown neighbourhoods in Grand Rapids, MI, in 1989. A dual focus is taken to raise questions about the patriarchal structuring of urban social space in relation to household-level modes of behaviour, and their differentiation across 'race', as well as the role played by social relations of sexuality in the creation of a lesbian residential area. In the conclusion I outline the need for contextual understandings of patriarchy to produce better interpretations of the ways in which the constitution of gender, 'race', class, and sexuality within specific localities contribute to the social geography of the city.

The city as a patriarchal site

Numerous debates have taken place among feminist academics on the nature of patriarchy concerning its definition, ranging through patriarchy as the power of the father; as a symbolic male principle; as an ideology that has arisen out of men's power to exchange women between kinship groups; as an expression of men's control over women's sexuality and fertility; as a description of the institutional structure of male domination; and as an abstract characterization of the structures and social arrangements within which women's oppression is elaborated (see Kramarae and Treichler, 1985). Debates have also occurred on the extent to which the definers incorporate these elements into their definitions. And, extensive exchanges have occurred on the role patriarchy plays in gender inequality; that is, is patriarchy an autonomous system, the primary form of social inequality? Is it intertwined with capitalist relations in one system of capitalist patriarchy or are there autonomous but 'substantially equal' systems of patriarchy and capitalism? (see Walby, 1986). Feminist geographers have contributed to this debate through the exchange in *Antipode* between Jo Foord and Nicky Gregson (1986; Gregson and Foord, 1987) and their critics (John Gier and John Walton, 1987; Louise Johnson, 1987; Linda McDowell, 1987).

Let me begin my argument by stating what it is not about. My argument is not with the notion of patriarchy,[1] rather it is with the unspoken assumptions of its universality which have permitted its ascendancy to what I would argue is a theoretically 'fatal abstraction' (Mancini Billson, 1991) which is now in danger of being hoisted on its own petard. I will preface my critique by investigating the ways in which patriarchy has been linked to studies of the urban. A wide range of studies exists which is framed by the concern to unpack the interrelationship between capitalism, patriarchy, and urbanization. In what ways, then, has the city been seen as a site embodying patriarchal principles?

First, feminists have argued, the city has been shaped to confine women to traditional roles within the family, the site from which women perform free domestic and child-care services (Garmanikow, 1978; Markusen, 1980). This process appeared to reach its zenith in the North America of the 1950s with the collusion of the ideology of domesticity and the form of the built environment expressed through the suburbanization of the nuclear family housing unit and the idealization of family life, epitomized in the phrase 'home as haven'. Symbolically, this expression was played out in the association of men with the public world of waged work in the city and of women with the family and community in the suburbs.[2] In other words, 'male control of, and access to the city are the result of patriarchal social relations imbedded in marriage and the family' (Garmanikow, 1978, page 398). Women's attempts to extricate themselves from this situation have been analyzed in terms of looking at the household not just as a site of oppression, but also as a resource system (Mackenzie, 1987; Wallman, 1984).

Second, women's responsibility for domestic labour circumscribes their mobility (especially given the transfer of the production of goods and services from the home into the public sphere and the decline in their home delivery, ranging from milk to doctors' visits) (Harman, 1983). A wide range of social and spatial inequalities which disadvantage women more than men has been documented. Access to transportation, jobs, services, and facilities is deemed to be more difficult both physically and socially, for women than it is for men. Segmentation of the labour market, for example, leads to differences in male and female job opportunities; indeed, jobs for women and men are often located in different parts of the city (see McLafferty and Preston, 1991; Pratt and Hanson, 1991). Such studies have differentiated among women to the extent that elderly women (Helms, 1974) and women with young children (Tivers, 1988) have received particular attention.

Third, these inequalities are deeply embedded in the design and organization of urban space, embodied, as they are, not only in the sexual division of labour in the family (and in waged employment) but also in the system of urban settlement planning and various state policies, such as the provision of child care. Women's responsibilities for the reproduction of labour power involve not only unpaid activities in the home and community but also maintaining client relations with agencies of collective consumption (such as health and welfare, and housing) (Bondi and Peake, 1988). Such agencies, whether they be state agencies, nongovernmental organizations, profit-making or cooperatives, tend to reproduce patriarchal relationships in that their structures and practices are constructed around the sexual division of labour in the family.

Fourth, there are a range of social and psychological pathologies that plague women in specific urban spaces such as the female suburban neuroses characterized by Betty Friedan (1963) as 'the problem with no name'. And women's curtailed use of public space is itself seen as a spatial expression of patriarchy (see Valentine, 1989), given the restrictions on women's behaviour resulting from fear of male violence and the state's apparent reluctance to protect them from it [see in particular Pain (1991), as well as studies on women's restricted use of transport (Wekerle, 1980; Wekerle and Rutherford, 1989)].

Furthermore, there is a debate over the extent to which it is possible to step outside the conceptual heritage bequeathed to us by patriarchal systems of thought. The dualisms which have peppered the pages of studies of the city (production versus reproduction; public versus private; city versus suburbs) have been rejected by some for being patriarchal themselves. Elizabeth Harman (1983, page 106), for example, claims we 'are adopting a conceptual apparatus which is itself patriarchal in that such categories not only describe the way in which cities divide men and women, they are the basis of an ideology which helps maintain and perpetuate sub-divisions'. Although few agree on the causal mechanisms of patriarchy, agreement with Harman's claim does not preclude that the form patriarchy takes shifts over time and space. The retraction of state-supplied services and their subsequent provision by the household or the market, for example, have obvious implications for the patriarchal bases of spatial organization in terms of its effects on women's activity patterns over time and space. Indeed, this is the area which has probably received more attention than any other by feminist urban geographers and is best exemplified in the work of Dina Vaiou (1992), Suzanne Mackenzie (1987), D. Rose (1989; 1990; also Mackenzie and Rose, 1983), Linda McDowell (1983), and Susan Hanson and Geraldine Pratt (1988; Pratt and Hanson, 1991) in their collective attempts to interpret the relationships between the restructuring of urban space, changing definitions of gender identity, and 'women's place in the city'.[3]

But, despite the tensions and contradictions highlighted by feminist geographers' studies, our current conceptions of 'women's place in the city' are at worst outmoded, barely advancing beyond the advanced industrial urban spatial divides of 'city centre versus suburbs', and at best they are confused. As Liz Bondi (1991, page 196) states:

> The linkages between urban change and the constitution of gender remain weakly elaborated for the contemporary period. With the benefit of hindsight dominant versions of femininity and masculinity may be identified, and their relationship to the organization of production and reproduction explained. In the present such patterning is harder to detect; indeed there may be more chaos than order.

This chaos, I would argue, is less a result of the inability to recognize empirically generalizable patterns of the 'fit' between gender identities and urban form. More importantly, it lies with geographers' apparent reluctance to address the heterosexist and 'white' cultural constructions that pervade our discourses on social and spatial relations.[4] Whereas, for example, class is a gendered relation, constituted and experienced through both masculine and feminine identities, which themselves are subject to negotiation, 'race' and sexuality remain largely untheorized. And it is not that race and sexuality are treated as fixed categories or essences or as individual attributes: they are simply, and conveniently, not considered (but see Jackson, 1989; Knopp, 1987; 1990; 1992; Lauria and Knopp, 1985). To a large degree this state of affairs can be understood in terms of the modes of analyses that have been employed in geography, ranging through imperialist narratives, ideographic studies of landscape and regions, statistically driven models, and structuralist concerns with class, none of which has had a sensitivity to social categories of power (outside of class) and all of which have attempted to divide 'the real into useful layers and manageable bits' (Stenstad, 1989, page 335). 'Race' and sexuality have been viewed as neither useful nor manageable. Postmodernist modes of discourse, while explicitly encouraging understanding of differences, do not appear to have substantially advanced the treatment of race and sexuality. Why is this so?

Although the postmodern form of comparison of self/subject and 'other' encourages us to realize that our own identities are constituted relationally to 'others' it does not break down these differences. Indeed, Tessie Liu (1991, page 266) claims, 'maintaining the divide between *us* and *them*, I suspect, is one way of distancing the uncomfortable reality of unequal power relations, which come to the fore once we include those previously excluded'. Postmodernism's claims to be critiquing patriarchal discourse appear to be compromised by its complicit patriarchal practices (see Bondi, 1990, page 165). And, uncomfortable though it may be, if, as feminists, we want to rise above this dilemma we need to take on board the critiques of the other 'others' in particular the critiques currently being voiced by women of colour and by radical feminist and lesbian women demand our attention on two levels. Understanding the 'raced' and sexualized nature of social and spatial relations affects not only the ways we interpret our empirical findings; implicatively, it also requires we address the terms on which we conduct our research. Each of these matters will now be addressed, starting with the latter.

Patriarchy, 'race', and sexuality

Specifically, I wish to address the question of the extent to which white feminists can do research 'on' or 'with' black women and, similarly, heterosexual women can do research 'on' or 'with' lesbian women. Let me start by outlining what I think cannot be done. Two arenas are closed to me. First, as a feminist, I have abandoned the notion of a social science built upon concepts of objectivity, detachment, and value-free research; choices about what to study and how to study are politically laden. Second, I would agree with certain black feminists that there are spaces that some theorists, for example, white feminist academics, cannot occupy (or at least their occupation is only possible with the agreement of black feminists). I am not putting forward this argument on the grounds of some essential notion of whiteness or white colour, but on the recognition that we all have identities forged out of experiences of domination and oppression. Not being able to identify with those struggles, as bell hooks (1990) claims, being too distanced from the pain of that domination and oppression, denies any meaningful entry into that space.

Certain black feminists, however, now appear to dominate the terms of this debate to the extent that other feminists have to elaborate their positions in relation to them.[5] In their efforts to wrest control of developments in feminist theory they are delivering a potent rhetoric of political correctness that can strike panic in feminists who are sympathetic to their concerns.[6] Their argument, as articulated by its most radical proponents, appears to be based on the assumption that work conducted by white, First World feminist academics consciously or nonconsciously serves to increase and extend racist ways of thinking. They are deeply sceptical that any of this work can be turned to good purpose. Feminists appear to stand divided again, though now it is a division characterized by the 'bad white feminists' versus the 'good black feminists'. Moreover, it increasingly appears to be a divide that is recognized in terms of skin colour.

Although black feminists are right to insist upon us all being aware of our ethnocentric biases and the gendered and 'raced' construction and, often, purpose of knowledge, I would question the argument that skin colour is the primary determinant of one's suitability to conduct research.[7] The reasoning underlying this proposition appears to be based on a historical, and undifferentiated and absolute, notion of 'blackness' and 'whiteness'. Even within a racist society are not some black/white women far removed from and having interests variant with other black/white women? As well, what are the implications of such a view for internationalism? To counterpose black and white feminists in such a way leads to an impasse difficult to resolve. And, by pitting category against category, the discussion has effectively masked the real women behind these debates, depriving itself of a human and humane face, as if the emotional realm lay outside the research process. It removes from consideration the many successful research projects entered into jointly by black and white feminists (see Albrecht and Brewer, 1990; Bunch, 1987). To deny that a number of white feminists have entered into research projects 'on' or 'with' Third World women for dubious reasons, such as their supervisor's whim or uncritical, romanticized, and racist beliefs that they are 'doing good', would be foolish. To deny white feminists and black feminists the opportunity to work jointly, albeit from a conscious recognition that material inequalities in positions of empowerment and disempowerment exist, would be even more so.

In contrast, a more positive way to go forward would be to recognize the contextually specific nature of our understandings – both intellectual and emotional – of each other and of the social world, which form the bases of our 'ontological distinctiveness' from each other (Stanley and Wise, 1990). To build upon these individualist perspectives requires a process of intersubjectivity, that is, a recognition that we all share experiences, in that we 'recognize ourselves in others and they in us' (Stanley and Wise, 1990, page 23). Indeed, it is the very process of intersubjectivity that can allow for a progressive union of the tensions – political and epistemological – between feminist discourses and allow us to relate to women who have not been reproduced in our own image. My relationship towards the women who spent time with me doing research is one of, what Maria Mies (1988) refers to as, 'conscious partiality'. That is, I have some degree of identification with these women, not on the grounds of us all belonging to some essentialist category of 'woman' but because as Dorothy Smith (1987, page 142) asserts: 'The sociological inquirer herself [is] a member of the same world she explores, active in the same relations for whom she writes. Like Jonah, she is inside the whale'. White feminists, then, need to accept that they are simultaneously privileged and oppressed.

The cornerstone here is that, although these subjective experiences of domination and oppression are authentic starting points for research, they alone are insufficient. The political does not reduce to the personal. As Pratibha Parmar asserts, we have to be willing to work across all our differences (1989, page 64). The methodological implications of this view are that our work must necessarily address the issue of intersubjectivity with the explicit recognition of two points. The first is that we are engaged in an interpretative act, what Clifford Geertz (1973) refers to as 'thick description': the double hermeneutic of interpreting others' interpretations of what they are doing (Jackson, 1989). And second, we enter into this interpretative act from what has been variously referred to as a 'politics of position' (Jackson, 1991) or 'politics of location' (Borsa, 1990), that is, 'those places and spaces we inherit and occupy which frame our lives in very specific and concrete ways, which are as much a part of our psyches as they are a physical or geographical placement' (Borsa, 1990, page 36).

Identifying these positions allows an enhanced awareness of the epistemological shifts that enable us to 'identify the spaces where we begin the process of re-vision' (hooks, 1990, page 145).[8] These spaces, in relation to a critique of patriarchy, have started to be opened up by black feminists and lesbian and radical feminist academics who have pointed out the roles of racism and heterosexism in contributing to material inequalities between women. The analytical contributions they have made to this process are outlined below.

A number of black feminists claim that the (largely) Eurocentric assumptions embodied within patriarchy and its use as an analytical tool have to be qualified within the context of racist society (Tang Nain, 1991, page 5). For example, a key element of women's subordination central to an understanding of patriarchy is that of reproduction. Black feminists argue that this concept becomes problematic when we take into account the role that black women and women of colour have played and still play in reproducing white labour, both through slavery and domestic service (Carby, 1982; Tang Nain, 1991). These are processes in which white women, as much as white men, have played a complicit part. Furthermore, a related point is the substantially different roles played by white and black men in the reproduction of the family unit. White women's economic dependence or subordinate position to men is not echoed in black families in which black women have higher rates of labour-force participation and more experience of being the sole income provider for their dependants (Stasiulis, 1990).

Second, black feminists have attacked white feminist critiques of the family, pointing out that the family is a contradictory site, embodying oppressive elements to women but also being an arena of solidarity and resistance against racism (Carby, 1982; Jones, 1985). Furthermore, Hazel Carby (1982, page 215) claims that black family structures are seen as deviations from the norm of the white nuclear family and are being pathologically constructed within white feminist theory as being more oppressive, that is, the extended family structure is associated with Third World 'backwardness' and precapitalist oppressive structures. Furthermore, the concentration on the nuclear family as *the* site of women's oppression negates the range of household and kinship relations in black cultures. It also serves to obscure from analysis the racist nature of citizenship and immigration laws which have often served to separate members of black families (Stasiulis, 1990).

Last, they argue that the conventional portrait of black fathers as irresponsible and absent also feeds into a construction of black masculinity in which black men are portrayed as failures 'who are psychologically ''fucked up'', dangerous (especially to white women), violent, sex maniacs' (hooks, 1992, page 89; see also Mama, 1989; Wallace, 1990). It follows that black men have largely been denied the benefits of white patriarchy. The use of the term patriarchy must be modified then to recognize that black women have been dominated patriarchally in different ways by different groups of men. Similarly, it follows that other relations of patriarchal oppression, such as those of waged labour and the state, are 'raced' and have differential impacts on white and black women. Quite obviously, any analysis of patriarchy which fails to place it within a broad political and economic framework may well serve to obscure the racialized nature of, for example, state relations, and the intertwining of relations of patriarchy and racism.

Critiques of patriarchy have also come from lesbian and radical feminist academics and can be applied to redefinitions of the meaning of urban space, 'through the subversion of the heterosexual assumption' (Davis, 1992, page 1).[9] Current work by lesbians and radical academics centres on a discussion of patriarchy in terms of its links to male-dominated sexuality (see Ettore, 1980; Ferguson, 1989; Rich, 1984; Stenstad, 1989; Walby, 1990). The concern is that patriarchal institutions and ideologies are understood not only in terms of sexual inequality, but also in relation to enforcing a compulsory heterosexuality. The two are intertwined, although the degree to which and the extent to which they are characterized as having causal properties are questions open to dispute (see Mackinnon, 1979). But most would agree that 'sexuality is socially organized and critically constructed by gender inequality' (Walby, 1990, page 121). Their argument is not based though on an essential female sexuality: sexuality manifests itself in widely different forms from place to place.

The heterosexism and homophobia of many heterosexual feminists has been highlighted through, for example, lesbian baiting within 'mainstream' feminist organizations. Other practices in which heterosexual women have often been complicit include attempts by the state to exclude lesbians from certain reproductive activities such as child rearing. These practices, like the criticisms of black feminists, problematize the notion of reproduction. But, because sexuality is a patriarchally constructed institution, lesbianism, which questions the structure of power relations and which explores contradictions about beliefs between biology and culture, between the sexual and the ideological, provides the basis for a serious challenge to patriarchy (Ettore, 1980; Ferguson, 1989; Kitzinger, 1987). This is discussed below in relation to a lesbian politics of place construction.

Challenging the constitution of 'race' and sexuality in Grand Rapids

To build on this analytically expanded conception of patriarchy, an investigation was conducted into the everyday activities of women in two low-income downtown neighbourhoods in the city of Grand Rapids, MI, with the aim of ascertaining whether empirical challenges are being made to the patriarchal structure of urban social space. Two aspects of this research are reported: one investigating activities mediated by 'race', and the other concerned with sexuality. These will be prefaced by a brief description of the Grand Rapids area.

Grand Rapids is a prime example of the uneven urban development resulting from economic restructuring and changes in state policies. Its context is of a buoyant local economy, though one in which a decline in real wages for a substantial sector of the population can be depicted. The last fifteen years in particular have witnessed a major restructuring of the local economy, with its transition to a service economy and an emphasis on headquarters functions and business services. The downtown corporate business complex now dominates the city and has generated a major spin-off of low-level service jobs, predominantly for women (Michigan Employment Security Commission, 1990).

Data were collected in the period January–April 1989, from in-depth interviews with ninety-seven women, fifty-three of whom were Anglo-American and forty-four African-American, living in two low-income downtown neighbourhoods. The selection of neighbourhoods was based on a number of factors. A group interview was conducted with neighbourhood organizers from the Grand Rapids Council of Neighbourhood Associations to identify those willing to have research conducted in their neighbourhoods. Although this process left over half of the neighbourhoods agreeing to participate, the small size of the majority of the neighbourhoods required that two be selected. Contiguous areas which utilized the same spatially delineated welfare services and transportation systems were needed. Primarily, the study required a mix of both Anglo-American and African-American women residing in the neighbourhoods. In total these considerations left open only a handful of neighbourhoods suitable for study. I will now turn to interrogate the role of the household, that is, the site of social reproduction in which living arrangements such as modes of behaviour and resources are mediated and experienced.

'Race', low-income women, and living arrangements

As argued earlier, patriarchy, as a primary determinant of urban spatial structure, is expressed and reproduced primarily through the patriarchal form of household organization and the single-family form of housing. The patriarchal structuring of social reproduction adopted here follows that of Eva Garmanikow (1978, page 390) 'in which the labour power of women within a capitalist system is employed partially or wholly in the service of men in the household and where the returns to both women's and men's labour are contained in the family wage'. First, then, it is necessary to ask to what extent the households in the study fitted this partiarchal pattern. Approximately 63 per cent of African-American women and 49 per cent of Anglo-American women were heads of households; that is, only about a third of African-American women and a half of white women lived in households with men.[10] This situation was not necessarily out of choice; the majority of Anglo-American heads of household were elderly women on their own, and African-American head of household were young women who through a combination of the economic marginalization of the African-American male and state policies [they were only eligible for Aid for Families with Dependent Children (AFDC) if no men were residing in the home] have had to separate arrangements for child rearing from the institution of marriage.[11] This finding was further emphasized by the fact that no African-American women were living with men who were unemployed. Whether through choice or constraints, well over half of the women in the study were adapting to and struggling against a new form of patriarchal structures by gaining access to residence in the city neither through marriage nor through relations with men.

Moreover, when we look at income coming into the household, in only 13 per cent of households (in six African-American households and seven Anglo-American ones) were men responsible for providing the sole income and in 64 per cent of African-American households and 45 per cent of Anglo-American households women provided the only income.[12] More significantly, the data on who was responsible for making decisions about money that came into the household revealed that in only 5 per cent of cases was it the male alone, in 34 per cent of households decisions were taken jointly, but in 58 per cent they were taken by women alone; that is, in only 10 per cent of households where a male was present was he alone responsible for making financial decisions. It would appear that the patriarchal ideology of the male breadwinner was not a daily reality in the majority of these women's lives.

It is not merely household composition that is of import to understanding the patriarchal reproduction of social space: the spatial organization of households is also relevant. This is probably best illustrated in relation to child care. In the thirty-five households with preschool age children requiring day care, fully 78 per cent used family resources alone to cope, a practice which possibly arose from the large number of these households living in extended family arrangements or living 'round the corner' from relatives. Indeed, one third (eleven) of these households had moved into these neighbourhoods solely to be near family, that is, female relatives, for day-care purposes. That this use of the family as a resource was a low-income household strategy appeared to be exemplified in the finding that only ten households paid for day care (eight with sitters and two in formal day care) and these all had higher incomes than the other households concerned.

Can we surmise that modes of behaviour such as these leading to the dismantling of patriarchal household arrangements are 'restructuring' the social geography of cities? Although patriarchal household forms and urban spatial patterns are being transformed, there is little evidence that patriarchal relations per se are breaking down. They are particularly evident in the allocation of activities within the household.

Although there were male spouses present in 44 per cent of the households, they systematically exhibited a very low level of involvement in housework. The only significant areas of their involvement were repairs and maintenance, and yardwork, which have traditionally been viewed as 'man's work'. Perhaps, even more significant is that even in households with no spouse present the only tasks not undertaken solely by women were washing dishes, which was also done by children, and once again yardwork, and repairs and maintenance which were either undertaken by male landlords, other male relatives not living in the home, or paid help, which was again, without exception, male. There is little evidence of equality in power relations between men and women in the division of labour in the household: women are overwhelmingly organizing household activities in a way which allows them to bear the costs of the reproduction of labour power.

Furthermore, economic independence from men appears only to change the nature of vulnerability for low-income women. For example, when we examine the source of income which is used for day care, rather than its level, the most severe lack of choice in terms of day-care options was for single African-American women on AFDC, of whom there were nine (plus one Anglo-American woman) with no relatives living nearby. Their only option was to stay at home. Although mothers on AFDC are given child care allowances, these were only $1.05 per child per hour. The average national cost of day care is $3,000 a year, or approximately $2 an hour; that is, AFDC payments are insufficient to cover day-care costs. Moreover, because AFDC is paid on a reimbursement schedule, some family day-care providers will not accept children of AFDC clients because of the time lag between service provision and reimbursements. The decision to stay at home appears to be a rational choice given the insufficient AFDC subsidy and the low wages these women could command in the marketplace. With AFDC the state is taking over the role of the absent father, thus keeping intact the gendered power structure of the family; in these cases the patriarchal ideology of the male breadwinner persists even when the reality recedes (Acker, 1988).[13] Indeed, the state was a major resource for nearly half the women in the survey. Over half of the African-American women and a third of Anglo-American women received transfer payment and for all the African-American women (bar three) and for ten of the seventeen Anglo-American women these were their main sources of income. Patriarchy thus does not appear to be breaking down so much as changing its form from private to public: from the control of individual patriarchs to an arena of collective control by men to whom they may not be related (see Walby, 1990).

In this brief review, I have attempted to illustrate how the social impact of the patriarchal reproduction of urban social space varies across different types of households. It appears that it is the most marginalized groups that are bearing the brunt of patriarchal forces, expressed in the many references by these women which indicated their despair: 'there's a lot of pain out there'. Although the nuclear form of the household no longer appears to be the major site of patriarchal oppression, patriarchal relations outside the household help to shape dependency on men. As shown, the state has taken over as a major source of subordination in many of the women's lives, effectively preventing many of them from full participation in the public arenas. Lack of family support, inadequate public transit systems, reductions in state welfare payments, etc. contribute to the constraints on these women's activity spaces, restricting them to the confines of their neighbourhood. Most importantly the raced and classed outcomes of these are manifest in that it is low-income, single African-American women with children who are most confined, and single, Anglo-American women without children who have the greatest mobility.

Sexuality, lesbian women, and urban space

Although the social relations of sexuality have received increasing attention from geographers (see Bell, 1991; Jackson, 1989; Knopp, 1987; 1990; Lauria and Knopp, 1985; Lloyd and Rowntree, 1978; McNee, 1984; Weightman, 1981; Winchester and White, 1988), the majority of studies focus on descriptive accounts of (male) homosexual geographies investigating the role of institutions, such as housing and leisure services of the homosexual community, in particular, the role of gay bars in emerging gay subcultures. More sophisticated studies examine the role and impact of gay communities on the urban fabric at the neighbourhood level. However, these also concentrate almost extensively on gay males' territories. The dearth of geographical (and other social science) research on lesbian residential areas (but see Adler and Brenner, 1992; Davis and Lapovsky Kennedy, 1986; Ettore, 1978; 1980; Goleman Wolf, 1979; Valentine, 1993a; 1993b; Winchester and White, 1988) has been 'explained' by lesbians' low visibility, which in turn has been viewed partially as a function of their 'lower average disposable income' (McNee, 1984, page 2); of the possible perception of a greater need for territory by male gays (Johnston, 1978, referenced in Lauria and Knopp, 1985); to lesbians' lack of the 'innate male territorial imperative' (Castells, 1983, referenced in Lauria and Knopp, 1985, page 158).

Such empirical and conceptual generalizations smack of an inability to rise above the level of the patriarchal mire, of being unable to unpack the heterogeneity of class, 'race', and other relations that characterize the lesbian community. As Ross (1990, page 88), in her analysis of lesbian organizing in Toronto points out,

> In large urban centres across Canada and other Western countries, the 1980s have heralded the subdivision of activist lesbians into specialized groupings: lesbians of colour, Jewish lesbians, working class lesbians, leather dykes, lesbians against sadomasochism, older lesbians, lesbian youth, disabled lesbians and so on.

It is not that spatial expressions of sexuality are unimportant for lesbians, or that territorial bases have no significance for their political identity. Rather, the form these spatial expressions take are different from those that have so far been studied. Indeed, it is highly improbable that the studies of the high-visibility gay ghettoes of the Village (New York) and Castro (San Francisco) encapsulate the territorial experiences of most gay men. The few studies conducted so far reveal that lesbian residential areas usually appear to be low profile, not easily identifiable, nor necessarily permanent.[14] But this is not to argue that lesbian sexuality has no part to play in defining the social character of space at the neighbourhood level, as is described in the following brief outline.

In Grand Rapids, one lesbian, Karen (not her real name), has been responsible, over the last fifteen years, for establishing a lesbian and women-only environment, covering six or so houses in two short adjacent streets in one of the neighbourhoods under investigation in this study. Up until the mid-1970s she had been living in a rented house in the suburbs, accumulating the capital to buy her own house. She liked the look of the solid houses on one of these streets in this neighbourhood so she waited until a duplex came up for sale and bought it (for $23,000). Soon after this she was made redundant from her job, and using her severance pay she was able to purchase two commercial stores (for $26,500) which she rented out to friends to open up a neighbourhood restaurant. Using her rental income she was able to buy more houses in the area. She only bought houses she would like to live in herself or houses she could see from her own home. She renovated these and rented them out to both lesbian and heterosexual women. Her next goal is to purchase a nearby commercial property in order to rent it out to a neighbourhood business.

What does this short synopsis reveal about the spatial basis of gay identity and politics? First, it shows that sexuality and gender are entwined in our understanding of 'gay ghettoes'. Unlike many male gay ghettoes, which are characterized by their high visibility, this lesbian residential area had a low visibility and was, therefore, less likely to be subject to police harassment and other forms of repression (such as hate male and graffiti). Indeed, the study area can hardly be defined as a ghetto or enclave given the intermixing of lesbian and heterosexual households and the fact that Karen believed that few local people were aware of the existence of such a community.[15] Further evidence of this overlap between gender and sexuality is in their bases for women's oppression. Apart from Karen, all other women occupying her houses were renters. Whether by choice or because of lack of capital, these women, lesbian and heterosexual alike, in their gendered restricted access to private property, were experiencing the same patriarchal forms of exploitation (see Jackson, 1989). This experience stands in direct contrast to that of gay men who tend to benefit economically, although not all to the same degree, from processes of gay ghetto formation (see Knopp, 1990).[16]

Second, it illustrates that the public space of lesbians cannot be understood purely in terms of the public versus private, production versus reproduction conceptualization of space; the creation of lesbian public space is motivated by the need to provide a location for the nurturing of lesbian identities. The specific act of renting to sympathetic straight and lesbian friends whom Karen could trust has led to the emergence of a space in which lesbian friendships and networks can function. One of her own statements best testifies to this: 'It's also important to have fun events – street parties – otherwise you're only ever asking people for help and that gives the impression that nothing good is happening and that's not the case'. Moreover, she views her relationship to her tenants as one of 'partnership'. Her desire to only buy houses she can see from her own home is based on the fact that she can then 'stay in touch with it'. She feels that her responsiveness gives 'security to the tenant'. Quite emphatically, she was not gentrifying the neighbourhood but 'improving' it, providing a physically attractive, economically affordable, secure residential area for her, primarily lesbian, women friends. Karen, then, had a vision. She is transforming the local area in such a way as to reflect her cultural politics and serve the needs of (lesbian) women. Sexuality and sexual identity were not incidental to this endeavour but rather its linchpin; the formation of a lesbian residential area represents a political act aimed at securing access to residential areas of the city which are not mediated through relations with men.

Third, this synopsis reveals a close intermeshing of lesbian politics with the new urban politics. The political agenda of gay liberation in this case study is concerned not only with sexuality; it also has an economic dimension. The availability of affordable and renovatable housing in an area which, because of its location and social history of 'white flight', holds little appeal to gentrifiers provided Karen with the opportunity to 'develop a territorial *and* economic base for . . . developing community resources' (Knopp, 1992, page 665). This is illustrated through the cultural coding in which increases in both use and exchange values were generated by (what Karen referred to as) this 'slow and sensitive' project. The revalorization of physical structures in the neighbourhood created both an income for Karen and workplaces and homes for others. In addition, she also promoted local employment by using only local people from the neighbourhood to work with her on major renovations. Her plan to buy a property on the nearby commercial strip is part of a project of the South East Economic Development Corporation. This is a nonprofit community development agency with its origins in a grass-roots neighbourhood association, which is aiming to revitalize the businesses on the strip that serve the neighbourhood by renting out properties to business owners sensitive to, and willing to cater to, the needs of local residents.

These activities, however, can be only partially understood as contributions to the process of community economic development, for they are both constituted and experienced within the cultural construction of lesbian politics. This is most vividly realized in the statements by Karen about the commercial property. She knew she would be 'taking a chance as to whether she'd get her investment back', and that her prime motivation was not economic but to regain the territory (from the graffiti gangs, and drug dealers), 'to be in practically on what's going on in . . . Street' so that she could help control the aesthetics of redevelopment and 'keep the streetscape intact, i.e. no demolitions'. This concern with aesthetics was further evidenced by the fact that her house repairs and renovations were always carried out in relation to the long-term goal of improving the physical attraction of the street. More importantly, for her, she feels her work has attracted women 'who care about property instead of just using it up. Visually, it's an uplifting experience to see a place renovated.' Moreover, she argues that, as she has acquired the skills to do these renovations along the way, she can carry out the bulk of the work herself and, hence, keep rents for her women friends at a 'reasonable' level, further testimony to the argument that economic considerations were mediated through her lesbian identity.

Last, these residential activities and choices have to be considered in light of the constraints imposed by dominant ideologies of heterosexism and homophobia. The austere morality of the 1980s, the concern with 'family values' and the 'work ethic' (often cited as Grand Rapids' greatest assets), are reemphasized by the local hegemonic controls upon acceptable behaviour constructed through Calvinism and the Dutch Reformed Church.[17] As Jeff Swanson, president of the Lesbian/Gay Community Network of Western Michigan stated, 'To get along here in Grand Rapids you have to be hidden' (Kyle, 1988, page 74). And Gwen DeJong, their female president, believes lesbian women are more likely than gay men to keep closeted. In the claustrophobic atmosphere of a city where the mayor refused to sign a proclamation in celebration of a gay rights day, likening gays and lesbians to Nazis, the low-level visibility of lesbian residential areas takes on yet another meaning. As Tim Davis (1992, page 2) asserts, 'heterosexism and homophobia are social constructions with spatial impacts that are not always clearly visible in the physical landscape'.

Conclusion

The aim of this study has been to reveal how dominant conceptualizations of patriarchy in urban studies have been contested, most specifically by the arguments of African-American and lesbian and radical feminists. Their rejection of an abstracted conceptualization of patriarchy arises from their insistence on a politics of location. This development allows us to understand better and interpret empirical evidence, simultaneously requiring us to interrogate constantly our work on both political and aesthetic grounds (hooks, 1990).[18] Most importantly, the acceptance of a politics of location means going beyond the recognition that structures of power privilege certain voices and silence others to reviewing categories and concepts that lack applicability and relevance to nonwhite and lesbian women, thus uncovering mechanisms of racism and heterosexism. This not only enables us to see the connections among the different ways in which women are oppressed (as opposed to the 'sameness' of women's oppression) but also the ways in which these are specific to gender, 'race', sexuality, and class.

With these considerations in mind the dominant concern is not whether patriarchy is being dismantled by the social practices of urban life. Rather rejection of abstracted conceptions of patriarchy requires focusing understanding on the differential experiences and negotiation of social relations of gender, 'race', class, and sexuality as sites for the contesting of patriarchal structures. This shift necessitates adopting a contextual approach to the ways in which space is being used to challenge the hegemonic discourses of patriarchy. This requires investigation of the range of meanings through which patriarchy's cultural representations and material bases are expressed.

Both case studies demonstrate the attempts by women to insulate themselves from patriarchal oppression and adapt themselves to the oppression while retaining dignity and identity. But, whereas women have always adapted to the constraints and limitations of patriarchal social relations, the spatial outcomes of strategies vary within and between 'race', sexuality, class, and family household status, in a complex manner. Obviously building up a social geography of the city from this microscale juxtaposition of social groups is a difficult task; but it is only at this local scale that the degree and character of modes of behaviour can be exemplified (Mingione, 1987) and meanings of different resources, of the family, of social networks, the state, and the formal and informal economy, can be understood. The various combinations of these resources form the determinants of spatial variations in social patterns of production and reproduction: the processes whereby they come to be differentially manifested between gendered, 'raced', and classed groups appear to hold most promise for the ways we conceptualize urban places and understand whether, and how, patriarchal structures are being undermined or reinforced.

Notes

1 The contested nature of the term patriarchy is, however, recognized. The argument I am propounding here is that its analytical usefulness has been hampered by its broad treatment which has masked the variable nature of the social sites and institutions through which patriarchal practices are played out. Furthermore, although it operates as a system of domination, it is also a system of conflict; women, although not all women at all times and places, have resisted male tyranny (Rich, 1984). Such recognition highlights the contestable nature of patriarchy, as a system of bargaining and compromises, rather than one of absolute male domination (Gordon, 1992). It is my contention that there is more to be gained than lost by continued, but reinvigorated, use of the term.

2 But see Margaret Marsh's (1990) work on femininity and the suburbs which richly documents the complexity of the patterning of events that led to the ideology of domesticity.

3 The reference to collective attempts should not be interpreted as indicative of a unified body of thought or beliefs on this topic. Since the late 1980s differences and dissension among feminist geographers have been increasing as witnessed by the debate on feminism(s) in the academy by members of the Women and Geography Study Group of the Institute of British Geographers (see Penrose *et al.*, 1992).

4 I am referring to whiteness here as sign and symbol, as bell hooks (1990, page 166) affirms, 'as a concept underlying racism, colonization and cultural imperialism'. White refers not only to skin colour, but to the norm from which all 'others' deviate.

5 See Elshtain (1990) for a similar argument in relation to ecofeminists and reproductive technology.

6 See Valerie Amos and Pratibha Parmar (1984, page 11), for example, who state, 'White feminists beware – your unquestioning and racist assumptions about the black family, your critical but uninformed approach to "Black Culture" has found root and in fact informs state practice'.

7 See Parmar (1989) who argues that certain constructions of blackness by black feminists have been based on essentialist notions. Although skin colour is an important element in racist ideologies I agree with Daira Stasiulis that 'arguments most commonly expounded by Black feminists . . . have the effect of linking racism to skin colour, rather than to the structural location of particular groups of women in concrete and historically specific social relations and to the accompanying discourses that aid in the processes of denigration, subordination and exploitation' (1990, page 290). This is not to deny, however, that black women by virtue of their skin colour have something in common that white women cannot experience: 'A poor woman may have a wildly different life from an upper class black woman but the upper class woman is still black – she cannot get away from that no matter how many privileges she has. She will always be black in a "white dominant" society' (written communication with Alison Grant).

8 According to hooks (1990, page 205), these epistemological shifts involve us separating out the useful from oppressive knowledge gained from the dominating group in order to refrain from participating in ways of knowing that can lead to estrangement, alienation, assimilation, and co-optation. In addition it requires that we start to dismantle the 'whiteness' of our conceptual frameworks, and recognize the worth of values and traditions that come from outside the academic tradition.
9 The terms lesbian and radical feminist are used to denote their differences; whereas some lesbians may categorize themselves as radical feminists, not all radical feminists are lesbians or choose to identify themselves as such.
10 Of these 63 per cent of African-American women, two thirds had children under the age of 18 years old living with them, a sixth lived with relatives other than children, and a sixth lived on their own. The corresponding figures for the 46 per cent of Anglo-American women were a fifth, none, and four fifths.
11 See Susser (1982, chapters 3 and 4). Ida Susser's ethnography of a low-income urban area in Williamsburg – Greenpoint, Brooklyn in the mid-1970s during the period of New York's fiscal crisis highlights the way in which the welfare system further contributed to the economic marginalization of low-income women.
12 Over 90 per cent of the women had annual household incomes of less than $30,000 and a half of the African-American women and a third of the Anglo-American women had annual incomes of less than $15,000.
13 This is especially so when we consider that US state programs directed to women are linked to marital and parental status, as childbearers or caretakers of children, whereas those directed to men (who are the majority recipients of the state welfare payments) are allocated on the basis of having had periods of employment or military service and are often linked to past earnings (see Acker, 1988).
14 The best known of these studies is probably that of Elizabeth Ettore (1978). She documents the rise of a lesbian feminist ghetto in Lambeth, London, in the early and mid-1970s which survived for six years and housed fifty lesbians on two streets in a working-class neighbourhood. See also the studies by Adler and Brenner (1992), Davis and Lapovsky Kennedy (1986), Goleman Wolf (1979), Valentine (1993b), Winchester and White (1988).
15 Indeed, many residents would be unaware because it is considered normal (in Western societies) for women to live together. The same is not true for men who, therefore, when they do live together have a higher, although unintentional, visibility (Alison Grant, written communication).
16 This case study appears to contrast sharply with that of Mickey Lauria and Larry Knopp (1985) in which the contradictions between use value and exchange value resulted in tensions in the male gay community, although their conclusion is certainly apposite to the study: 'Gays, in essence, have seized an opportunity to combat oppression by creating neighbourhoods over which they have maximum control and which meet long-neglected needs' (page 161).
17 Although in terms of the number of churches in the standard metropolitan area the Dutch Reformed Church has a declining proportion, its influence is out of all proportion to its members. In a city in which institutionalized religion plays a significant role – there are more churches per 1,000 population than in any other US city – the ultraconservative ideology espoused by the Dutch Reformed Church has a deep and widening hold.
18 This integration applies on two levels, both for our work and ourselves: as Joan Borsa states: 'Part of our struggle is to be able to name our location, to politicize our space and to question where our particular experiences and practice fit within the articulation and representations that surround us . . . by articulating our

specific experiences and representing the structural and political spaces we occupy, we offer concrete accounts of how and where we live, what is significant to our experience of cultural identity, how we have been constructed and how in turn, we attempt to construct (and reconstruct) ourselves' (1990, pages 36–37). Without understanding these things about ourselves, how can we hope to understand them?

References

Acker, J. 1988: 'Class, gender and the relations of distribution'. *Journal of Women in Culture and Society* 13, 473–97.

Adler, S. and Brenner, J. 1992: 'Gender and space: lesbians and gays in the city'. *International Journal of Urban and Regional Research* 61, 24–34.

Albrecht, L. and Brewer, R. (eds), 1990: *Bridges of Power, Women's Multicultural Alliances.* Philadelphia, PA: New Society Publishers.

Amos, V. and Parmar, P. 1984: 'Changing imperial feminism'. *Feminist Review* 17, 3–19.

Bell, D. 1991: 'Insignificant others: lesbian and gay geographies'. *Area* 23, 323–9.

Bondi, L. 1990: 'Feminism, postmodernism, and geography: space for women?' *Antipode* 22, 156–67.

Bondi, L. 1991: 'Gender divisions and gentrification: a critique'. *Transactions of the Institute of British Geographers* 16, 190–8.

Bondi, L. and Peake, L. 1988: 'Gender and the city: urban politics revisited'. In Little, J., Peake, L. and Richardson P. (eds), *Women in Cities.* Andover, Hants: Tavistock Publications, 21–40.

Borsa, J. 1990: 'Toward a politics of location. Rethinking marginality'. *Canadian Women Studies/Les Cahiers de la Femme* 11(1), 36–9.

Bunch, C. 1987: *Passionate Politics.* New York: St Martin's Press.

Carby, H. 1982: 'White woman listen! Black feminism and the limits of sisterhood'. In Centre for Contemporary Cultural Studies (ed.), *The Empire Strikes Back.* London: Hutchinson, 212–35.

Castells, M. 1983: *The City and the Grassroots.* Berkeley, CA: University of California Press.

Cox, K. 1991: 'Questions of abstraction in studies of the new urban politics'. *Journal of Urban Affairs* 13, 267–80.

Davis, M. and Lapovsky Kennedy, E. 1986: 'Oral history and the study of sexuality in the lesbian community: Buffalo, New York, 1940–1960'. *Feminist Studies* 12(1), 7–26.

Davis, T. 1992: 'Social processes and political activism: alternative lesbian and gay spatial strategies for social change', paper presented at the Annual Meeting of the Association of American Geographers, San Diego, April: copy available from author at Graduate School of Geography, Clark University, Worcester, MA 01610.

Elshtain, J. 1990: *Power Trips and Other Journeys. Essays in Feminism and Civic Discourse.* Madison, WI: University of Wisconsin Press.

Ettore, E. M. 1978: 'Women, urban social movements and the lesbian ghetto'. *International Journal of Urban and Regional Research* 2, 499–519.

Ettore, E. M. 1980: *Lesbians, Society and Women.* Andover, Hants: Routledge and Kegan Paul.

Ferguson, A. 1989: *Blood at the Root: Motherhood, Sexuality and Male Dominance.* Winchester, MA: Pandora.

Foord, J. and Gregson, N. 1986: 'Patriarchy: towards a reconceptualization'. *Antipode* 18, 186–211.

Friedan, B. 1963: *The Feminine Mystique*. New York: W. W. Norton.

Garmanikow, E, 1978: 'Introduction. Special issue on women and the city'. *International Journal of Urban and Regional Research* 2, 390–403.

Geertz, C. 1973: *The Interpretation of Cultures*. New York: Basic Books.

Gier, J. and Walton, J. 1987: 'Some problems with reconceptualizing patriarchy'. *Antipode* 19, 54–8.

Goleman Wolf, D. 1979: *The Lesbian Community*. Berkeley, CA: University of California Press.

Gordon, L. 1992: 'A right not to be beaten: the agency of battered women, 1880–1960'. In Helly, D. and Reverby, S. (eds), *Gendered Domains. Rethinking Public and Private in Women's History*. Ithaca, NY: Cornell University Press, 228–43.

Gregson, N. and Foord, J. 1987: 'Patriarchy: comments on critics'. *Antipode* 19, 371–5.

Hanson, S. and Pratt G. 1988: 'Reconceptualizing the links between home and work in urban geography'. *Economic Geography* 64, 299–321.

Harman, E. 1983: 'Capitalism, patriarchy and the city'. In Baldock, C. and Cass, B. (eds), *Women, Social Welfare and the State in Australia*. Sydney: Allen and Unwin, 104–29.

Helms, J. 1974: 'Old women in America: the need for social justice'. *Antipode* 6(3), 26–33.

Holcomb, B. 1986: 'Women in the city'. *Urban Geography* 5, 247–53.

hooks, b. 1990: *Yearning*. Toronto: Between the Lines.

hooks, b. 1992: 'Reconstructing black masculinity'. In *Black Looks: Race and Representation*. Boston, MA: South End Press, 87–114.

Jackson, P. 1989: *Maps of Meaning*. London: Unwin Hyman.

Jackson, P. 1991: 'The crisis of representation and the politics of position'. *Environment and Planning D: Society and Space* 9, 131–4.

Johnson, L. 1987: '(Un)realist perspectives: patriarchy and feminist challenges in geography'. *Antipode* 19, 210–15.

Johnston, G. 1978: *Which Way Out of the Men's Room?* South Brunswick, NJ: A. S. Barnes.

Jones, J. 1985: *Labour of Love, Labour of Sorrow*. New York: Basic Books.

Kitzinger, C. 1987: *The Social Construction of Lesbianism*. London: Sage Publications.

Knopp, L. 1987: 'Social theory, social movements and public policy: recent accomplishments of the gay and lesbian movements in Minneapolis, Minnesota'. *International Journal for Urban and Regional Research* 11, 243–61.

Knopp, L. 1990: 'Some theoretical implications of gay involvement in the urban land market'. *Political Geography Quarterly* 9, 337–52.

Knopp, L. 1992: 'Sexuality and the spatial dynamics of capitalism'. *Environment and Planning D: Society and Space* 10, 651–69.

Kramarae, C. and Treichler, P. 1985: *A Feminist Dictionary*. London: Pandora Press.

Kyle, C. 1988: 'Grand Rapids: applause amidst angst'. *Grand Rapids Magazine* September, 62–76.

Lauria, M. and Knopp, L. 1985: 'Towards an analysis of the role of gay communities in the urban renaissance'. *Urban Geography* 6, 152–69.

Liu, T. 1991: 'Teaching the difference among women from a historical perspective: rethinking race and gender as social categories'. *Women's Studies International Forum* 14, 265–76.

Lloyd, B. and Rowntree, L. 1978: 'Radical feminists and gay men in San Francisco: social space in dispersed communities'. In Lanegran, D. and Palm, R. (eds), *Invitation to Geography*. New York: McGraw-Hill, 78–88.

McDowell, I. 1983: 'Towards an understanding of the gender division of urban space'. *Environment and Planning D: Society and Space* 1, 59–72.

McDowell, L. 1987: 'Beyond patriarchy: a class-based explanation of women's subordination'. *Antipode* 19, 311–21.

Mackenzie, S. 1987: 'Neglected spaces in peripheral places: homeworkers and the creation of a new economic centre'. *Cahiers de Géographie du Québec* 31, 247–60.

Mackenzie, S. and Rose, D. 1983: 'Industrial change, the domestic economy and home life'. In Anderson, J., Duncan, S. and Hudson R. (eds), *Redundant Spaces in the Cities and Regions*. London: Academic Press, 155–200.

Mackinnon, C. 1979: *The Sexual Harassment of Working Women: A Case of Sex Discrimination*. New Haven, CT: Yale University Press.

McLafferty, S. and Preston, V. 1991: 'Gender, race and commuting among service sector workers'. *Professional Geographer* 43, 1–14.

McNee, B. 1984: 'It takes one to know one'. *Transition* 14(3), 2–15.

Mama, A. 1989: 'Violence against black women: gender, race and state responses'. *Feminist Review* 32, 30–47.

Mancini Billson, J. 1991: 'The progressive verification method. Towards a feminist methodology for studying women cross-culturally'. *Women's Studies International Forum* 14, 201–15.

Markusen, A. 1980: 'City spatial structure, women's household work, and national urban policy'. In Stimpson, C., Dixler, E., Nelson, M. and Yatrakis, K. (eds), *Women and the American City*, Chicago, IL: University of Chicago Press, 20–41.

Marsh, M. 1990: *Suburban Lives*. New Brunswick, NJ: Rutgers University Press.

Michigan Employment Security Commission 1990, 'Annual Planning Information, Program Year 1989', mimeo, Michigan Employment Security Commission, Grand Rapids, MI.

Mies, M. 1988: *Women: The Last Colony*. London: Zed Books.

Mingione, E. 1987: 'Urban survival strategies, family structure and informal practices'. In Smith, M. and Feagin, J. (eds), *Capitalist City: Global Restructuring and Community Politics*. Oxford: Basil Blackwell, 297–322.

Pahl, R. 1984: *Division of Labour*. Oxford: Basil Blackwell.

Pain, R. 1991: 'Space, sexual violence and social control: integrating geographical and feminist analyses of women's fear of crime'. *Progress in Human Geography* 15, 415–32.

Parmar, P. 1989: 'Other kinds of dreams'. *Feminist Review* 31, 61–87.

Penrose, J., Bondi, L., McDowell, L., Kofman, E., Rose, G. and Whatmore, S. 1992: 'Feminists and feminism in the academy'. *Antipode* 24, 218–37.

Pinch, S. and Storey A. 1991: 'Social polarization in a buoyant labour market: the Southampton case – a response to Pahl, Dale and Bamford'. *International Journal of Urban and Regional Research* 15, 453–60.

Pratt, G. and Hanson, S. 1991: 'On the links between home and work–family–household strategies in a buoyant labour market'. *International Journal of Urban and Regional Studies* 15, 55–74.

Rich, A. 1984: 'Compulsory heterosexuality and lesbian existence'. In Darty, T. and Potter, S. (eds), *Women-identified Women*. Palo Alto, CA: Mayfield, 119–48.

Rose, D. 1989: 'A feminist perspective on employment restructuring and gentrification: the case of Montreal'. In Wolch, J. and Dear, M. (eds), *The Power of Geography*. Winchester, MA: Unwin Hyman, 118–38.

Rose, D. 1990: '"Collective consumption" revisited: analysing modes of provision and access to childcare services in Montreal, Quebec'. *Political Geography Quarterly* 9, 353–80.
Ross, B. 1990: 'The house that Jill built: lesbian feminist organizing in Toronto, 1976–1980'. *Feminist Review* 35, 75–91.
Smith, D. 1987: *The Everyday World as Problematic: A Feminist Sociology*. Boston, MA: Northeastern University Press.
Stanley, L. and Wise, S. 1990: 'Method, methodology and epistemology in feminist research processes'. In L. Stanley (ed.), *Feminist Praxis*. Andover, Hants: Routledge, Chapman and Hall, 20–60.
Stasiulis, D. 1990: 'Theorizing connections: gender, race, ethnicity and class'. In Li, P. (ed.), *Race and Ethnic Relations in Canada*. Toronto: Oxford University Press, 269–305.
Stenstad, G. 1989: 'Anarchic thinking: breaking the hold of monotheistic ideology on feminist philosophy'. In Garry, A. and Pearsall, M. (eds), *Women, Knowledge and Reality*. Winchester, MA: Unwin Hyman, 331–40.
Susser, I. 1982: *Norman Street: Poverty and Politics in an Urban Neighbourhood*. New York: Oxford University Press.
Tang Nain, G. 1991: 'Black women, sexism and racism: black or antiracist feminism?' *Feminist Review* 37, 1–22.
Tivers, J. 1988: 'Women with young children: constraints on activities in the urban environment'. In Little, J., Peake, L. and Richardson, P. (eds), *Women in Cities: Gender and the Urban Environment*. Andover, Hants: Tavistock Publications, 84–97.
Vaiou, D. 1992: 'Gender divisions in urban space: beyond the rigidity of dualist classifications'. *Antipode* 24, 247–62.
Valentine, G. 1989: 'The geography of women's fear'. *Area* 21, 385–90.
Valentine, G. 1993a: '(Hetero)sexing space: lesbian perceptions and experiences of everyday spaces'. *Environment and Planning D: Society and Space* 11, forthcoming.
Valentine, G. 1993b: 'Negotiating and managing multiple sexual identities: lesbian time–space strategies'. *Transactions of the Institute of British Geographers New Series* 18, 237–48.
Walby, S. 1986: *Patriarchy at Work*. Cambridge: Polity Press.
Walby, S. 1990: *Theorizing Patriarchy*. Oxford: Basil Blackwell.
Wallace, M. 1990: *Invisibility Blues: From Pop to Theory*. London: Verso.
Wallman, S. 1984: *Eight London Households*. London: Tavistock.
Weightman, B. 1981: 'Commentary: towards a geography of the gay community'. *Journal of Cultural Geography* 1, 106–12.
Wekerle, G. 1980: 'Women in the urban environment'. In Stimpson, C., Dixler, E., Nelson, M. and Yatrakis, K. (eds), *Women and the American City*. Chicago, IL: University of Chicago Press, 185–211.
Wekerle, G. and Rutherford, B. 1989: 'The mobility of capital and the immobility of female labor: responses to economic restructuring'. In Wolch, J. and Dear, M. (eds), *The Power of Geography*. Winchester, MA: Unwin Hyman, 139–72.
Winchester, H. P. M. and White, P. E. 1988: 'The location of marginalised groups in the inner city'. *Environment and Planning D: Society and Space* 6, 37–54.
Wolch, J. 1990: *The Shadow State*. New York: The Foundation Centre.

6 G. Pratt and S. Hanson, 'Gender, Class, and Space'

Excerpts from: *Society and Space* **6**, 15–35 (1988)

A central concern of social geography has been the study of the residential structure of the city. Much activity from the 1950s to the 1970s was devoted to establishing the dimensions of urban social structure and the spatial patterning of those dimensions in metropolitan areas. Since the mid-1970s, social geographers have shifted the focus of their interest toward the processes that explain urban residential structure and the implications of that structure; the patterns themselves seem no longer to be questioned but now merely to be taken for granted. Recent sweeping social and demographic changes, however, suggest that the dimensions of urban social structure and the spatial patterning of that structure have probably changed fundamentally and that therefore urban residential patterns need to be reexamined. Moreover, the current widespread acceptance, in theoretical arguments, of the force of existing residential patterns in social reproduction (for example, Harris, 1984; Harvey, 1975; Scott, 1980; 1985; 1986) underlines the need to question the nature of these long taken-for-granted residential patterns and to assess the implications of contemporary change for the social reproduction argument.

Basic to both social area analysis and factorial ecology was the assumption that residential areas were relatively homogeneous with respect to class; it is this fundamental belief that has been picked up and used in the theoretical arguments on the social reproduction of class. The increase (especially since the 1950s, when the earliest social area analyses were first being done) in female labor-force participation renders the validity of this homogeneity assumption questionable: the existence of high levels of sex segregation in the workplace (Reskin and Hartmann, 1986) reflects the fact that men and women hold very different positions in the work force. Most important for the issue at hand, women's employment opportunities tend to be restricted to a relatively limited number of low-paid, low-status, and lower-class occupations. This leads us to suspect that the geography of social class is also different for men and for women; if it is, then the social geography of the city is a great deal more complex than the standard models of residential structure would lead us to believe, and census tracts that may be homogeneous with respect to male social class will not be homogeneous when both men's and women's social classes are considered. This point holds major implications for the validity of the social reproduction thesis.

Our purposes in this paper are, first, to lay out our reasons for expecting changes in urban residential structure and to outline the nature of those expected changes; second, to present some empirical evidence from the Worcester, MA SMSA (Standard Metropolitan Statistical Area) against which to judge the validity of our arguments; and, third, to sketch out the implications of the changes we document. Our purpose extends beyond the goal of 'setting the record straight' at a descriptive level. Given the current theoretical interest in the effects and implications of residential structure, a more adequate description of the geography of class has considerable theoretical significance.

Background

We develop our argument through a number of steps. First, we outline the gender bias implicit in the many factorial ecologies that form the basis of accepted descriptions of urban residential pattern. This bias raises doubts about the validity of the resulting descriptions, and leads us to question the descriptive base upon which recent theories of social reproduction have been constructed. We then review contemporary demographic trends and spell out our expectations about the influence of these changes on the residential structure of the North American city.

Descriptive studies of residential pattern

Literally hundreds of studies have been carried out to determine the dimensions along which residential differentiation is organized and to identify the spatial patterning of each of these dimensions (for a review see Timms, 1971). In the majority of the studies a factorial ecological approach has been employed, which involves the factor analysis of a large number of indicators in order to uncover the underlying structure of residential differentiation. Although such studies are often presented as 'objective', the results of these analyses reflect a myriad of decisions on the part of the investigator, made at various junctures in the analysis; as a result, the outcome can actually be quite subjective in that it reflects the investigator's own biases, views, and preconceptions. For example, because the structure that is discovered in a factorial ecology is largely determined by the variables included, the investigator makes important theoretical decisions in choosing which variables to enter into the factor analysis. The conventional solution to the problem of supporting theoretical propositions through the choice of indicators has been to include as wide-ranging and as representative a set of variables as possible (Timms, 1971).

Even when this precaution has been taken, many factorial ecologies describe how the social class of male but not female workers varies across residential areas in the city. In the most extreme cases, this is because information on occupation type and income is restricted to male workers. For example, in his analysis of the Auckland urban area in 1966, reported in his influential book *The Urban Mosaic* (1971), Timms includes the following variables to measure socioeconomic status of households in each census tract (page 70): percentage of the male work force in professional and managerial occupations, percentage of the male work force in nonmanual occupations, percentage of males earning $3,000 or more per annum, percentage of males classified as self-employed, and a standardized social rank index. This last index was constructed from the preceding occupational and income variables.

Not all factorial ecologies have relied exclusively on male occupation and income data. Even when variables on women's labor-force participation are included, however, the investigator's expectation about the appropriate factor structure (established from years of absorbing dozens of prior factorial ecologies), leads the investigator to ignore women. A striking example of this is Patterson's factorial ecology of the Vancouver region, cited extensively by Ley in his textbook *A Social Geography of the City* (1983). Patterson factor analyzes a comprehensive range of 192 variables, including many that measure characteristics of the labor-force participation of women. Seven factors emerge, two of which are closely related to the social area analysis construct of economic status. One factor appears to measure characteristics of male labor-force participation whereas the other measures that of women. Variables that are strongly positively associated with the first of these factors include percentages of the labor force in professional, technical, and managerial occupations, in the education, finance, insurance, and wholesale trade industries. Negatively correlated with this first factor are percentage of employees in the industries of wood finishing, construction, food and beverages, hotels, restaurants, and transportation and storage. Variables that load on the second, areally independent factor clearly measure women's labor-force participation. They include proportion of the labor force in sales and clerical occupations and in the industries of retail trade and telecommunications.

Patterson labels the first factor socioeconomic status and concludes that the subdivision of the socioeconomic construct into two factors is an artifact of the spatial resolution of the data, specifically the use of small areal units. An alternative explanation is that the spatial distribution of female clerical and retail workers differs from that of workers in traditionally male-dominated jobs. This issue is implicitly suppressed when Patterson reserves the label of 'socioeconomic status' for the first factor. The effects of the labelling process are clearly evident in Ley's presentation of the 'socioeconomic status' factor as evidence of the sectoral patterning of socioeconomic status in Vancouver. The first factor is taken to represent the entire construct, and the spatial distribution of clerical and sales workers is ignored.

Of course, women were not totally ignored in traditional social area analyses and factorial ecologies, but the point we wish to make is that women's employment characteristics were usually not considered in the definition of the social rank (socioeconomic status) dimension. From the very beginning (for example, see Shevky and Bell, 1955), female labor-force participation was one of the three key variables used to define the 'familism' or 'urbanization' dimension (the other two were fertility and number of single-family dwelling units). In other words, these studies *did* take into account *whether* women worked or not – as a dichotomous variable – but ignored the question of women's occupational or social *status*. Perhaps this reflects the gender-role stereotypes prevalent at the time these studies were conducted: women either worked (an indication of the 'urban' status of a census tract) or they were home with the family (an indication of the 'familism' status of the census tract), but working and having a family certainly did not occur together for women. It is ironic that in most US metropolitan areas now, female labor-force participation rates are higher in the suburbs – traditionally the bastions of 'familism' – than they are in the central cities – traditionally the bastion of 'urbanism'. As we demonstrate below in our study of Worcester, the fact that women constitute a substantial proportion of the labor force throughout the metropolitan area renders the dichotomous view of women's labor-force participation relatively useless as an indicator of any dimension of contemporary social life.

Theoretical implications of residential structure

Recent theoretical work builds from the empirical foundations laid by factorial ecologists; it is assumed that residential areas are well defined and homogeneous with respect to class. 'In the modern American city, a particularly sharp and deeply entrenched line of disjunction can be drawn between white-collar occupations and blue-collar occupations' (Scott, 1985, page 488). Such a residential structure is considered not only to mirror but also to reproduce class relations in capitalist societies (Harris, 1984; Harvey, 1975; Scott, 1980; 1985; 1986). 'Occupationally defined neighborhoods assist in the smooth social reproduction of the specific human attributes, attitudes, behavioral patterns, and so on, that are an essential ingredient of labor market capacity' (Scott, 1985, page 490). Residential segregation is seen as fragmenting the working class into factions defined by skill level and as reinforcing workers' identification with a class faction. It encourages blue-collar workers to see themselves as blue-collar workers, separated spatially, socially, and politically from workers in white-collar occupations.

This argument also holds that social reproduction facilitates the intergenerational transfer of skills; it therefore implies that the class composition of neighborhoods determines the shape of intraurban labor markets. Educational resources are allocated unevenly across the city, varying directly with the wealth of the school board jurisdiction (Cox, 1979). Children living in white-collar neighborhoods tend to receive a disproportionate share of educational resources. Neighborhoods are also social milieux in which varying educational priorities are established; children of blue-collar workers are encouraged by parents and peers to acquire skills that prepare them for manual labor, whereas children of white-collar workers learn the value of skills that prepare them for white-collar jobs (Bourdieu, 1974; Bowles and Gintis, 1977; Huckle, 1985; Scott, 1980).

The simplicity of the reproduction argument is readily acknowledged by its advocates: 'we must be careful that we do not claim too much for the basic structuring effects of the production system on urban social life' (Scott, 1985, page 490). The qualification typically takes the form of admitting ethnicity and 'life-style' into the conceptual schema, as additional factors that structure residential segregation. The basic social homogeneity of residential communities, however, goes unquestioned, and without the assumption of homogeneity the reproduction argument founders.

Yet a view of the city in which neighborhoods are structured into homogeneous enclaves of white- and blue-collar workers rests on the suppression of the issue of gender. It is assumed that women who work can be classified in terms of their husbands' occupational standing, and the fact that the conditions of women's labor-force participation are radically different from men's is ignored. Typical of class-based theories, the issue of gender is ignored. In doing this, these theorists reproduce the social reality of patriarchy at a theoretical level (Delphy, 1981).

Gender and residential structure

A neglect of gender leaves factorial ecologies and the reproduction argument on shaky ground. Contemporary demographics do not support the stereotype of a nuclear family maintained by a single male breadwinner.

The 'typical American family' consisting of a bread-winner father, a home-making mother, and two children now constitutes only 7 percent of US households. A number of factors have conspired to erode the predominance of this typical American household. First is the well-known and well-documented increase in female labor-force participation since World War 2. Whereas in 1950 only 28.3 percent of US women aged 14 to 64 were in the labor force (DOL, 1983, page 14), by 1982 the female labor-force participation rate had risen to 53 percent. In 1984 about 50 million women were employed, constituting 43 percent of the US work force (Reskin and Hartmann, 1986, page 1). Moreover, the labor-force participation rate is considerably higher among younger women than it is among older women, indicating that the overall female labor-force participation rate is likely to increase, not decrease, over time: among women aged 25 to 34, 67.4 percent were employed outside the home in 1981, despite the fact that women in this age group are in their peak childbearing years (DOL, 1983, page 13). This high labor-force participation rate for younger women reflects the dramatic increase in employment among married women and among mothers of young children, both preschool and school aged. For example, in 1982, 66 percent of US mothers with children aged 6 to 13 worked outside the home (DOL, 1983, page 15), and in 1985, 49.4 percent of married women with children less than one year old were in the labor force (Hayghe, 1986).

A second factor that has contributed to an altered view of what constitutes the 'typical American family' (and is a factor perhaps related to the increase in female labor-force participation) has been the increasing divorce rate. Whereas in 1960 only 2.9 percent of the female population over 18 years of age in the USA was divorced, by 1984, 8.3 percent of that population was divorced. In the USA the divorce rate has risen from 35 per one thousand married persons in 1960 to 121 in 1984 (BOC, 1986, page xvii). A third source of change contributing to the demise of the 'typical American family' has been the growing proportion of single-person households. In 1960, 13.1 percent of all households had only one person; by 1984 the figure had risen to 23.4 percent of all households (BOC, 1986, page 47). A fourth factor has been the increase in female-headed households and single-parent families, most of which are headed by women. The proportion of all family households that are headed by women in the USA has increased from 9.3 percent in 1960 to 14.6 percent in 1980 (BOC, 1986, page 45), and in 1985 nearly one quarter (22.2 percent) of all families with children under 18 were single-parent families. Also increasing (but more difficult to document with census data) has been the number of two-earner households. The proportion of wives in married couple households that worked in 1950 was 27 percent (BOC, 1986, pages 39 and 398); by 1983 that proportion had risen to 68 percent (BOC, 1986, page 455).

Despite the current high level of female labor-force participation documented by these figures, women still work in a relatively small number of occupations. It is this concentration of men and women in distinctly different occupations that is known as occupational segregation. The 1980 US Census recognized 503 different occupational categories (for example, hairdresser, accountant, bookkeeper, elementary school teacher). For 187 of these categories, at least 90 percent of the workers were members of one sex; in 275 of these categories at least 80 percent of the workers were of one sex (Reskin and Hartmann, 1986, page 7). About half of all employed women and more than half of all employed men work in occupations where at least 80 percent of the workers are of one sex (Reskin and Hartmann, 1986, page 7).

These demographic trends are well documented, and their theoretical implications have been actively explored for a number of years (Acker, 1973; Crompton and Jones, 1984; Garnsey, 1977). Certainly our concern with the relationship between gender and class is not a new one. These dramatic demographic changes have not yet, however, been systematically related to space. We wish to highlight the significance of these trends for a reconceptualization of urban residential structure.

Implications of demographic changes for urban spatial structure

In spelling the demise of the 'typical American family', the demographic changes outlined in the previous section have also probably obliterated the utility of traditional models of urban residential structure. Certainly these changes prod us to question the long-accepted bases of residential patterning in cities and lead us to expect that contemporary patterns will themselves be quite different from those predicted by the traditional models. We test this notion not by conducting yet another factorial ecology, but by examining the validity of several expectations posed by the changes described above. These expectations are as follows.

First, it seems unlikely that women's labor-force participation is still a valid discriminator of social areas. As it is no longer accurate to assume that child-rearing and paid employment are mutually exclusive activities and as female labor-force participation is now so widespread, our first expectation is that there is relatively little spatial variation in female labor-force participation across the city.

The fact of occupational segregation by sex leads us to our second, third, and fourth expectations. The second one is that male income and occupational class[1] will, on average, be 'higher' than women's but that the variance in women's income and class will be less than the variance in men's. Moreover, we anticipate that measures of male occupational class and income will show more variation over space (by residential area) than will measures of female class and income; this is our third expectation. Fourth, in view of the second and third propositions described above, we expect a high degree of within-household and therefore within-neighborhood heterogeneity, in terms of both income and occupational class. We reason that because of occupational segregation by sex (with women crowded into a small number of female-oriented, low-paying, lower-class jobs), there should be relatively little spatial variation in women's income or occupational class across the city. But precisely because of this, together with the fact of higher male income and occupational class, there should be considerable variation in income and occupational class within households and hence within neighborhoods: we expect high-income households to include low-income women, blue-collar suburbs to include a substantial proportion of white-collar (that is, female clerical) workers, and high-status white-collar suburbs to house a good number of workers (again, women) employed in low-status white-collar jobs.

Fifth, the increase in the number of two-earner households is expected to have led to more class heterogeneity between households living within the same neighborhoods. Household income reflects not only the occupational status of employed members but also the number of such workers. A household with two lower-income workers may be able to afford a house in the same residential area as a household supported by one higher-income worker. Variability in the employment patterns of women (for example, whether or not they work for pay and whether they are the household's primary or secondary wage earner) may loosen the link between individual incomes, occupational class, and residential segregation. This trend is likely to be reinforced by other social and demographic changes. For example, high divorce rates create a sizeable new subpopulation of divorced women who own homes but have virtually no income and who are not adequately classified within the white-collar–blue-collar occupational schema (Rein *et al.*, 1980).

Finally, new bases for residential segregation may have developed, reflecting the interaction between gender and class. The segregation of women into low-paying jobs means that families headed by women are extremely vulnerable to poverty. In 1986, one third (34.6 percent) of female-headed households in the USA were poor, as compared with 6.1 percent of two-parent households (US Bureau of the Census, 1986, page 23, table 15). If female-headed families are concentrated in particular parts of the city, this is a significant new basis for income-related residential segregation. Moreover, the spatial concentration of female-headed households would represent a new gender-related basis for residential segregation.

Some evidence from Worcester

We evaluate the validity of each of these expectations with data for the Worcester, MA, SMSA. Worcester was chosen because it is a city we are both familiar with and because its female labor-force characteristics are typical of the nation as a whole.[2]

Data

To test our ideas, we need small-area data (for example, for census tracts) on measures of class and of income by gender. Unfortunately data of this sort are not routinely provided in the US Census of Population; although figures are published on the occupational makeup of the population within each tract, these are not broken down according to sex. We use the Census of Population to examine the spatial patterns of female-headed households, and of female labor-force participation, but for evidence of the other patterns that concern us here, we had to find alternative data sources. We used two such sources, which together allow us to test the viability of the ideas we have outlined above: (1) the 1980 Public Use Micro-data 1 percent B sample (PUMS) for the Worcester SMSA, and (2) special runs from the 1980 Census Journey-to-Work file for the Worcester SMSA.

Containing records for a sample of housing units, the PUMS data set provides information on the characteristics of each housing unit and the people in it. Of particular importance for this study is the fact that in the PUMS data each person's occupation is coded according to the detailed three-digit census SOCs (Standard Occupation Codes – for example, accountant, bar manager, postal clerk, high school teacher), which allows us to identify each individual's occupational class. The PUMS data are, therefore, extremely useful for comparing male and female occupational class in the Worcester SMSA; the PUMS data are, however, limited by the fact that they do not provide geographic information at a scale finer than 'central city' and 'rest of the SMSA'. We can therefore use these data to compare male and female class and occupational patterns in city versus suburb, but not at the census-tract level.

Special runs from the 1980 Census Journey-to-Work files enable us to describe at the census-tract level the residential location patterns of men and women in households of different types. Unfortunately, income is the only surrogate for class available in these data (though we also have information on occupations in terms of their gender composition, for example, female-dominated, male-dominated, or gender-integrated).[3] By aggregating these journey-to-work flows at the origins, we can describe the residential location patterns of specified population subgroups.

Analyses

Using these data sources, we examine evidence for each of our expectations in turn.

To investigate our first proposition – that female labor-force participation will exhibit relatively little spatial variation – we constructed maps of (1) female labor-force participation rates (that is, percentage of the female population aged 16 and above in a census tract which is in the paid labor force) and (2) percentage of resident workers in a tract who are female (that is, percentage of employed individuals who are women). The maps in Figures 6.1 and 6.2 indicate support for our proposition. Although there is of course some tract-to-tract variation in female labor-force participation rates (Figure 6.1), it is not high; the rates range from 26.4 percent to 65.8 percent, with a mean ($\bar{X}$) of 51.4 percent and a standard deviation (s) of only 7.0. The generally higher suburban rates are evident, but there are also many urban tracts with rates above the SMSA average. Similarly, Figure 6.2 shows that everywhere women constitute a significant proportion of the labor force; in no census tract does their proportion fall below 34 percent. In 93 of the 94 tracts women form between 34 percent and 57 percent of the work force (the proportion in the 94th tract is 66 percent!), and again the variance in the tract-to-tract distribution is relatively low ($\bar{X}$ = 44.1, s = 4.9).

To show the relative dearth everywhere of 'typical' households (typical in having a single male breadwinner) we include Figure 6.3, which reveals that there is only one census tract in the whole SMSA where more than half of the households are typical in this sense.

We tested our second expectation with data from the 1980 Worcester PUMS. Recall that we anticipate men's income and occupational class to be higher than women's, and we expect the variance in each of these variables to be lower for women than for men. The analyses were conducted on the 1,002 employed men and 789 employed women in the 1980 Worcester PUMS sample. The results consistently confirm our expectations. First, men's incomes (for 1979) are significantly higher than women's, and there is less variation among women's incomes than there is among men's ($\bar{X}$ for men = \$15,249, s = \$10,197; $\bar{X}$ for women = \$7,501, s = \$5,490, $t = 19.9$, $p < 0.001$). Second, class was measured by assigning each occupation in the data set to

Figure 6.1 Proportion of women over 16 years of age, employed in labor force. (a) Worcester City, (b) Worcester Metropolitan Statistical Area

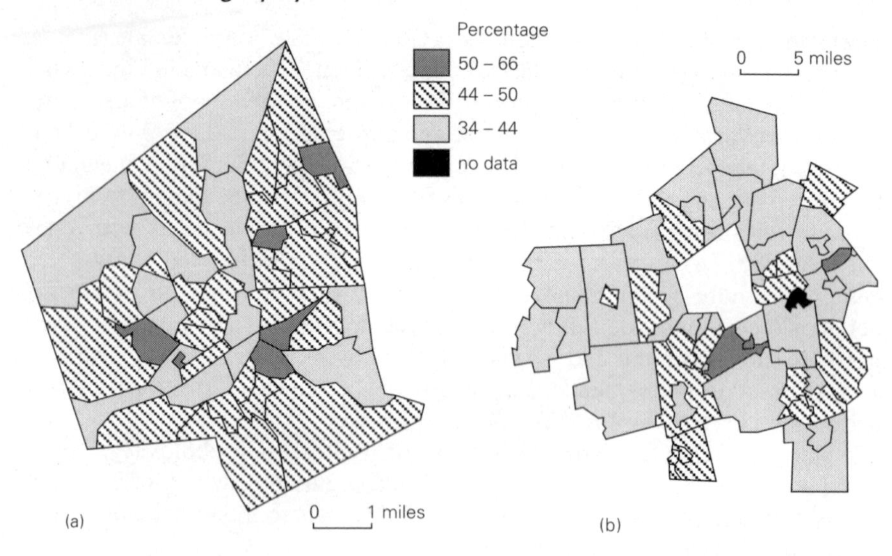

Figure 6.2 Percentage of resident workers in each tract who are female. (a) Worcester City, (b) Worcester Metropolitan Statistical Area

Figure 6.3 Incidence of dual-headed households with single male breadwinner. (a) Worcester City, (b) Worcester Metropolitan Statistical Area

one of the five classes shown in Table 6.1. The data in Table 6.1 clearly show how differently men and women are distributed over these class categories: the row percentages document the occupational segregation of men and women, and the column percentages confirm that men are more evenly spread than are women over these five categories; nearly three quarters (72 percent) of all employed females work in nonmanual skilled and nonmanual

Table 6.1 Occupational class of women and men in Worcester Standard Metropolitan Statistical Area (SMSA)

Occupational class	*Men*			*Women*		
	no.	*row %*	*col. %*	*no.*	*row %*	*col. %*
Unskilled manual	296	72.0	29.5	115	28.0	14.6
Unskilled nonmanual	104	26.8	10.4	284	73.2	36.0
Skilled manual	251	82.8	25.0	52	17.2	6.6
Skilled nonmanual	172	37.6	17.2	285	62.4	36.1
Managers, professionals, owners	179	77.2	17.9	53	22.8	6.7
Total	1002	55.9		789	44.1	

$\chi^2 = 370.2$, degrees of freedom = 4, $p < 0.001$.

Source: Calculated from the PUMS sample for the Worcester SMSA, 1980.

unskilled occupations. The existence of occupational segregation is further demonstrated in Table 6.2, where the preponderance of women in female-dominated occupations and of men in male-dominated occupations is very much evident. In sum, the evidence strongly supports the existence of deep cleavages between men and women in their occupations, and these differences have associated with them important income and status differences as well. Moreover, it is clear that women in the Worcester labor market are, like women throughout the USA and elsewhere, crowded into fewer occupations than are available to men and that there is therefore less variability in women's income and occupational status than is the case for men.

We test our third expectation – that male income and occupational class measures will show more spatial variation than will such measures for women – with the PUMS data as well as with data from the journey-to-work file. With the PUMS data, as mentioned earlier, the only spatial patterns we can examine are those for city versus suburbs, as no finer level of spatial resolution is available in those data.

Table 6.2 Gender composition of occupations for male and female workers in Worcester Standard Metropolitan Statistical Area (SMSA)

Gender compositions of occupations	*Men*		*Women*	
	no.	*%*	*no.*	*%*
Female-dominated	59	5.9	461	58.4
Gender-integrated	252	25.1	240	30.4
Male-dominated	691	69.0	88	11.2
	1002	100	789	100

$\chi^2 = 763.30$, $p < 0.001$.

Source: Calculated from the PUMS sample for the Worcester SMSA, 1980.

When the city–suburb patterns are examined, the evidence supports our contention that women's class standing does not vary over space as much as men's does. For men, income, class, and occupational status are all significantly higher in the suburbs than in the city; for women there is no significant city–suburb difference on any of these variables. Average income for men living in the city ($N = 400$) is \$12,462 ($s = \$8,248$) and for suburban men ($N = 602$) is \$17,062 ($s = \$10,917$); the difference is highly significant ($t = 7.32, p < 0.001$). Women's incomes, however, are not significantly higher in the suburbs: average income for women workers living in the city ($N = 358$) is \$7,218 ($s = \$4,985$) and for suburban women ($N = 431$) is \$7,735 ($s = \$5,868$); $t = 1.30$, p is not significant. Table 6.3 shows that the same pattern (significant city–suburb differences for men but not for women) exists for the class measure.

The only one of the measures of social status that is available in the journey-to-work file is income; therefore income is the only one of these indices whose spatial distribution we can examine at the census tract level. The maps in Figures 6.4 and 6.5 reveal not only the overall higher male (than female) income levels – in both city and suburban tracts – but also the generally 'flatter' distribution of women's (than of men's) incomes. The distribution of women's income is clustered in the lowest two categories; in fact in no census tract does the mean female income reach into either of the two highest income categories. Yet in only nine tracts does mean male income fall *below* the two highest categories.

To test our fourth expectation – that because women's jobs are very different from men's there will be considerable heterogeneity in social status within households – we selected from the PUMS data the 387 households in

Table 6.3 Occupational class of men and women for central city and suburbs in Worcester Standard Metropolitan Statistical Area (SMSA)

Occupational class	*Men*				*Women*			
	central city		*suburbs*		*central city*		*suburbs*	
	no.	*%*[a]	*no.*	*%*	*no.*	*%*	*no.*	*%*
Unskilled manual	143	48.3	153	51.7	55	47.8	60	52.2
Unskilled nonmanual	52	50.0	52	50.0	140	49.3	144	50.7
Skilled manual	87	34.7	164	65.3	20	38.5	32	61.5
Skilled nonmanual	62	36.0	110	64.0	122	42.8	163	57.2
Managers, professionals, owners	56	31.3	123	68.7	21	39.6	32	60.4
Total	400	39.9	602	60.1	358	45.4	431	54.6

Men $\chi^2 = 22.6$, degrees of freedom = 4, $p < 0.001$.
Women $\chi^2 = 4.5$, degrees of freedom = 4, not significant.

[a] Row percentages within sex.

Source: Calculated from the PUMS sample for the Worcester SMSA, 1980.

Figure 6.4 Mean female income. (a) Worcester City, (b) Worcester Metropolitan Statistical Area

Figure 6.5 Mean male income. (a) Worcester City, (b) Worcester Metropolitan Statistical Area

which both husband and wife were in the labor force. Through a series of contingency tables in which wives' class and status characteristics are arrayed against husbands', we attempt to evaluate the validity of our notion. Again, the evidence supports our contention.

First, as expected, husbands' incomes are, on average, significantly higher than those of their wives: $\bar{X}$ for husbands = \$16,952, s = \$9,436; $\bar{X}$ for wives = \$7,622, s = \$6,486; $t = 17.6$, $p < 0.001$. When the wife's income is arrayed

Table 6.4 Spouses' incomes in dual-earner households in Worcester Standard Metropolitan Statistical Area (SMSA)

Husband's income ($)	*Wife's income ($)*				*Totals*
	< 10,000	*10,001–15,000*	*15,001–20,000*	*> 20,000*	
< 10,000	52	9	3	1	65
10,001–15,000	53	27	3	0	83
15,001–20,000	78	18	14	0	110
> 20,000	86	24	9	10	129
Totals	269	78	29	11	387

$\chi^2 = 64.1$, degrees of freedom = 9, $p < 0.001$.

Source: Calculated from the PUMS for the Worcester SMSA, 1980.

against her husband's, as is shown in Table 6.4, it is clear that in the bulk of dual-earner households wives earn less than their husbands; in only 16 of the 387 households did the wife's earning outstrip her husband's. Moreover, as the husband's income increases, the likelihood decreases that the wife's earnings will equal her husband's.

Table 6.5 shows the distribution of wives' occupational class (in the columns), given the distribution of the husbands' class (in the rows). Clearly there is intrahousehold heterogeneity with respect to occupational class. A number of interesting points emerge from this table. First, it is only in rare cases that the wives of men in the 'highest' occupational class share their husbands' class standing. Second, if a man is in a skilled manual occupation (the category with the largest number of male workers), his wife is twice as likely to be nonmanual as a manual worker, raising real questions about the class homogeneity of blue-collar households.

In sum, these results on dual-earner households support our argument that, because so many women are now in the labor force and because they are segregated into a few occupations, one cannot assume that urban neighborhoods – or even households – are homogeneous with respect to class. We conclude this section with a look at the ways in which occupational segregation by sex produces heterogeneity within two types of neighborhoods: (1) traditional working-class neighborhoods (where a high percentage of the male labor force hold blue-collar jobs) and (2) traditional upper middle-class neighborhoods (where a high percentage of the male labor force hold white-collar jobs).

To examine this, we chose one census tract in the city of Worcester where many workers are employed in blue-collar jobs and another where there is a high proportion engaged in professional and managerial occupations.[4] Data constraints force us to assess the occupational class of residents through a number of indirect measures: gender composition of occupation, industrial sector, and personal income. These are reasonable proxies for the construct of interest. Female-dominated occupations tend to be white- (or pink-) collared; blue-collared occupations tend to be concentrated in the manufacturing and

Table 6.5 Occupational class of wife and husband, within household in Worcester Standard Metropolitan Statistical Area (SMSA)

Husband's occupational class	*Wife's occupational class*[a]					*Row total*
	unskilled manual	*unskilled nonmanual*	*skilled manual*	*skilled nonmanual*	*managers*[b]	
Unskilled manual	24 27.6 40.7	30 34.5 24.6	11 12.6 39.3	21 24.1 13.2	1 1.1 5.3	87 22.5
Unskilled nonmanual	7 21.2 11.9	15 45.5 12.3	1 3.0 3.6	9 27.3 5.7	1 3.0 5.3	33 8.5
Skilled manual	20 19.2 33.9	35 33.7 28.7	12 11.5 42.9	37 35.6 23.3	0	104 26.9
Skilled nonmanual	4 5.2 6.8	22 28.6 18.0	2 2.6 7.1	46 59.7 28.9	3 3.9 15.8	77 19.9
Managers[b]	4 4.7 6.8	20 23.3 16.4	2 2.3 7.1	46 53.5 28.9	14 16.3 73.7	86 22.2
Column total	59 15.2	122 31.5	28 7.2	159 41.1	19 4.9	387 100

$\chi^2 = 87.4$, degrees of freedom = 16, $p < 0.001$.

[a] The entries are: top, number; middle, row percentage; bottom, column percentage.

[b] Managers, professionals, owners.

Source: Calculated from the PUMS sample for the Worcester SMSA, 1980.

construction industrial sector; and income provides an index of occupational class (for example, professionals and managers tend to earn more than those in nonskilled white-collar occupations).

There is a tremendous amount of heterogeneity within each census tract, though the occupational differences between men and women are remarkably consistent across both tracts. Within the 'blue-collar census tract', 68 percent of women in the labor force hold jobs in female-dominated occupations, an indication that they are *not* blue-collar workers. This impression is confirmed when jobs are broken down by industrial sector: 46.3 percent of women in the blue-collar tract work in the education, health, and welfare sector. Examining the 'white-collar census tract', working women are clearly in lower-status occupations than the men. The majority of women (59.3 percent) hold female-dominated jobs in the education, health, and welfare sector and earn considerably less than men living in the census tract. This analysis strengthens our point that it is inappropriate to conceive of neighborhoods as homogeneous with respect to occupational class. This may be the case within each gender, but the differences between the occupational classes of men and women are such that the generalization is not true for the population as a whole.

Our fifth expectation concerns the impact of two-earner households on intra-neighborhood heterogeneity. Does the presence of both two- and single-earner households in the same neighborhood increase the level of class heterogeneity within an area? Our expectation will be supported if, in high-income tracts for example, we find a mix of (a) single-earner high-income households and (b) high-income dual-earner *households* in which each *individual* worker's income is, say, only mid-range. Similarly, middle-income tracts that have a large number of two-earner *households* consisting of low-income *individuals* will support the notion that the growing number of dual-earner households is an important source of intra-neighborhood heterogeneity. Our data do not permit us to tackle this issue as solidly as we would like; we do not have data on household income (by whether a household has one or two earners) at the tract level. We can, however, use the PUMS data to probe the idea at the SMSA level.

We examine the income of individual workers in single-earner and in dual-earner households within each of three household income categories (derived by dividing households into top, middle, and lower thirds on family income).[5] The results lend support to the idea that dual-earner households are indeed a potential source of heterogeneity within neighborhoods: among dual-earner households whose *family* income is in the top third for the SMSA (> $29,045), there are many workers whose *personal* income is in the lower and middle categories. Moreover, more than 80 percent of the highest-income households had two earners, and although it is clear that having both earners in the highest personal income quartile is a significant factor in putting a household's income in the top third, three fifths of the individual workers in dual-earner households had personal incomes below the 75th percentile. A similar pattern holds for the middle-income households: three quarters of middle-income households have two earners, but nearly one quarter of the individuals in two-earner households earn incomes in the lowest quartile. Insofar as family income is important in sorting households across urban space, then, we would expect to find dual-earner households creating heterogeneity within neighborhoods even if all households in an area had identical family incomes.

Our final expectation is that female-headed households constitute a new gender-related basis for residential segregation. The map in Figure 6.6 shows that female-headed households constitute at least 6 percent of households in *all* census tracts. There is, however, a clear concentration of female-headed families within the inner city of Worcester, where such families tend to make up at least 25 percent of resident households. In one striking instance, 66.4 percent of households in the census tract are headed by women. This extreme degree of concentration reflects the fact that the majority of households in this tract live in public housing. It is recognized that public housing is increasingly dominated by female-headed households (Freeman, 1981). To the extent that public housing is spatially concentrated, so too are female-headed households. In general, Figure 6.6 demonstrates a remarkable degree of gender-based segregation within inner-city neighborhoods.

Figure 6.6 Families headed by women. (a) Worcester City, (b) Worcester Metropolitan Statistical Area

Implications

Our expectations about changes in urban residential structure have been supported. Residential areas are no longer clearly distinguishable on the basis of whether or not female residents hold paid employment. 'Familism' has a new meaning as a factor that structures urban residential space: female-headed households tend to be concentrated in different parts of the city from dual-headed households. As well, the conventional wisdom about the distribution of social classes throughout the city needs qualification. Once gender has been highlighted, considerable within-household and within-neighborhood heterogeneity is evident.

These empirical findings lead us to question the adequacy of current claims about the functionality of neighborhoods for the social reproduction of classes. At the very least, the processes are a great deal more complex and much less automatic than presently specified.

Men and women experience class with different degrees of consistency. In general, men experience the same class standing at work, at home, and within the family. For men, residential status reinforces perceptions of class. This uniformity of experience may be changing for men also, if dual-earner lower-status households are able to purchase houses in the same neighborhoods as single-earner higher-status ones. But it is nevertheless arguable that, in general, men interact within a narrow range of class experience, because home and work tend to be mutually reinforcing.

For many working women, however, there is a real discontinuity between their class position, as defined in the workplace, and both their husband's class and their residential status. In the PUMS sample, of the 122 women in dual-earner families who were classified as unskilled nonmanual, over 16 percent were married to men who were classified as managers or professionals. As a family, they may live in a middle-class residential suburb. Conversely, 36 percent of the women married to skilled blue-collar workers did skilled white-collar jobs. This disjunction in women's experience may have some interesting consequences for politics at work and at home.

Considering first the effects at work, one may draw upon the reproduction argument. The diversity in residential status and 'family' class positions among women with the same occupations may hinder collective workplace action. This parallels the general claim that residential segregation has fragmented working-class consciousness (Harris, 1984; Harvey, 1975). Women with common work experiences may differ in their perceptions of class, social status, and the legitimacy of workplace action. They may feel a significant social distance between themselves and their coworkers, with implications for the politics of the workplace.

This is precisely what Cho (1985) reports from her participant observation study of women working at Microtek, Inc. in Silicon Valley. Women occupied about 95 percent of the assembly jobs at Microtek. Though the assembly jobs were unskilled, the female workers came from diverse class backgrounds: 'One Korean woman worker's husband was a medical doctor, while one American woman's boyfriend was a fabrication operator at a firm in Silicon Valley. . . . In general, a significant number of married women's husbands had high-income jobs, such as chemical engineering. There were also quite a few single women who had two jobs to meet their living expenses' (Cho, 1985, pages 200–1). This diversity reinforced the extreme atomization of the work force. The married women tended to show less concern over wages. For example, all of the Korean married women reported that they 'never read [their] pay checks with care' (page 201). It is therefore not surprising that it was a single mother who discovered that the company was violating its own rules and regulations by not paying the workers 150 percent for their overtime work. Cho argues that the diversity of cultural, economic, and class backgrounds hindered the Microtek workers from discovering and pursuing their common interests as assembly workers.

We cannot assume, however, that women's work experiences have no effect in the other sphere of their lives, at home. Drawing out these influences tends to lead one away from the reproduction argument. We consider two influences.

First, working women possibly bring home with them some knowledge of the conditions under which women who reside in other parts of the city and in other family circumstances conduct their lives. It is not unlikely that a secretary who owns her own home will hear the concerns of a coworker who rents. The clerk who drives her car to work may come to know the circumstances of another who relies on public transportation. We are suggesting that working women tend to have a greater opportunity to build up knowledge of a more diverse array of life experiences. This may influence their views on a number of issues: for example, rent controls, tenant rights, 'snob' zoning, transportation policy. Women who are middle-class in terms of their husband's class standing may have more liberal attitudes because of their exposure to a broader range of experience at work.

Second, because the circumstances of women's employment are different from those of men, women tend to be more supportive of government provision of social services than are men. Women's employment opportunities are tied to the public provision of social services in two ways: many women are employed in the public sector and the employment of many is dependent upon such services (for example, subsidized child care). Edgell and Duke (1983) have outlined the impact of government cutbacks on women's employment opportunities in Britain: as an index of this, between 1975 and 1982, male unemployment in the United Kingdom has increased by nearly 200 percent whereas female unemployment has increased by over 350 percent (Edgell and Duke, 1983, page 363).

Women's reliance on social services leads them to be more supportive of government provision of them. A 1980 election study in the USA indicated that only 28 percent of women were inclined to reduce government services, as compared with 38 percent of surveyed men: 'These differences in support of social services provided the potential basis for a woman's vote' (Klein, 1984, page 159). (A similar attitudinal tendency has been found in Britain: Edgell and Duke, 1983; Mark-Lawson *et al.*, 1986.)

The significance of these attitudes for electoral politics has been recognized in the USA since 1980. It was in this year that the 'gender gap' was first noted (Klein, 1984). As an example of the continuing importance of this gap, in the recent US midterm elections on 4 November 1986, it is estimated that 50 percent of men and 56 percent of women voted for the Democrats: 'It was women who ended the Republican grip on the Senate . . . "CBS News polls showed that nine Democratic senators owed their victory to the women's vote"' (Landsberg, 1986, page 2). Working women tend to support the provision of community-based social services and this tends to draw their vote to liberal or leftist political parties. This means that women, particularly working women (Klein, 1984), living in the higher-status suburbs are likely to vote differently from the men in their household and community. The 'gender gap' indicates a divergence of political opinion, within the household and within the community, that is inconsistent with the reproduction thesis.

Theories of social reproduction that draw upon outmoded characterizations of urban residential structure simplify and distort the processes through which political consciousness and social identity are defined. Neighborhoods and families often are not homogeneous in class terms; even individuals (particularly women) may experience a variety of class positions in different facets of their lives. The occupational segregation of women and men, along with the increased participation of women in the paid labor force, effectively break down any occupation-based residential segregation that may have existed in North American cities in the past. The implications of this for social reproduction remain to be explored more closely. Our focus on gender, class, and space effectively opens up these implications for closer scrutiny.

Notes

1 We focus on occupational class (for example, blue-collar and white-collar workers) because these occupational distinctions are highlighted in the reproduction argument. For a description of the operationalization of occupational class, see Pratt (1986).
2 For example, the female labor-force participation rate in the Worcester metropolitan area (51.4 percent) is identical to the US metropolitan rate and so is the percentage of women in the metropolitan labor force (43.9 percent).
3 We define female-dominated occupations as those in which at least 70 percent of the workers are women; male-dominated occupations are those occupations in which at least 70 percent of the workers are male. The remainder of occupational categories are defined as gender-integrated.
4 Although this is a selective presentation, we did examine the characteristics of residents living in a number of census tracts selected on this basis. We are convinced that two highlighted census tracts are not unique or extreme cases.
5 The cutoff point between lower and middle thirds was $20,415; between middle and upper thirds it was $29,045.

Selected references

Acker, J. 1973: 'Women and social stratification: a case of intellectual sexism'. *American Journal of Sociology* 78, 936–45.
BOC, 1986: *Statistical Abstracts of the US*, 106th edition. Washington, DC: US Bureau of the Census.
Bourdieu, P. 1974: 'Cultural reproduction and social reproduction'. In Karabel, J. and Halsey, A. H. (eds), *Power and Ideology in Education*. New York: Oxford University Press, 487–511.
Bowles, S. and Gintis, H. 1977: *Schooling in Capitalist America*. New York: Basic Books.
Cho, S. K. 1985: 'The labor process and capital mobility: the limits of the new international division of labor'. *Politics and Society* 14, 185–222.
Cox, K. R. 1979: *Location and Public Problems*. Chicago, IL: Maaroufa.
Crompton, R. and Jones, G. 1984: *White-collar Proletariat: Deskilling and Gender in Clerical Work*. London: Macmillan.
Delphy, C. 1981: 'Women in stratification studies'. In Roberts, H. (ed.), *Doing Feminist Research*. Andover, Hants: Routledge and Kegan Paul, 114–28.
DOL, 1983: *Time of Change: 1983 Handbook of Women Workers*. US Department of Labor Bulletin 298. Washington, DC: US Government Printing Office.
Edgell, S. and Duke, V. 1983: 'Gender and social policy: the impact of public expenditure cuts and reactions to them'. *Journal of Social Policy* 12, 357–78.
Freeman, J. 1981: 'Women and urban policy'. In Simpson, C. R., Dixler, E., Nelson, M. J. and Yatrakis, K. B. (eds), *Women and the American City*. Chicago, IL: University of Chicago Press, 1–9.
Garnsey, E. 1977: 'Women's work and theories of class stratification'. *Sociology* 12, 223–43.
Harris, R. 1984: 'Residential segregation and class formation in the capitalist city: a review and directions for research'. *Progress in Human Geography* 18, 26–49.
Harvey, D. 1975: 'Class structure in a capitalist society and the theory of residential differentiation'. In Peel, R., Chisholm, M. and Haggett, P. (eds), *Processes in Physical and Human Geography*. London: Heinemann Educational Books, 354–69.

Hayghe, H. 1986: 'Rise in mothers' labor force activity includes those with infants'. *Monthly Labor Review* 109(2), 43–5.

Huckle, J. 1985: 'Geography and schooling'. In Johnston, R. J. (ed.), *The Future of Geography*. Andover, Hants: Methuen, 291–306.

Klein, E. 1984: *Gender Politics: From Consciousness to Mass Politics*. Cambridge, MA: Harvard University Press.

Landsberg, M. 1986: 'Gloomy analysis ignores women's breakthrough in US vote'. *The Globe and Mail* 22 November, Section A, page 2.

Ley, D. 1983: *A Social Geography of the City*. New York: Harper and Row.

Mark-Lawson, J., Savage, M. and Warde, A. 1986: 'Gender and local politics: struggles over welfare policies, 1918–1939'. In The Lancaster Regionalism Group (eds), *Localities, Class and Gender*. London: Pion, 195–215.

Patterson, J. 1974: 'Factorial urban ecology of Greater Vancouver: characteristics of the data base', unpublished MA thesis, Department of Geography, University of British Columbia, Vancouver.

Pratt, G. 1986: 'Housing-consumption sectors and political response in Canada'. *Environment and Planning D: Society and Space* 4, 165–82.

Rein, M., Bane, M. J., Frieden, B., Rainwater, L., Coleman, R., Anderson-Khlief, S., Clay, P., Pitkin, J. and Bartlett, S. 1980: 'The impact of family change on housing careers', unpublished manuscript available from Joint Centre for Urban Studies of MIT and Harvard University, Cambridge, MA.

Reskin, B. and Hartmann, H. 1986: *Women's Work, Men's Work: Sex Segregation on the Job*. Washington, DC: National Academy Press.

Scott, A. J. 1980: *The Urban Land Nexus and the State*. London: Pion.

Scott, A. J. 1985: 'Location processes, urbanization, and territorial development: an exploratory essay'. *Environment and Planning A* 17, 479–501.

Scott, A. J. 1986: 'Industrialization and urbanization: a geographical agenda'. *Annals of the Association of American Geographers* 76, 25–37.

Shevky, E., and Bell, W. 1955: *Social Area Analysis*. Stanford, CA: Stanford University Press.

Timms, D. 1971: *The Urban Mosaic: Towards a Theory of Residential Differentiation*. Cambridge: Cambridge University Press.

US Bureau of the Census. 1986: 'Current population reports' series P-60, number 157. Washington, DC: US Bureau of the Census.

7 Hilary P. M. Winchester,

'Women and Children Last: The Poverty and Marginalization of One-parent Families'

Excerpts from: *Transactions of the Institute of British Geographers* NS **15**(1), 70–86 (1990)

Introduction

This paper aims to provide a preliminary geographical analysis of one-parent families, drawing on evidence from Great Britain and Australia. One-parent families constitute one of the most rapidly-growing types of household in the industrialized developed world; however, the geographical distribution of one-parent families and their socio-economic characteristics remain largely unknown and unconsidered by geographers. The geography of one-parent families is important both at the empirical and the theoretical levels. At the empirical level, a study of one-parent families provides insights into the geographical implications of household change; it also provides evidence of the socio-economic and spatial marginalization of a growing group of households. At the theoretical level, the marginalization of one-parent families exemplifies the analysis of the feminization of poverty, related to women's dual role in contemporary western society. The relative deprivation and segregation of one-parent families has significance for socialist and feminist theory and policy. Furthermore, it has immediate material significance for the well-being of a large number of women and children. The remainder of this introductory section outlines the issue of the feminization of poverty; examines general patterns of household change in Britain and Australia; and reviews the small group of geographical studies of one-parent families.

Studies of marginalization have indicated that certain groups are alienated or excluded from capitalist–patriarchal society (Merlin, 1986; Vant, 1986; Winchester and White, 1988). The marginalization of these groups is related in general to the structural characteristics of power relations within society. However, more specific explanations are required for groups as diverse as punks, prostitutes, pensioners and lone parents. The analysis of the marginalization of lone parents in this paper draws heavily on a recent body of literature concerned with the feminization of poverty (Pearce, 1978; Stallard, Ehrenreich and Sklar, 1983; Holcomb, 1986). The analysis of the feminization of poverty has contributed to an understanding of the causes of socio-economic polarization, and its unequal impact on women and men.

A significant polarization is occurring between groups of women. On the one hand is the group described by Holcomb (1986) as 'new urban women'; skilled, professional, independent women who develop successful careers and are financially secure. At the other extreme are dramatically increasing numbers of women below the poverty line, mainly households with no resident adult male, including lone female parents and lone elderly women.

Stallard, Ehrenreich and Sklar (1983) showed that in 1969, 37 per cent of adult poor in the United States of America were women; by 1979 this proportion had jumped to 67 per cent.

The analysis of female poverty has stressed that it is different to male poverty; male poverty is generally due to a lack of jobs, while female poverty is due to occupational segregation into low-paying jobs. The occupational segregation of women reflects an underlying societal assumption that women will be engaged in caring and child-rearing, but will be financially supported by men. This assumption is reflected in the results of divorce and custody proceedings, which tend to leave children in the care of their mothers rather than their fathers. The particular problems of lone mothers make them particularly vulnerable to poverty. Sarvasy and Van Allen (1984) have argued that these women bear not only a dual role, but an unjust dual role made up of unpaid domestic labour with underpaid wage labour.

The Welfare State and public policy have played a contradictory role in the feminization of poverty. Positive support for women in poverty has been provided by direct financial benefits, and indirectly by the provision of numerous job opportunities in the public services, which have been predominantly filled by women. Negative impacts of public policy have occurred when solutions to female poverty have been based on the ideology of the traditional family. Such an ideology directs attention towards male employment rather than to female financial self-sufficiency. Sarvasy and Van Allen (1984) propose a socialist policy package which is centrally feminist to overcome the injustice of the dual role.

The geographical study of households has tended to fall down the gap between population geography and social geography. The traditional concerns of population geographers in the industrialized developed world have been with migration, and, to a lesser extent, with patterns of fertility and mortality. As migration flows have diminished in intensity, and as trends in fertility and mortality have become increasingly predictable, so the dynamic nature of household change has become clearer. The study of household change as one of the most rapid and significant forms of population change has come to the fore in geography only since about 1985 (see, for example, Gober, 1986; Jones, 1987; Hall, 1988). The relative lack of attention given to households and families by population geographers may also be related to the difficulties of definition; this is the case, for example, in Italy, where the household is deemed to be synonymous with the family (Hall, 1986).

Social geographers, although concerned with welfare issues and with the spatial and socio-economic characteristics of various sub-groups, have rarely paid attention to particular household types; an exception to this is the geographical study of the elderly. More commonly, social geographers have studied population sub-groups defined by race, class, or occupational stratification. Even the techniques of social area analysis, which nominally included the study of familism, rarely provided the opportunity for the analysis of household types as such. More recently, the increasing literature on the interaction of women with the built environment has emphasized the importance of urban social differentiation by gender (see, for example,

Holcomb, 1984; Lewis and Foord, 1984; Wagner, 1984; Institute of British Geographers, Women and Geography Study Group, 1984; Allport, 1986), and has shown the inadequacy of traditional household definitions (Pratt and Hanson, 1988). Despite the welfare tradition in social geography and the recent focus on marginalization of particular sub-groups within the city, the geographical study of households in general, and of one-parent families in particular, remains limited.

The structure of households and families has changed very rapidly in the last 20 years in western Europe (Hall, 1986) and in other industrialized developed societies. The most significant change has been an increase in the proportion of smaller households (Table 7.1). In Great Britain, the average size of household fell from 2.9 people in 1971 to 2.6 in 1984 (*Social Trends*, 1986). In Australia, the average size of household fell from 3.3 people in 1971 to 3.0 in 1986 (*Social Indicators Australia*, 1984). The larger average household size in Australia may be explained by the younger age structure of the population.

The trend to smaller households in both Great Britain and Australia results from a substantial diminution both in large families, and in complex households consisting of more than one family. The number of households in Britain with five or more people fell from 14 per cent to 9 per cent between 1971 and 1984. In Australia, households of five or more people have not only decreased as a proportion of total households since 1976, but the actual number of such households has also declined (Hugo, 1986). Much of this reduction can be accounted for by a general decrease in fertility, and by the ageing of the population. In western Europe, a range of other social and economic factors have also been considered to be important by Hall (1988; p. 22): these include the increasing 'individualism and independence' of both the young and the old, and the decline of traditional economies.

At the same time as large households have declined, there has been a substantial increase in one-person households. In Britain, these have

Table 7.1 Changing household composition, Great Britain and Australia, 1971–1986 (percentages of all households)

	Great Britain			*Australia*		
	1971	*1981 (%)*	*1984*	*1971**	*1975 (%)*	*1986*
Number of people per household						
1 person	18	22	25	14	18	19
2 people	32	32	32	27	29	31
5+ people	14	11	9	25	18	15
Mean number of persons per household	2.9	2.7	2.6	3.3	3.1	3.0
One-parent families	2.8	4.1	8.9†	3.1	4.3	5.6

Notes: *Some data for 1969 (Source: *Social Indicators Australia*, 1984). † Data for England only. (Source: *Regional Trends* (1988); see text for discussion).
Sources: *Social Indicators Australia*, 1984; *Census of Population* (Australia), 1986; Haskey, 1986; *Social Trends*, 1986; *Regional Trends*, 1988.

increased from 18 per cent in 1971 to 25 per cent in 1984. In Australia, the trends are similar, from 14 per cent in 1971 to 19 per cent in 1986 (Table 7.1). The proportion of one-person households is lower in Australia because of significant differences in the age structure of the population. The increase in one-person households is primarily caused by greater numbers of the elderly living alone, but there have also been increases in single-person households of other age groups, including the young after leaving the parental home, and older divorced or separated people. Most single-person households, both young and old, are female. There are only a small number of middle-aged people living alone, but the majority is male, often the divorced partners of female lone parents. Other changes in household types include a reduction in the number of family households (parents with children), and an increase in the number of households made up of one parent with child(ren). This paper focuses on the distribution and characteristics of this latter group.

This paper offers a preliminary examination of the extent of need, poverty and marginalization of one-parent families in Great Britain and Australia. The evidence presented in the following sections shows the rapid growth of one-parent families, their national and intra-urban locations, and their economic marginalization in terms of income, employment, housing and mobility. The evidence for economic and spatial marginalization of one-parent families has wide-ranging social implications, such as the need for support networks in particular areas; these social consequences can be inferred, but they require further detailed study which is in progress, but which is beyond the scope of the secondary data used in this paper.

This paper also examines briefly some of the housing and welfare policies directed towards one-parent families in Great Britain and Australia. A number of state policies impinge on lone parents, especially policies for child care, family income supplements, housing, and the creation of equal employment opportunities for women. The location and housing of deprived groups are particularly important issues in affluent home-owning societies, as marginal groups concentrated in other tenure forms may carry a certain social stigma. Similarly, welfare policies may stigmatize their recipients, and can only be seriously evaluated by examination of their underlying ideology. Such policies in Britain and Australia will be examined in the final section of this paper in relation to the socialist feminist agenda proposed by Sarvasy and Van Allen (1984). It must be emphasized that detailed empirical study of policy uptake is beyond the scope of this present paper, but that further work is in progress.

The number of geographical studies of one-parent families is very small. Thumerelle and Momont (1988), analysing the distribution of one-parent families in France from the 1982 census, attribute the dearth of earlier studies both to data difficulties and to more deep-rooted causes. They argue that although there have always been one-parent families, mainly because of widowhood, their existence did not challenge the norm of the nuclear family (Thumerelle and Momont, 1988). In the case of the black population of the United States of America, this norm, if it ever existed, has been shattered since the 1960s: by 1984, over half of black children were being raised by the mother only (Jones, 1987). In this instance, the poverty of lone parents is exacerbated by racial as well as gender segregation in the job market.

Locational studies of one-parent families have emphasized the importance of the structure of the local housing market. In Montreal, Rose and Le Bourdais (1986) have shown that one-parent families have increased in number and have become increasingly youthful since the early 1970s; this change in composition has brought about a more even distribution across the city as the younger families located in central zones. On the other hand, a British-based study found that lone parents were particularly concentrated in peripheral estates (Robertson, 1984). More general studies of female-headed households have emphasized patterns of segregation in the inner city (Cook and Rudd, 1984) and the inappropriate nature of much of the existing housing stock (Birch, 1985; Watson, 1986).

The data available for the study of one-parent families are derived mainly from national censuses, and will vary depending on definition of household and family types. Households are defined by residence rather than kinship, and therefore include people living alone, and groups of unrelated people living together. Wherever possible, in this paper, one-parent families are given as a proportion of total households rather than as a proportion of families. However, some of the cross-tabulations of the Australian census compare one-parent families with other family types. The cross-tabulations of one-parent families with other indicators vary between Britain and Australia. For example, the British census provides information on housing tenure and car ownership, while the Australian census provides cross-classifications of family type with income. However, the census data are sufficiently comparable to enable a useful general study to be undertaken. The reliance on census data is necessary for a full geographical coverage, especially for small areas.

One-parent families in Great Britain and Australia

This section examines at a national level the growing numbers and changing location of one-parent families within Great Britain and Australia; indicative evidence is also presented on income and poverty. The national trends are used as a context for more detailed studies of the two medium-sized cities of Plymouth and Wollongong.

The basic data sources for family and household types are the national censuses and statistical surveys but the data on the number of one-parent families can be conflicting and confusing. In Great Britain, census data indicate that the proportion of one-parent families increased from 2.8 per cent of all households in 1971 to 4.1 per cent in 1981 (Table 7.1). Data from the General Household Survey for the combined years 1983–85 indicate that one-parent families as a proportion of all families with dependent children had increased from 8 per cent in 1981 to 14 per cent in the early 1980s (*Social Trends*, 1987). However, *Regional Trends* (1988) suggests that in 1985 in England alone, one-parent families formed 8.9 per cent of households, constituting over one and a half million such households; this level is substantially higher than that indicated by the Census. In the absence of recent census data, these latter figures by county are used in the construction of Figure 7.1. More recent data are only available at an aggregate scale, and therefore census data for 1981 are used for the detailed study of Plymouth.

In Australia, the more frequent occurrence of censuses obviates some of the problems of reliance on sample surveys. It is clear that the number of one-parent families has increased in Australia, as it has in other advanced countries. One-parent families formed the fastest-growing family type between 1966 and 1981 (Hugo, 1986), and their numbers have increased even more dramatically since 1982 (Brownlee, 1988). In 1986, the Australian Census recorded 215,000 one-parent families, making up over 5 per cent of households. However, the estimates from the Social Security Review were of 316,000 such families in 1985; 14 per cent of families (Raymond, 1987). A small proportion of the Census underenumeration may result from the concealment of one-parent families in complex households of three or more generations, but these households are in any case increasingly uncommon (Hugo, 1986). The data used in this study, both nationally and within Wollongong, are derived from the Australian Census of 1986. Some of the cross-tabulations of these data compare one-parent families with all households,

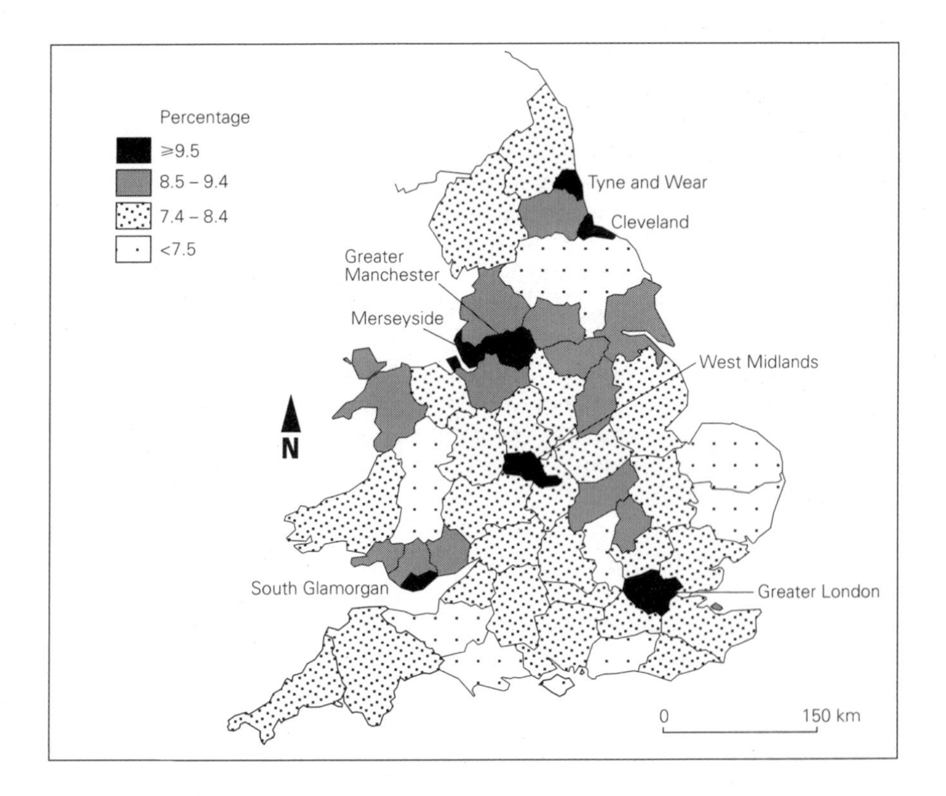

Figure 7.1 One-parent families in England and Wales, 1985, as a percentage of total households. Source: *Regional Trends*, 1988

some with all families; where comparisons are only available with all families this is clearly stated on the table.

Although there are difficulties in establishing the precise numbers of one-parent families, it is clear that there has been a substantial increase in their number since 1971. The reasons for this increase in Great Britain and Australia in particular, and in western industrialized societies in general, are predominantly related to the rising divorce rate, and to the period of separation before divorce is finalized (Thumerelle and Momont, 1988). Although the marriage rate has fallen in Britain and Australia since the early 1970s (*Social Trends*, 1986; National Population Council, 1987), the divorce rate has increased between four and fivefold. In 1960, the divorce rate in the United Kingdom was 0.5 per 1,000 inhabitants; by 1980, the rate had increased to 2.8 per 1,000 (Commission of the European Communities, 1984). In Australia, marriage rates are at an all-time low (Anon., 1988), and divorce rates per 1,000 married people have increased substantially from 2.8 in 1961 to 12.1 in 1981 (*Social Indicators Australia*, 1984). In both cases, the increase in divorce can be related to legal changes which facilitated divorce. In Britain, a marked rise in the divorce rate occurred in 1971, after the 1969 Divorce Reform Act came into effect; and another sharp increase occurred in 1984, when further changes in the law allowed petitions for divorce after only one year of marriage. In Australia, the passing of the Family Law Act in 1975 produced a sharp upward trend in divorce statistics, reaching 20 per 1,000 married people in 1976, a rate which has subsequently declined (*Social Indicators Australia*, 1984).

Twenty years ago, the most common reasons for the existence of one-parent families were illegitimacy and widowhood. While rates of widowhood have declined, there is evidence that the rate of non-marital births is rising. It is suggested that the character of these births is gradually changing, from the clandestine and largely unwanted to members of low-status groups of the population, to the belated but positively chosen for higher-status groups (Deville and Naulleau, 1982). Many of these recent non-marital births may be to cohabiting couples rather than to lone mothers; evidence in Britain for this comes from the increasing practice of registering the birth by both parents rather than by the mother alone. Indicative evidence in Australia also comes from the increase in ex-nuptial births to women aged 30 and over, and from the high proportion of children recorded as living with couples in *de facto* relationships, especially to couples of this age group. However, births to unmarried teenagers, although relatively less common, may be more likely to result in a one-parent family in the 1980s than was the case in the 1960s, as there is now less social pressure for young mothers to give up their babies for adoption or to rush into marriage. In both Britain and Australia, between 21 and 22 per cent of lone mothers have never been married. Lone parenthood therefore arises from a number of causes, and affects a wide range of socio-economic groups. Despite the increase in non-marital families, only a small number of lone parents can be considered to be so by choice.

The geographical distribution of one-parent families is complex. The most recent data available for England and Wales (*Regional Trends*, 1988) show a marked metropolitan concentration, and a less marked core/periphery distinction (Figure 7.1). The highest concentration of one-parent families is found in Merseyside, where they form 11 per cent of households; all the other metropolitan counties also exhibit concentrations of 9 per cent or more. The lowest concentrations of one-parent families are found in the Isle of Wight (6.3 per cent) and in parts of East Anglia, the South West, and mid-Wales, together with more rural areas of the Home Counties and Yorkshire. It is likely that these geographical differences reflect structural differences in the age, socio-economic, and racial composition of the population.

In Australia, there is some evidence of concentration of one-parent families in large isolated cities, such as Darwin and Perth, although not in the major metropolitan centres (Figure 7.2). The regional pattern shows areas of concentration in New South Wales and in parts of the outback; a dual pattern complicated by socio-economic structures, the Aboriginal composition of the population, and by the distorted structure of frontier populations. The highest concentrations of one-parent families are found in the cities of Darwin (8.4 per cent) and Canberra (7.1 per cent). The contrasting socio-economic and

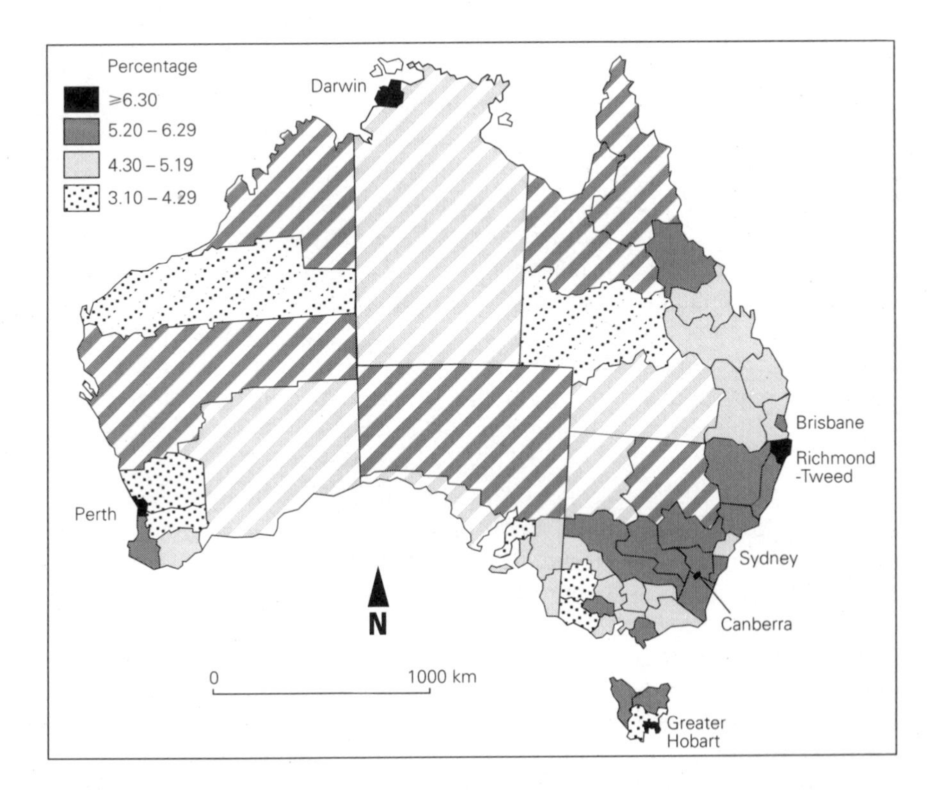

Figure 7.2 One-parent families in Australia, 1986, as a percentage of total households. Source: *Census of Population* (Australia), 1986. Note: intermittent shading denotes areas of very low population density

racial composition of the two cities exemplifies the dual nature of the regional pattern. Canberra, the national capital, contains a large number of professional people and is characterized by a sophisticated lifestyle and high divorce rates; Darwin, on the other hand, is a tropical frontier city with a substantial Aboriginal population, lower socio-economic status, and low rates of marriage.

The increase in number of one-parent families has affected men and women unequally. Most lone parents with dependent children are women, whatever the cause of lone parenthood. In particular, the operation of divorce proceedings still tends to give custody of the dependent children to the mother rather than the father, a pattern heavily influenced by the expectation that women will naturally fill the caring role. In Britain, 89 per cent of one-parent families are headed by a woman, and this proportion has increased since the 1970s (Haskey, 1986); while in Australia the proportion is 86 per cent (*Census of Population* (Australia), 1986).

The two types of household which are growing most rapidly, single-person households, and one-parent families, are both predominantly female. It is noteworthy that these two groups of households are particularly vulnerable to poverty. It has been cogently argued that poverty is increasingly a female and urban phenomenon. The lone elderly are impoverished because of their dependence on welfare benefits, but lone elderly men are often better off than women in similar circumstances, as men are more likely to have occupation-related pension entitlements. Similarly, lone fathers with children experience less poverty than lone mothers with children as they have greater access to better-paid jobs. Young mothers are particularly disadvantaged in this respect, as they are likely to have limited work skills and experience; Holcomb (1986; p. 453) has characterized teenage motherhood as a 'fast track to poverty'. The difficulties of female occupational segregation are compounded for young black women (Jones, 1987). The urban concentration of female poverty may arise either from marginal groups such as the poor elderly and blacks being trapped in a particular segment of the housing stock, or from the selective migration of the service-dependent poor (Wolch, 1980) to central areas where specialist services and jobs are relatively accessible.

Despite the variety of causes of one-parent families, their general vulnerability to poverty has caused them to be considered as a group which is economically marginal (Merlin, 1986; Winchester and White, 1988). Their economic marginality is demonstrated by national data. In Britain, half of all one-parent families in the early 1980s were found to be living below the poverty line. They were 6 times more likely to be within the definition of poor, and twice as likely to have incomes in the bottom 20 per cent of the income scale than were married couples with children (Family Policy Studies Centre, 1986). In Australia, more than half of one-parent families in 1986 had an income which was less than $12,000; although one-parent families constitute 5 per cent of all families in Australia, they make up over 21 per cent of all families in this lowest income band. This poverty is brought about in large part by the requirements of the dual role of child care and poorly-paid waged employment, exacerbated by the time and mobility constraints experienced particularly by women.

The occupational segregation of women results in low average female earnings; in Australia women earn only two-thirds of male income, and this discrepancy is exacerbated by lone parenthood. Recent data on family incomes in Australia indicate that married couples with one wage-earner and two children earned $460 per week on average, whereas female wage-earning lone parents earned $302 (66 per cent of the 'married' wage). However, the majority of female lone parents are in receipt of benefit, providing an income of only $160 per week. The average weekly income for female lone parents is therefore $186, compared with the average weekly income for male lone parents of $283 (the female income again forms 66 per cent of the male income) (Brownlee, 1988). In Britain, the employment difficulties for lone mothers result in half of this group relying on state benefits, whereas only 30 per cent of lone fathers rely on benefits, reflecting their better access to employment opportunities and higher earnings (Popsay, Rimmer and Rossiter, 1983). In the mid-1980s, however, male lone parents' reliance on social security had increased as employment opportunities diminished during the recession. Harrison (1988) recorded that in Australia, the proportion of male lone parents who are unemployed doubled between 1981 and 1986, from 17 to 33 per cent.

One-parent families in Plymouth, Great Britain, and Wollongong, Australia

The two cities of Plymouth and Wollongong provide more detailed case-studies of the location and status of one-parent families than the national statistics of the previous section. Plymouth is a city of a quarter of a million people in the south west of England, which has traditionally had a strong naval presence, and is a service centre for much of Devon and Cornwall. The naval dockyards remain the major employer, but these have suffered substantial job losses as a result of retrenchment and privatization (Maguire, Brayshay and Chalkley, 1987). Wollongong, 80 km south of Sydney, is slightly smaller, with a population of 168,000 in the Local Government Area in 1986. It is a coastal city set in a spectacular physical environment, but dominated economically by traditional manufacturing industries which are now in decline (Haughton, 1989). The years of booming steel, coal and textile industries brought many immigrants from different ethnic backgrounds, but neither Wollongong nor Plymouth contain a significant black population. In these cities, therefore, the marginalization of one-parent families is essentially uncomplicated by racial factors. Both cities are predominantly working class, and adjusting with some difficulty to economic restructuring.

Plymouth occupies a congested site and has a history of housing overcrowding which has been remedied, at least in part, by an intensive programme of peripheral council house building in the post-war period (Gill, 1979). The majority of remaining inner city housing consists of terraced properties, but much of the inner city was bombed during the Second World

War, and further gutted by post-war reconstruction. The replacement housing is mainly two- or three-bedroomed terraced or semi-detached; there are only three high-rise blocks in the city (Winchester, 1987). Wollongong's site is also physically restricted by the dominating escarpment, but housing is nonetheless predominantly single-family detached housing.

Wollongong contains a higher proportion of public housing and of medium-density housing than is usual in non-metropolitan Australia (Keys and Wilson, 1984). However, the Australian zoning of medium-density and high-density is lower density than its British counterpart, and includes both three-storey blocks of flats and town houses. A first wave of Australian public housing in the 1950s catered for the expanding population on relatively large estates of individual fibro-clad iron-roofed houses; subsequent additions to the public housing stock have been more varied in style and scale. Both cities have a relatively large stock of council housing (and a significant proportion of the population in low-income groups) (Keys and Wilson, 1984; Maguire, Brayshay and Chalkley, 1987).

Given the nature of the economies, social composition, and housing stock of the two cities, it would not be surprising to find a relatively high proportion of one-parent families in these environments. However, the two cities differ strongly in this regard. One-parent families are relatively under-represented in Plymouth, where they constituted only 2.5 per cent of households in 1981, considerably less than the proportion in the metropolitan areas of the country. Although the Census appears to underestimate the number of one-parent families, and although their number has increased since 1981, this relatively low proportion emphasizes the importance of the metropolitan areas either as a form of trap for the impoverished or as a centre of potential jobs, services and cheap housing. By contrast, Wollongong conforms almost exactly to the national average, with one-parent families forming 4.8 per cent of households in 1981 and 5.6 per cent by 1986. Although there are differences in the proportion of one-parent households in the two cities, these households are predominantly female-headed, with 89 per cent in Plymouth, and 87 per cent in Wollongong.

The Census data of 1981 for Plymouth and 1986 for Wollongong give some indication of the marginality of one-parent households in relation to all households in terms of housing, mobility, employment and income. In both Plymouth and Wollongong, most one-parent families are only on the fringes of the job market (Table 7.2). Lone fathers are much more likely to be in employment than lone mothers; and many more women than men are economically inactive. Employment levels are generally higher in Wollongong, but in both cities over half the lone mothers either considered themselves unable to enter the labour force or were unemployed. In Plymouth, the gender differential is very marked for lone parents of very young children; furthermore, more women than men are engaged in part-time employment. Data broken down in this way are not available for Australian cities, but data on incomes provide further indications of the marginal occupational status of one-parent families. In Wollongong, 70 per cent of one-parent families received an income of less than $12,000 in 1986, compared with 13 per

Table 7.2 Employment status of lone parents in Plymouth, 1981 and Wollongong, 1986

Plymouth (percentage)	*In full-time employment*	*In part-time employment*	*Not economically active or unemployed*
Lone fathers with children aged 0–4	53.3	0.0	46.7
Lone fathers with children aged 5–15	65.9	2.6	31.5
Lone mothers with children aged 0–4	3.2	9.9	86.9
Lone mothers with children aged 5–15	18.4	25.9	55.7
*Wollongong (percentage)**	*In employment*	*Not in labour force*	*Unemployed*
Lone fathers	66.5	20.2	11.7
Lone mothers	24.5	64.4	10.1

Note: *These do not add to 100; 1–2 per cent do not state their employment status.
Source: *British Census*, 1981; *Census of Population* (Australia), 1986.

Table 7.3 Housing type and income for one-parent families and all families in Wollongong, 1986 (percentages)

Income	*$0–12,000*	*$12–32,000*	*$32,000+*	*Not stated*
All families	13.0	45.6	22.8	18.6
One-parent families	69.6	22.0	1.8	6.7
Housing type	*Owner/buyer*	*Renters*	*Detached houses*	*Medium density*
All families	72.4	24.3	67.3	8.7
One-parent families	28.1	71.2	58.2	34.3

Note: This information is available only for families, not households; single-person households and households of unrelated adults are therefore excluded.
Source: *Census of Population* (Australia), 1986.

cent of all families; less than 2 per cent of one-parent families earned $32,000 or more, compared with 23 per cent of all families (Table 7.3). The concentration of poverty in this group of households is shown by the fact that a quarter of the city's low-income families were headed by a lone parent, a proportion even higher than the national average.

Low incomes are a particularly important factor constraining residential choice (Birch, 1985), and it has been argued that low-income women are further disadvantaged by urban revitalization (Holcomb, 1984). It would be expected therefore that one-parent families would be concentrated in certain parts of the city. The exact location is likely to be severely constrained by the structure of the housing market; in North America, the location of lone parents appears concentrated in city centres, whereas in Great Britain there is some evidence of concentration on peripheral estates. In Plymouth, one-parent families are not particularly concentrated in any one part of the city, whereas other marginal groups such as the lone elderly are clearly concentrated in older housing in the city centre (Furnival, 1987). Only in two wards of the city, Budshead and St Peter, is the concentration of one-parent families greater than 4 per cent of the total; these districts are both on the western edge of the city in areas of naval estates and council housing (Figure 7.3). The lack of spatial concentration is counterbalanced by the over-concentration of one-parent families into particular categories of housing tenure. Over half the city's one-parent families are accommodated in council housing, compared with 27 per cent of households in the city as a whole (Table 7.4). This is because one-parent families easily accumulate points in the council house allocation system, and have high priority on housing lists, but may be allocated a house in any of the city's council estates. For these households in Plymouth, this location means that the housing amenities available to one-parent families are generally good, as the peripheral housing estates are relatively new. In Plymouth, the rates of overcrowding for one-parent families are worse than for households as a whole, but their small family

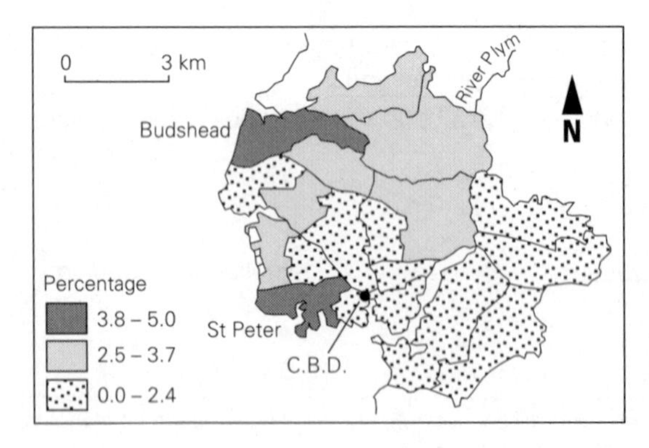

Figure 7.3 One-parent families in Plymouth, 1981, as a percentage of total households. Source: *British Census*, 1981

Table 7.4 Housing amenities, housing tenure and car ownership for one-parent families and all households in Plymouth, 1981 (percentages)

			Housing amenities (%)		
Household type	*One or more persons per room*	*Lack bath*	*Lack inside wc*	*Not self-contained*	*No car*
All households	2.4	1.9	2.6	5.5	41.4
One-parent families	9.3	0.6	1.0	2.9	73.0
Household type	*Owner-occupied*	*Housing tenure (%) Council rented*	*Rented unfurnished*	*Rented furnished*	*Other*
All households	54.8	27.1	6.7	6.1	5.3
One-parent families	28.7	56.1	4.5	3.8	7.0

Source: *British Census*, 1981

size and eligibility for public housing results in less overcrowding than for other families with children (Furnival, 1987). This concentration of one-parent families in public housing in Britain would help explain their peripheral location in many smaller British cities; however, more detailed analysis is required of the intra-urban location and housing conditions of one-parent families in metropolitan areas of Britain.

The location of one-parent families in Australia is less metropolitan-based than in Britain, but the spatial concentration of one-parent families is greater within the city. In Wollongong, for example, in one collector's district in Bellambi, 25 per cent of households were one-parent families (Figure 7.4). This greater concentration reflects in part the finer spatial scale of analysis, but also reflects the more limited availability of public housing in Australia. However, even on a neighbourhood scale (which approximates to the wards used in Plymouth), 16 per cent of households in the five collector's districts which make up Bellambi were one-parent families.

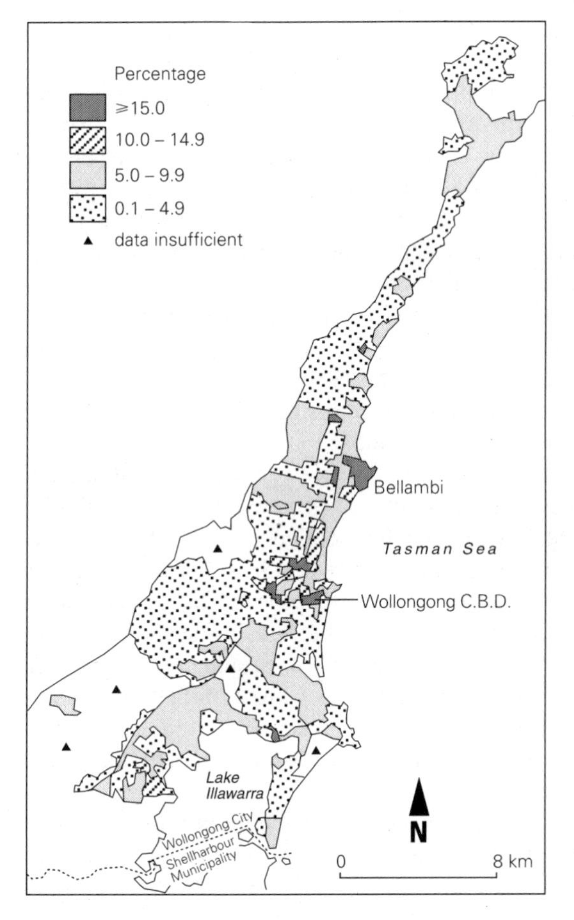

Figure 7.4 One-parent families in Wollongong, 1986, as a percentage of total households. Source: *Census of Population* (Australia), 1986

Although information by tenure is not cross-classified by household type, it is nonetheless apparent that one-parent families are over-represented in rented and medium-density accommodation (Table 7.3). Seventy-one per cent of Wollongong's one-parent families live in rented accommodation and 34 per cent live in medium-density blocks (compared with all families of which 24 per cent rent and 9 per cent live in medium-density blocks). A large proportion of the rented and medium-density housing is public housing, although some is privately rented. Public medium-density housing consists either of three-storey walk-up blocks built in the 1950s, or newer and more attractive town-house developments such as those at Bellambi. Concentrations of one-parent families occur particularly in newer public housing, because of the relationship between age of housing and age of occupier. Australia therefore stands in contrast to countries such as Britain and France where public housing forms a major segment of the housing market (Winchester and Furnival, 1989). Although both Britain and Australia exhibit a concentration of lone parents in rented public housing, in Britain this does not bring about a marked spatial concentration, whereas in Australia the more limited supply of public housing induces locational segregation.

The rented and medium-density accommodation in Wollongong does not only consist of public housing, however. Medium-density redevelopment has taken place in much of the central area, close to the shops, and close to other amenities such as the beach, with one or two expensive high-rise private developments boasting both ocean and mountain views. Less expensive redevelopment has also occurred in the older residential suburbs, to the north and west of the Central Business District, replacing older detached housing on large blocks. This new development has provided housing niches for one-parent families. In the six collector's districts immediately to the west of the Central Business District, over 10 per cent of the families are one-parent families, whereas at the 1981 Census, this was an area with one of the 'lowest levels' of this family type (Keys and Wilson, 1984: p. 62). This change of location of one-parent families, facilitated by housing market changes, also implies an attraction to central services and jobs.

In the centre of Plymouth, there is no comparable privately rented accommodation, with the exception of a limited amount of new housing association property which is likely to be beyond the financial resources of most lone parents. The privately rented housing stock in Plymouth is largely run-down and subdivided for student use, although in the process of gentrification (Winchester, 1987). Nonetheless, this sector of the housing stock does provide temporary accommodation for many groups of people, particularly in college vacation periods; a significant proportion of one-parent families has spent some time in temporary accommodation, hostels, or rented unfurnished property, especially in the period immediately after marriage breakup and before being rehoused (Furnival, 1987).

Restrictions on mobility for one-parent families in Plymouth are shown by the high proportion which do not have a car. In Plymouth, over 70 per cent of one-parent families do not have a car, whereas only 30 per cent of other households with children lack a car. The inability to own a car is less often a matter of choice than of economic necessity, particularly for women who are not economically active. Furnival (1987) in interviews with lone parents in Plymouth found that a car was the first 'luxury' to be disposed of in times of hardship. The relatively compact structure of Plymouth, and its frequent bus, minibus, and train services does mean, however, that mobility and access is not totally restricted, and that one-parent families are no more deprived than many other low-income households.

However, in Wollongong, the linear structure of the city and the relative lack of public transport severely handicap households without a car, especially as dependence on the private motor vehicle is part of the Australian way of life. In a country in which 90 per cent of households own a car, to be without one is a matter of harsh economics, physical inability, or occasional idiosyncrasy. Although there are no data on car ownership by household type, the low income levels of lone parents suggest that they are more likely to be without a car than are two-parent families. For families close to central Wollongong, the deprivation may be less severe than for those in the suburbs, and indeed a central city location may have been chosen to minimize the need for and to obviate the expense of a car. However, the lack of a car may still inhibit participation in activities in the evenings and at weekends when public transport is restricted, although low income is likely to be a more significant primary factor in limiting access to a variety of facilities.

Marginalization, policy and the dual role of lone parents

Winchester and White (1988) have outlined a number of factors leading to socio-spatial polarization in society and to the marginalization of different population groups. One-parent families have been recognized as a group that is marginalized and vulnerable to poverty (Merlin, 1986; Vant, 1986; Winchester and White, 1988). This study now uses the evidence presented above to examine the various types of marginality experienced by one-parent families, and the special problems they face in coping with their dual role.

One-parent families may be considered to form a group which is marginal in economic terms (Winchester and White, 1988). Economic marginality is seen most clearly in relation to the labour market and the housing market. Marginality in relation to the labour market is seen clearly in the national evidence of poverty, low incomes, limited employment opportunities and receipt of state benefits presented in this study; this evidence is replicated at the local level in Plymouth and Wollongong.

A further important aspect of economic marginality is access to the housing market (Merlin, 1986; Winchester and White, 1988). One-parent families are often disadvantaged in this respect, in particular because of their low income (Birch, 1985). They are also disadvantaged because of the structure of the housing market, which normally subsidizes the house-buyer and which is still geared to the construction of housing for 'conventional' family units (Darke, 1986; Morrow-Jones, 1986; Watson, 1986). Britain and Australia both have sharply rising proportions of home-owners, but one-parent families are economically excluded from this pattern. In both countries, one-parent families are less likely than other types of family to own their own home, and are more likely to rent housing, to share accommodation, to lack basic amenities, and to suffer overcrowding in higher-density accommodation or even homelessness. They will also spend a disproportionately large amount of their income on housing, but are economically excluded from some accessible areas of the central city because of the process of gentrification (Holcomb, 1984).

The housing provision for one-parent families can be deemed marginal in a limited sense. Both Britain and Australia encourage the ideology of home ownership (Headey, 1982; Kemeny, 1986), and people who cannot aspire to that are stigmatized (Stretton, 1986). The class stigma associated with the occupation of council housing in Britain is increasingly paralleled by the 'welfare' nature of Australian public housing (King, 1986). In Australia more than in Britain, public housing provides accommodation of reasonable quality, but it may be relatively peripheral and isolated from services and facilities; privately rented accommodation in the central city may provide poorer housing amenities but gives better access to facilities within walking distance. Both options are remote from the norm of owner-occupation, the detached house on the quarter-acre block in Australia. In Britain, the more widespread council tenure is less removed from the 'normal' owner-occupation of the three-bedroomed semi-detached suburban house.

A third form of economic disadvantage arises from the restriction of activity spaces (Parkes and Thrift, 1980). Activity spaces of all economically marginal groups tend to be restricted because of financial constraints on mobility; women tend to be less mobile than men, especially as they are more likely to be without private transport (Coutras, 1983; Allport, 1986). Lone parents are further restricted because of the problems of travelling with children, and the need to be in specific places at specific times, such as at the school gate to meet children at the end of the school day.

The structure of suburban areas, the zoning of residential uses, and the separation of home from work and from other facilities and services all serve to exacerbate the isolation of many women in the suburbs: these problems are compounded for lone parents (Lewis and Foord, 1984; Wagner, 1984; Institute of British Geographers, Women and Geography Study Group, 1984). The limitations of urban spatial structure, built to reflect a patriarchal familistic social structure, restrict activity spaces, social interaction, and the search for job opportunities.

Other types of marginality which may be just as significant for the daily lives of one-parent families include social and legal marginality (Winchester and White, 1988). One-parent families are not socially marginal in Britain and Australia in the same way as would be the case in Ireland or Italy. Nonetheless, there is still an element of social stigma and blame attached to the status of the lone parent. This is particularly the case for women, who are deemed to have failed in their primary role, whereas men may be offered more social support by friends and family, in that they are deemed to have been deserted. Social stigma is accentuated by low income, the receipt of benefits, and the occupation of rented housing, especially government housing (Headey, 1982; King, 1986). Social marginalization may be particularly acute in the case of black women, who are over-represented as lone parents and further disadvantaged in terms of employment and income.

One-parent families are not illegal, but the legal structures of society have not yet fully adapted to changing demographic realities. In particular, the structure of divorce law, child support, and tax and benefit systems, may embody the ideology of the nuclear family with a male breadwinner and female (formally unpaid or underpaid) carer. The public policy response may therefore even contribute to the feminization of poverty.

The support for the ideology of the family is most clearly shown in Britain, where the structure of benefits differentiates between widows and other lone parents, and between women and men. The majority of one-parent families were entitled to an extra benefit of £4.60 per week in 1986, and low-earning lone parents could claim a Family Income Supplement (Department of Health and Social Security, 1986). Widows are entitled to a Widow's Allowance which, unlike other benefits, is not means-tested, and so allows widows to work without loss of income. Other one-parent families, brought about by non-marital births or by separation and divorce, are only entitled to ordinary rates of supplementary benefit plus the one-parent allowance, and earned income is deductible from benefit. Lone parents, especially women, are also adversely affected by the 'cohabitation rule' whereby a man deemed to be cohabiting is liable to support the woman and children. The social security (and taxation) system is therefore based on the expectation of traditional gender roles, that men will support their wives and families. This expectation also exists in Australia, but it is recognized that it is increasingly difficult to identity 'marriage-type' relationships; furthermore, under existing law there is no obligation for a person to support their new partner's children (Raymond, 1987). Changing social realities are bringing about pressure for changes to the family basis of assessment.

In Britain, the continuation of the archaic Widow's Allowance implies that widows are deemed to be 'deserving' of support, whereas other one-parent families who have 'chosen' their lifestyle are 'non-deserving'. As such, the legal situation, based on the historical assumption that the nuclear family is normal and the one-parent family abnormal (Robertson Elliot, 1986), marginalizes most one-parent families and limits their participation in the job market, in the housing market and in all activities which require personal mobility and disposable income. In Australia, the taxation and social security system is less divisive by sex or marital status and provides supporting parents' allowance for all classes of lone parents; the major distinction between widows and other lone parents was removed in 1980 (Raymond, 1987).

The evidence from both British and Australian case-studies supports the view that one-parent families are marginalized in relation to employment and income. This marginalization derives particularly from the overwhelmingly female composition of the group, their segregation into low-paid employment, and the competing needs of child care and paid employment. It is perpetuated by the structure of the benefit system, which supports lone parents to about the poverty line, but then reduces benefit as earned income increases. Economic marginality limits other choices, notably housing; the limitations of cost preclude owner-occupation in many cases, so that one-parent families are predominantly found in cheap rented accommodation. This type of accommodation is spatially concentrated in particular areas of the city; this is especially the case in Australia where the norm is owner-occupied detached housing. In Britain, cheap rented housing is extensively provided by the local authority, and forms a significant (but reducing) component of the total housing stock, so residential distributions of one-parent families are less spatially concentrated; they are merely one element in the large low-income group of council tenants. The evidence from both Plymouth and Wollongong further indicates that one-parent families headed by men are less likely to suffer from occupational marginality, and are therefore likely to have more disposable income to choose housing and to purchase mobility.

The policies affecting one-parent families may be considered in relation to the feminist socialist perspectives outlined by Sarvasy and Van Allen (1984). They propose five long-term aims for public policy strategies designed to reduce vulnerability to poverty and to alleviate the injustices of the dual role (Sarvasy and Van Allen, 1984). These aims are: economic self-determination for women; equalization of men's and women's shares of paid labour, unpaid labour and caring; sensitivity to class and racial differences among women; democratic planning and control of social programmes, and the expansion of the public conception of social responsibility. The most intractable of the aims is that of economic self-determination for women. In Australia, a new policy known as JET (Jobs, Education, Training) has recently been put into operation for lone parents; this is in fact a combination of existing programmes but specifically directed towards lone parents. Improvement of women's prospects in the labour market has also been assisted by affirmative action for women and racial minorities in the government service and higher education. Although it is too early to offer a full assessment of these policies, it would appear to have begun to influence job segregation and public opinion in these employment sectors, and not just by creating poorly-paid dead-end jobs to 'get women off welfare'.

The aims of equalizing the gender division of labour, and of increasing sensitivity to class and racial differences have been addressed in part by the operation of race and sex discrimination legislation, operative in both countries. In Australia, class and racial divisions are less pronounced than in Britain, and programmes of affirmative action for employment have progressed further. Neither Britain nor Australia can claim much progress in the democratic planning of social programmes, as in both countries these are dominated by professionals rather than by clients. Nor has the public conception of social responsibility moved strongly towards a socialization of the costs of child rearing, which are presently borne largely by the individual parent or parents. Family allowances, family income supplements and supporting parents allowances move some way to alleviating the private burden, but many of these benefits and child care schemes are means-tested, often leaving the lone parent in a poverty trap. The emphasis on JET in Australia is still dependent on the availability of existing child care facilities. The public policy response to the poverty and marginalization of one-parent families has therefore gone some way towards a feminist socialist welfare solution, but in Britain even more than Australia the costs of child care are sustained privately, and solutions to the poverty of women are more often offered which emphasize and reinforce economic dependence on individual men.

However, the policies as yet in place have had little time to influence the patriarchal societal structures evident in law, the job market and the family. The problems of poverty and marginalization of one-parent families are extremely pressing. The dramatic rise in divorce rates since the early 1970s and the accompanying increase in one-parent families has changed the distribution of urban poverty. Female lone parents now form one of the most deprived groups in Britain, Australia and other western industrialized countries. It is projected that this group will increase in absolute terms, although the rate of increase is likely to diminish. The implications of this changing family structure are at least as dramatic and far-reaching as the well-publicized decline in fertility and the ageing of western populations. As with the elderly, the vulnerability to poverty requires policies with immediate impact. Policies directed at the improvement of the occupational status of women are inevitably long-term in nature; more immediate impact on the related issues of poverty, child care and employment would be derived from an extension of subsidized child care facilities, and of non-taxable benefits which could be supplemented by earned income.

The issue of housing policy is also of fundamental importance in overcoming marginality. Low-income groups, such as one-parent families, constitute a problem in situations where normal (i.e., owner-occupied) housing is beyond their economic resources. Possible options which operate within the constraints of owner-occupation include a number of methods of housing subsidy. In Britain, housing subsidy for owner-occupiers operates through tax relief, and in Australia through grants to low-income families and a deferral of stamp duty. However, families on the poverty line are not in a position to be able to take up these options, and so they are heavily concentrated in public housing. In Australia, there is pressure for continued monitoring of the income levels of tenants of public housing; high-income tenants would have their rents raised to market levels to encourage them to vacate public housing for low-income tenants. This policy would maintain the supply of public housing but would not reduce the stigma attached to its occupation. The stigma of occupying 'welfare' housing suggests that there is a need to move back to a position where welfare housing is built to be sold cheaply to sitting tenants, to bring low-income groups into 'normal' owner-occupied housing, and to provide a continued turnover of capital for construction. These types of housing policy, combined with child care, income supplements, and equal employment policies, could act together to reverse the marginalization of this group in terms of housing and employment. These are policies which could be implemented in the short term given the political will. They are basically aspatial policies, although with spatial implications.

The spatial segregation of one-parent families into relatively peripheral public housing estates in Australian cities is geographically significant. Segregation into such estates appears to cause a number of social problems, notably difficulties of access to facilities, and a lack of social support. Further research is being undertaken into the operation of community facilities, the dissemination of knowledge about policy entitlements and the uptake of benefit. Further research is also needed on the location and housing conditions of one-parent families in other areas such as the metropolitan areas of Britain, and the isolated cities of Australia.

This paper has shown that there is substantial evidence of the marginalization of one-parent families. It is argued that this marginalization from the norms of capitalist–patriarchal society occurs primarily through the expectations of the dual role which lone parents undertake, as unpaid domestic workers and low-paid waged workers. This dual role brings about a vulnerability to poverty, which can be alleviated by long-term feminist socialist policies aimed at the full and equal participation of women and men in paid and unpaid labour, and the socialization of the costs of child care.

References

Allport, C. 1986: 'Women and suburban housing: post war planning in Sydney 1943–61'. In McLoughlin, J. B. and Huxley, M. (eds), *Urban planning in Australia: critical readings*. Melbourne: Longman Cheshire, 233–48.

Anon. 1988: 'Cohabitations', *Australian Society* 7(11), 54.

Birch, E. L. (ed.) 1985: *The unsheltered woman: women and housing in the 80's*. New Brunswick: Rutgers, The State University of New Jersey.

Brownlee, H. 1988: 'New data on family incomes'. *Family Matters* 20, 16–17.

Census of Population (Australia) 1986: Canberra: Australian Bureau of Statistics.

Commission of the European Communities 1984: *Women in the European Community*. Luxembourg: Commission of the European Communities.

Cook, C. C. and Rudd, N. M. 1984: 'Factors influencing the residential location of female householders', *Urb. Affairs Qu.* 20, 78–96.

Coutras, J. 1983: 'La ville au féminin'. In Noin, D. (ed.), *Géographie Sociale*. Paris: CNRS, 432–43.

Darke, J. 1986: 'Some recent writings on women and the environment: a review essay', *Antipode* 18, 16–29.

Department of Health and Social Security 1986: *Social Security benefit rates*. Leaflet S39. London: HMSO.

Deville, J.-C. and Naulleau, R. 1982: 'Les nouveaux enfants naturels et leurs parents', *Economie et Statistique* 145.

Family Policy Studies Centre 1986: *Family trends and Social Security reform*. London: Family Policy Studies Centre.

Furnival, R. E. 1987: 'One-parent families in Plymouth: an example of a marginalized social group', unpubl. B.Sc. thesis, Dept of Environ. Sci., Plymouth Polytechnic.

Gill, C. 1979: *Plymouth: a new history*, vol. 2. Newton Abbot: David and Charles.

Gober, P. 1986: 'How and why Phoenix households changed: 1970–1980', *Ann. Ass. Am. Geogr.* 76, 536–49.

Hall, R. 1986: 'Household trends within western Europe 1970–1980'. In Findlay, A.

and White, P. E. (eds), *Western European population change*. London: Croom Helm, 18–34.

Hall, R. 1988: 'Recent patterns and trends in western European households at national and regional scales', *Espace, Populations, Sociétés* 1988(1), 13–32.

Harrison, M. 1988: 'Major changes to Family Law Act', *Family Matters* 20, 12–14.

Haskey, J. 1986: 'One-parent families in Great Britain', *Pop. Trends* 45, 5–13.

Haughton, G. 1989: 'Community and industrial restructuring: responses to the recession and its aftermath in the Illawarra region of Australia', *Environ. and Plann. A* 21, 233–47.

Headey, B. 1982: 'Housing conditions in the United Kingdom: who got what at what cost 1919–77?'. In Blowers, A., Brook, C., Dunleavy, P. and McDowell, L. (eds), *Urban change and conflict: an interdisciplinary reader*. London: Harper and Row, 98–102.

Holcomb, B. 1984: 'Women in the rebuilt urban environment: the United States experience', *Built Environ.* 10, 18–24.

Holcomb, B. 1986: 'Geography and urban women', *Urb. Geogr.* 7, 448–56.

Hugo, G. 1986: *Australia's changing population: trends and implications*. Melbourne: Oxford University Press.

Institute of British Geographers, Women and Geography Study Group 1984: *Geography and gender*. London: Hutchinson in association with the Explorations in Feminism Collective.

Jones, J. P. 1987: 'Work, welfare and poverty among black female-headed families', *Econ. Geogr.* 63, 20–34.

Kemeny, J. 1986: 'The ideology of home ownership'. In McLoughlin, J. B. and Huxley, M. (eds), *Urban planning in Australia: critical readings*. Melbourne: Longman Cheshire, 251–8.

Keys, C. L. and Wilson, M. G. A. (eds) 1984: *The urban Illawarra: a social atlas*. Wollongong: Illawarra Regional Information Service.

King, R. 1986: 'Housing policy, planning practice'. In McLoughlin, J. B. and Huxley, M. (eds), *Urban planning in Australia: critical readings*. Melbourne: Longman Cheshire, 274–87.

Lewis, J. and Foord, J. 1984: 'New towns and gender relations in old industrial regions: women's employment in Peterlee and East Kilbride', *Built Environ.* 10, 42–52.

Maguire, D. J., Brayshay, M. and Chalkley, B. S. (eds) 1987: *A social atlas of Plymouth*. Plymouth: Plymouth Polytechnic.

Merlin, P. 1986: 'Housing policies in the inner city and the development of ghettoes of marginal groups (the case of Paris)'. In Heinritz, G. and Lichtenberger, E. (eds), *The take-off of suburbia and the crisis of the central city*. Wiesbaden: Steiner Verlag, 228–34.

Morrow-Jones, H. A. 1986: 'The geography of housing: elderly and female households', *Urb. Geogr.* 7, 263–9.

National Population Council 1987: *What's happening to the Australian family?* Population Report 8. Canberra: Australian Government Publishing Service.

Parkes, D. N. and Thrift, N. J. 1980: *Times, spaces and places: a chronogeographic perspective*. Chichester: Wiley.

Pearce, D. 1978: 'The feminization of poverty: women, work and welfare', *Urb. and Soc. Change Rev.* 2, 28–36.

Popsay, J., Rimmer, L. and Rossiter, C. 1983: *One parent families: parents, children and public policy*. London: Study Commission on the Family.

Pratt, G. and Hanson, S. 1988: 'Gender, class and space', *Environ. and Plann. D: Soc. and Space* 6, 15–35.

Raymond, J. 1987: *Bringing up children alone.* Paper 3, Social Security Review Issues. Canberra: Australian Government Publishing Service.

Regional Trends 1988: London: HMSO.

Robertson, I. M. L. 1984: 'Single parent lifestyle and peripheral estate residence', *Town Plann. Rev.* 55, 197–213.

Robertson Elliot, F. 1986: *The family: change or continuity?* London: Macmillan Education.

Rose, D. and Le Bourdais, C. 1986: 'The changing conditions of female single parenthood in Montreal's inner-city and suburban neighbourhoods', *Urb. Resources* 3, 45–52.

Sarvasy, W. and Van Allen, J. 1984: 'Fighting the feminization of poverty: socialist-feminist analysis and strategy', *Rev. of Radical Political Economics* 16(4), 89–110.

Social Indicators Australia 1984: Canberra: Australian Bureau of Statistics.

Social Trends 1986: London: HMSO.

Social Trends 1987: London: HMSO.

Stallard, K., Ehrenreich, B. and Sklar, H. 1983: *Poverty in the American dream: women and children first.* Boston: South End Press.

Stretton, H. 1986: 'Housing – an investment for all'. In McLoughlin, J. B. and Huxley, M. (eds), *Urban planning in Australia: critical readings.* Melbourne: Longman Cheshire, 259–73.

Thumerelle, P. J. and Momont, J.-P. 1988: 'Eléments pour une géographie des familles monoparentales en France', *Espace, Populations, Sociétés* 1988(1), 128–32.

Vant, A. (ed.) 1986: *Marginalité sociale, marginalité spatiale.* Paris: CNRS.

Wagner, P. K. 1984: 'Suburban landscapes for nuclear families: the case of Greenbelt towns in the United States'. *Built Environ.* 10, 35–41.

Watson, S. 1986: 'Housing and the family: the marginalization of non-family households in Britain', *Int. J. of Urb. and Reg. Res.* 10, 8–28.

Winchester, H. P. M. 1987: 'Social areas of Plymouth'. In Maguire, D. J. and Brayshay, W. M. (eds), *Field excursions in the Plymouth region.* Northampton: Straw Barnes Press, 70–81.

Winchester, H. P. M. and Furnival, R. E. 1989: 'Les familles monoparentales: un groupe urbain marginalisé', *Espace, Populations, Sociétés* 1989(1): (forthcoming).

Winchester, H. P. M. and White, P. E. 1988: 'The location of marginalised groups in the inner city', *Environ. and Plann. D: Soc. and Space* 6, 37–54.

Wolch, J. 1980: 'Residential location of the service-dependent poor', *Ann. Ass. Am. Geogr.* 70, 330–41.

8 Richard Black,

'Livelihoods under Stress: A Case Study of Refugee Vulnerability in Greece'

Excerpts from: *Journal of Refugee Studies* **7**(4), 360–77 (1994)

Introduction

Protecting vulnerable people forms a cornerstone of modern welfare systems throughout the developed world. Central to the Victorian distinction between the 'deserving' and the 'undeserving' poor, vulnerability provides a crucial yardstick as to whether an individual is considered 'worthy' of public or private charitable assistance under even the most restrictive of welfare regimes (Goodin 1985). The same is true for refugee protection and assistance: it is the refugees' perceived vulnerable status that justifies the special responsibility felt by states to intervene on their behalf, and to provide refuge. Nonetheless, 'vulnerability' remains a nebulous concept, and refugees frequently appear to remain vulnerable in spite of, or indeed in some cases because of, government and NGO interventions.

This paper examines the meaning of vulnerability as it applies to refugee populations, and the livelihoods of refugee households and individuals. In the context of a developed world example – Greece – it is argued that vulnerability, when defined in terms of economic stress and dependence, represents a powerful issue in refugees' lives, negatively affecting their livelihoods. Targeting resources to the most vulnerable is a politically and socially desirable objective, but highly problematic in terms of deciding how to distribute limited resources. In turn, attacking the sources of vulnerability can provide an important lever to reduce poverty and disadvantage: in classic Victorian terms, 'to help refugees to help themselves'. However, such a policy may also involve refocusing attention away from 'vulnerable' people themselves, to the wider economic, social and political conditions that perpetuate vulnerability.

Definitions of vulnerability

There are two principal ways of defining vulnerability. A common-sense view of vulnerability in social analysis might identify as 'vulnerable' groups such as children and the elderly (Palmer *et al.* 1988), or those who are infirm or mentally ill, and who are unable to sustain a livelihood, or respond to threats to their well-being. For example, professionals concerned with care of the elderly have developed various precise indicators of vulnerability, such as a scale based on 'instrumental activities of daily living' (IADL), to determine individuals' abilities to carry out simple daily tasks (Morris *et al.* 1984; Atchley 1990). The notion of vulnerability here implies a dependence on others, usually involving either the state, or family or community arrangements designed to afford protection and assistance to the vulnerable person.

In addition to such groups or individuals with 'certain characteristics that render them in need of protection' (Kane 1990: 8), it is also possible to see vulnerability as deriving from circumstances external to the individuals concerned. In one sense, this relates to the division, identified by Dow (1992) in his discussion of vulnerability to global environmental change, between the coping abilities (resistance and resilience) of individuals, and their level of exposure to risk. Thus vulnerability can be seen to relate not only to the age, health, physical limitations and experience of an individual, but also to issues such as class, gender, ethnicity and market dependency that describe that person's place in the economy or society. A similar distinction is made by Watts and Bohle (1993: 45) in the case of vulnerability to famine in the Third World, between 'inadequate capacities to cope with stress, crises and shocks', and 'the risk of exposure to crises, stresses and shocks'.

External factors influencing vulnerability also relate to the extent to which particular groups are perceived and treated by the wider society. For example, Stevenson (1989) notes that the dependency (and therefore vulnerability) of the elderly is 'manufactured socially', through the imposition/acceptance of earlier retirement, the denial of rights to self-determination for the elderly in institutions, and the legitimation of low incomes. In this context, concern at the 'burden of dependency' of elderly people is misplaced, since the way that the elderly are treated increases that dependency. Similarly, Sokou (1987: 119) notes that in Greece, 'the handicapped . . . have been kept out of the labor force even though they are trainable', increasing their vulnerability and dependence on family support.

Turning to the case of refugees, it could be argued that in terms of their personal circumstances and experiences, all refugees could be considered as politically 'vulnerable', in the sense that they require protection, normally from the host state, against forcible repatriation to their country of origin. Indeed, the notion of a need for protection and/or assistance is used by the United States Committee for Refugees (USCR 1993) as a criterion for determining who is a refugee and should be included in annual statistics – such that those permanently resettled in developed countries and not vulnerable to forcible repatriation are not considered by this source at least as refugees. Zetter's (1988) comment in the first editorial of the *Journal of Refugee Studies* that 'the label (refugee) creates and imposes an institutionalized dependency', reinforces the point that vulnerability may be increased through the wider society's attitudes and practices towards a particular group. Thus, a major case is made by Harrell-Bond (1986) that refugee assistance regimes themselves promote dependency, by usurping the decision-making and organizational capacity of refugee individuals and communities, whilst Daley (1991) has argued that refugee assistance programmes and refugee status may further marginalize and create vulnerability for women-headed households.

Nonetheless, care must be taken not to assume that provision of assistance necessarily leads to dependency (Kibreab 1993), whilst similarly, it would be wrong to simply correlate women-headed households with vulnerability. In the context of a desire to allocate scarce resources to those most in need, it is important to go beyond a conception of vulnerability which classifies all refugees as vulnerable, whether because of their inherent characteristics as refugees, or because of the generalized treatment of refugees by the wider society. Thus in practical terms, consideration of the vulnerability of refugees generally has as its main aim distinguishing between different groups that are to receive greater or lesser assistance, and assessment of vulnerability is carried out by state or non-governmental agencies providing such assistance.

In making such a distinction, once again, two approaches might focus in turn on the characteristics of refugee individuals and households themselves, and on their relationship with the wider society. Thus in the context of vulnerability to famine, Anderson and Woodrow (1991) distinguish between the 'capacities' of individuals and households, such as capacity to work, skills, knowledge, decision-making capacity, and 'vulnerabilities', such as poverty, gender issues, class, race, and access to power, which describe the relationship with society. This 'capacities/vulnerabilities analysis' also distinguishes between physical/material, social/organizational and motivational/attitudinal issues, each of which have an impact on the level of vulnerability.

A more sophisticated method for analysing vulnerability to malnutrition has been developed by Pryer (1990), who considers the nature of livelihoods, based on a model of the household economy derived from field research in India and theoretical literature on other parts of the Third World. Pryer identifies a number of distinct elements of household economies, and derives from this a series of indicators, including assets, demography, labour participation, sources of income, debt and expenditure. These in turn can be used to classify households empirically using cluster analysis, with each cluster being cross-referenced to nutritional status. Distinguishing households on a range of economic indicators also forms the basis in practice for calculation of entitlement to social welfare benefits in many developed countries, although Pryer's system includes some social factors such as demography.

The discussion so far has remained relatively abstract, with the aim of introducing a conceptual framework for the analysis of vulnerability, and more detail will be provided on specific measures to identify vulnerable refugees in the Greek case below. However, one further theoretical point merits attention: that is, although 'vulnerability' represents a theme of the literature discussed, it is also important to draw a distinction in terms of vulnerability to what? Indeed, the *Oxford English Dictionary* definition of 'vulnerable' refers to ability 'to be hurt or wounded or injured', as well as being 'exposed to danger or attack', a definition which fits well with the notion of (all) refugees being in fear of persecution and violence. However, conceptions of vulnerability differ if, for example, the focus is on vulnerability to famine (Anderson and Woodrow 1991), natural hazards (Blaikie *et al.* 1994; Varley 1994), malnutrition (Pryer 1990) or simply to stress, isolation and job dissatisfaction (Earnshaw and Davidson 1991). In particular, there is a fundamental difference between the conception of vulnerability as something imposed by the context in which people find themselves, and the somewhat stricter conception of vulnerability as an *a priori* condition of certain groups, as with the potential frailty and dependence on others of people who are elderly, ill or incapable of helping themselves.

In this context, it is perhaps useful to think of vulnerability as associated with the need for protection, albeit relative to the resources available to an individual, with different types of protection, ranging from basic legal protection to social, economic and physical protection required for different but overlapping groups of people. In this context, legal protection might be seen as a minimum standard required or expected for all in society, and the absence of such protection for some – such as people who are gay or lesbian, illegal immigrants, and some refugees – constitutes a source of vulnerability. In contrast, other forms of protection may be needed to a greater or lesser extent by different groups depending on resources available to them: economic protection, for example, might be needed only by those in severe poverty, such that individuals are 'vulnerable' if they are very poor and have no economic protection (although all those without economic protection are potentially vulnerable, say, to the loss of employment, whilst those 'protected' by assistance might be vulnerable to the loss of assistance).

Refugees in Greece: background

The question of the vulnerability of refugees is particularly pertinent in a country such as Greece, where the welfare state, in comparison to much of the European Union, is relatively underdeveloped (Deleeck and Van den Bosch 1991; Petmesidou 1991), and where reservations made by the Greek state on signing the 1951 Geneva Convention place refugees from the outset in a relatively 'vulnerable' position. For example, in two important exceptions to the standard provisions of the 1951 Convention, Greece does not generally offer to refugees either the right to work,[1] or the right to permanent settlement.[2] In the absence of social welfare payments by the Greek state, many refugees become reliant on illegal employment, and limited financial assistance from UNHCR and other charitable organizations (Mestheneos 1988). In turn, this situation reflects the past assumption that eventually, most refugees arriving in Greece would be permanently resettled in third countries, especially the United States, Canada and Australia, an assumption that is becoming increasingly unrealistic (Black 1992a; 1994).

In this context, a situation clearly exists of a need for assistance (economic 'protection') to refugees, but also given pressure on resources, a need to discriminate in the allocation of assistance in favour of the most vulnerable. For example, UNHCR is unwilling to see significant sums spent on social assistance to refugees in a member state of the European Union, where subsistence costs per capita are significantly higher than in the Third World, and which in any case, as a member of this club of 'rich' nations, might be seen as a natural contributor to, rather than as a recipient of UNHCR funds. Thus whilst maintaining a budget for social assistance, UNHCR has sought to reduce its size. Meanwhile, the cost of assistance to refugees in Greece represented a contributory factor in the decision of another agency, the

International Catholic Migration Committee (ICMC) to withdraw its own programme of social support in the mid-1980s, although other factors were also involved.

In practice, the number of officially-recognized refugees in Greece is comparatively low. For example, from 1988 to 1991, official asylum applications (i.e. new arrivals) averaged only around 5,000 per year, with numbers falling to just 3,600 in 1991 (Black 1992a). UNHCR currently estimates the total number of refugees and asylum seekers resident in Greece to be around 6,000. Figures are complicated by the fact that the Greek government has an extremely low acceptance rate for refugees, standing at only around 2 per cent in 1991, with most of those accepted coming from Turkey. For this reason, many refugees prefer not to enter the asylum application procedure at all, applying instead to UNHCR for protection under their Mandate, or simply living illegally in the country. As in other publications (cf. USCR 1993; ECRE 1994), figures quoted here represent applications under this parallel UNHCR procedure, where around 33 per cent of applicants were 'accepted' in 1991, leading to the issuing of a UNHCR 'Blue Card'.[3]

Indeed, not only are applications for asylum low by European standards, they are also low in comparison to other groups of migrants entering Greece in recent years. For example, estimates of the number of illegal foreign workers resident in Greece range from around 300,000 to a recent, and probably inflated, figure provided by the Greek Trade Union Federation (GSEE) of one million (EIU 1993). In addition, Greece received over 52,000[4] new arrivals from the former Soviet Union of 'Pontic Greeks' between 1988 and 1993, a group who have fled situations of violence and persecution in their home country, but are not classified as refugees since they qualify for Greek citizenship by virtue of their Greek ethnicity (Baldwin-Edwards 1994). It is interesting to note that for this group, who are often referred to in the Greek press and popular discourse as 'refugees', the Greek state recognizes the need for social assistance and support, and has instigated programmes designed to assist integration into the Greek society and economy (Voutira 1991).

In spite of their low numbers, refugees recognized under the 1951 Convention or UNHCR's Mandate do nonetheless represent a significant group both for Greek public policy, and for UNHCR. In the case of the former, I have argued elsewhere that pressure to tighten immigration controls, both from northern European countries and from the effect of geopolitical changes in the Balkans on Greek public opinion, has resulted in a marginalization of the interests of recognized refugees, and a risk that Greece is not fulfilling its obligations as a signatory to the convention (Black 1992a). This marginalization of refugee interests has impacted on the vulnerability of refugees; meanwhile, although at the time of writing Greece had received few 'visible' refugees from the conflict in the former Yugoslavia, a significant number of young men are already 'hiding' in the country to avoid being drafted, whilst a southward movement of the conflict to Kosovo or even the 'Former Yugoslav Republic of Macedonia' could bring a large new group of refugees to Greece.

In turn, for UNHCR, the situation in Greece parallels that in other southern European states which have moved in the last decade from being countries of emigration to countries of large-scale immigration, but where consequent tightening of immigration controls has not been paralleled by the creation of adequate mechanisms to deal with the permanent settlement of refugees. Indeed, it can be argued that within Greece, not only asylum policy, but also immigration policy more generally is still underdeveloped: Mousourou (1991) for example argues that Greek immigration policies are still dominated by issues of 'repatriation' of Greeks living abroad rather than policies to deal with non-Greek migrants. Various measures implemented to assist the Pontic Greeks can be seen as an extension of this policy orientated towards the Greek 'diaspora', and can be contrasted with the lack of policy towards or assistance for 'foreign' migrants and refugees.

Iraqi and Iranian refugees in Greece

In the absence of official large-scale flows from the former Yugoslavia, the most significant flows of refugees to Greece in recent years have come from the Middle East, and specifically from Iran and Iraq (Table 8.1), although since 1993 arrivals from Somalia have also increased. Movement from Iran reached a peak in 1987, but despite a decline since that date, small numbers of opponents of the Tehran regime continue to arrive in Athens, many of them members of Protestant Evangelical Churches (which have suffered continual harassment in Iran), or members of the northern Kurdish minority. In contrast, movement from Iraq to Greece has been dominated by people who fled from Iraqi Kurdistan in the summer of 1988, in the face of large-scale attacks on villages by the Iraqi state using chemical weapons (Laizer 1991). Most of this group moved initially to Turkey, where there were allegations of mistreatment in refugee camps: the largest onward movement to Greece occurred in 1990, when nearly 3,000 refugees arrived either overland or on boats across the Aegean Sea. Subsequent but smaller-scale flows direct from Iraq occurred during and after the Gulf War, as attacks on Kurdistan intensified. Refugees from Iraq include both Kurds and Christians from the long-standing Assyrian and Chaldean communities living in Kurdistan and Baghdad. According to

Table 8.1 Applications for political asylum in Greece, 1987–91

Region of origin	*1987*	*1988*	*1989*	*1990*	*1991*
Middle East					
of which:	2815	2255	2343	4598	1903
Iran	1652	850	552	306	229
Iraq	323	570	536	2943	1056
Eastern Europe	2447	3828	689	365	1508
Africa	1440	1258	971	318	213
Total	6931	7930	4204	5502	3808

Source: Applications to UNHCR for Blue Card, plus applications to Greek Ministry of Public Order by asylum-seekers from Turkey.

UNHCR, as of 31 December 1991, there were 1,334 Iranian and 1,884 Iraqi refugees living in Athens, although available data does not distinguish between the separate Christian and Kurdish communities from Iraq, whilst these figures exclude those from the two countries which were refused, or did not apply for, recognition by UNHCR.

Field research was conducted by the author in Athens in 1992 with a sample of 90 refugee households, comprising 30 Iranian, 28 Iraqi Christian and 32 Kurdish households. In the absence of an adequate sampling frame, and in response to the need to win refugees' trust, respondents were selected by recommendation of translators, community leaders and previous respondents. Of these, 63 households were either in possession of a UNHCR Blue Card, or registered as 'of concern' to UNHCR, with a further six awaiting the outcome of their application. The inclusion of 21 households outside the UNHCR system altogether reflects the fact that a significant number of Iraqi and Iranian households in Greece consider themselves as refugees, and are considered as such by their compatriots, but have either failed to obtain UNHCR recognition, or have chosen not to apply for refugee status. The sample cannot be considered as statistically representative, but efforts were made to ensure a wide representation of ages, family situations and backgrounds, and locations within the Greater Athens area.

Interviews sought to compile basic demographic data; information on patterns of flight; past and current sources of income and expenditure; the deployment of household members in different locations and sectors of the economy (including the informal economy); the extent of social, economic, political and cultural relationships with other refugee households, refugee and other organizations, and with the host community; as well as strategies and aspirations for the future. Interviews were kept open-ended, to allow an opportunity for respondents to express their own problems and perceptions. Where possible, individual and family case histories of migration, employment, housing and experience of state policy or official assistance were examined in detail.

Identifying vulnerable households: policy perspectives

As noted above, UNHCR maintains a budget for financial assistance to refugees in Greece, although this is limited, and must therefore be targeted at those most in need, or in other words, at the most vulnerable households. Two principal criteria are used in practice for the targeting of financial assistance, namely the length of stay in Greece (newly arrived refugees receive priority), and secondly the size of the household. Financial assistance is implemented through two non-governmental organizations, the Social Work Foundation (SWF), and since 1989, the Greek Council for Refugees (GCR). It primarily takes the form of monthly payments made to refugee households during their first 12 months in the country, based on a sliding scale depending on family size: in 1992, this ranged in theory from 18,000 drachmas per month for individuals, up to a maximum of 36,000 drachmas per month for a family of five or more members. These payments, although clearly inadequate for survival – indeed, in most cases, the payment barely covered the cost of housing – represented a significant 'safety net' for those receiving them.

Some divergence from these guidelines was found in the author's survey in 1992. For example, some households reported receiving different amounts to those set in the guidelines, with a small number of very large households receiving more than was laid down, and some individuals receiving nothing at all. Meanwhile, although 39 of the 63 households 'recognized' in some way by UNHCR were receiving money, with most of the remainder having received payments which had since been stopped, 30 of those still receiving money had arrived in Greece prior to October 1990, 18 months before the survey was carried out (Table 8.2). This reflected the fact that decisions to discontinue payments were treated on a case-by-case basis, with informal assessments made by SWF or GCR social workers of the needs of the households concerned.

The judgement that refugee households are more vulnerable during the first year or more of their stay in Greece, and thus more in need of financial assistance, has some justification in reality. Initially, refugees often have little money, and although 85 per cent of those interviewed reported bringing savings or jewellery with them when they left their home country, only a third had assets of any kind on arrival after paying traffickers to bring or guide them to Greece, and usually then only a few hundred US dollars at most. In addition, most refugees face a considerable language barrier, and initially have less well-developed contacts and networks within the country that might help them to secure survival. This is reflected in the survey finding that rates of male unemployment were significantly higher amongst those in the country for under two years (34 per cent) than for those present for longer periods of time (19 per cent), whilst wages for men and women in employment were also generally lower for more recent arrivals (Table 8.3).

A further practical problem relates to the fact that financial assistance is only made available to those recognized by UNHCR, in a process that can

Table 8.2 Refugee status and payment of assistance by UNHCR to sampled households

Status Assistance	*Date of arrival in Greece*		
	Since Oct. 1990	*Before Oct. 1990*	*All*
Blue Card/'of concern' to UNHCR	14	49	63
of which:			
Receiving assistance	9	30	39
Assistance stopped	3	18	21
Never received assistance	2	1	3
Application pending	4	—	4
Application rejected	5	1	6
Total	30	60	90

Source: Field data, 1992.

Table 8.3 Wages of refugee sample by time spent in Greece

Daily wage (drachmas)	*Number present in Greece for* < *1 yr (%)*	*1–2 yrs (%)*	> *2 yrs (%)*	*All refugees in employment*
< 4,000	20 (59)	24 (37)	16 (34)	60 (41)
4,000–6,500	14 (41)	39 (60)	26 (55)	79 (54)
> 7,000	— (—)	2 (3)	5 (11)	7 (5)
Total households	34 (100)	65 (100)	47 (100)	146 (100)

Source: Field data, 1992.

take weeks or even months. In this context, it is immediately upon arrival that refugees in Greece are perhaps at their most vulnerable. Thus a small number of very recently arrived refugees interviewed were entirely dependent on the charity of other refugee households, or in some cases, of church or community-based charities, who provided them with basic accommodation and food. A number of households reported that they had spent their first weeks or even months sleeping rough, although in mid-1992 this situation had ameliorated. Worse, many refugees had spent their first months in the country in prison, after being arrested for illegal entry – the standard sentence being 14 months. This reflects the fact that this legal process takes precedence in Greek law over any claim to asylum (Black 1992a).

Identifying vulnerable households: alternative measures

In addition to duration of stay and size of family, a number of other criteria can be identified to distinguish vulnerable households. Most obviously, vulnerability can be associated with the means of households to sustain their livelihoods, a factor which in turn depends on the economic, social and political system within which they are operating. In this sense, vulnerability is not an absolute concept, but one which is relative to particular local circumstances. Such relativity is recognized in studies of poverty in general even within Europe: for example, the official definition of 'poverty-lines' in the European Community uses a relative standard of 50 per cent of average equivalent household income for single-person households in each member state (O'Higgins and Jenkins 1990); whilst relative standards of poverty and vulnerability are also likely to be different in a developed and Third World context. Such relativity also arguably extends to the political circumstances in which refugees find themselves, and these two aspects of vulnerability are considered in turn below.

Socio-economic vulnerability

In setting a 'poverty threshold' to measure socio-economic vulnerability, an obvious standard, at least in developed countries, is whether a household has sufficient income from legal employment, self-employment or welfare benefits to meet minimum household expenditure requirements. This forms the basis for a subjective measure of poverty thresholds in a study of seven EC countries conducted by Deleeck and Van den Bosch (1992), where respondents were asked essentially whether they could balance their household budgets for income and expenditure. However, in applying such a measure for Greece, it must be borne in mind that not only are refugees not legally allowed to work, with a few exceptions, but also that an unknown but highly significant proportion of employment of Greek workers in Athens is in the technically illegal 'informal sector' or 'paraeconomy' (Mestheneos 1991), with this sector accounting for as much as 30–35 per cent of GDP in 1990 (Baldwin-Edwards 1994). Thus an assessment of vulnerability based on the employment status of refugees in Athens needs to include illegal employment as constituting part of a 'legitimate' household strategy.

Taking this into account, each household surveyed was asked to estimate both their average current income, and their average minimum monthly expenditure on basic items such as housing, food, transport and clothing. Income sources were distinguished on the basis of whether they were 'permanent', in that employment was full-time and relatively secure, or whether they were temporary in the sense that employment was through daily contracts which depended on the availability of work on a particular day – although all of the employment recorded was technically illegal, and thus insecure in that the refugees concerned had no legal rights. Data presented in Table 8.4 show that amongst the 71 households able or willing to provide such estimates, only just over half (38) had incomes which met their estimated expenditure, whilst of these households, 20 would fall below this poverty line if their temporary employment were to be interrupted.

In practice, such interruptions to insecure work are common, depending on seasonal demand and economic fluctuations, especially in the construction industry and, outside Athens, in agriculture. Many refugees engaged in such daily work commented that frequently days or weeks would pass without any employment being offered. Meanwhile, as can be seen from Table 8.5, assistance from UNHCR is not particularly well-targeted at households which are vulnerable in terms of their current income sources, with only just over half of those UNHCR-recognized households currently earning insufficient income to meet expenses being in receipt of assistance, although

Table 8.4 Financial vulnerability of sample households

Financial situation	*No. households*	*(%)*
Income exceeds expenditure	18	(25)
Vulnerability to income loss*	20	(28)
Income less than expenditure	26	(37)
Reliant on charity	7	(10)
Total**	71	(100)

Notes: * Households where income would be less than expenditure if casual work ceased, or if assistance were withdrawn. ** Total excludes 19 households for which income/expenditure data are not available.
Source: Field data, 1992.

Table 8.5 UNHCR assistance to sample refugee households

Financial situation	*Total households*	*UNHCR recognized*	*Receiving assistance*
Income exceeds expenditure	18	11	5
Vulnerability	20	17	17
Income less than expenditure	26	20	12
Reliant on charity	7	0	0
Total	71	48	34

Source: Field data, 1992.

all of those in temporary employment did receive assistance. Jobs themselves were concentrated in construction and largely unskilled manufacturing work in very small companies that are characteristic of the Greek labour market (Leontidou 1993).

A more general view of the employment situation of the refugees interviewed as part of this study confirms the vulnerable situation of many. For example, unemployment in general is a problem, standing at 26 per cent overall for adult males, while 18 per cent of households have no family member currently employed. This compares with an official unemployment rate of 9.2 per cent amongst Greek men (EIU 1993), although higher rates are suggested by independent analysts (Leontidou 1993). Moreover, although it is not possible to compare average household income amongst the refugees interviewed with equivalent average incomes for Greek workers, it can be noted that wages are low, standing according to respondents at the equivalent of roughly 25–50 per cent below the average paid to Greek workers in similar jobs. For example, average daily wages found within the sample were just 3,350 drachmas for jobs in manufacturing, compared with over 7,000 drachmas reported for the economy as a whole (ESYE 1993).

A number of factors influence access to employment for refugees, including an apparent relationship with the length of time an individual has spent in the country, as noted earlier. In addition to this, though, the age of a refugee appears to exert a significant influence, with for example men aged 20–39 having unemployment rates half (20 per cent) the level for those over 40 (39 per cent) or aged 15–19 (40 per cent). The employment situation for women is more complicated, given the competing demands of domestic and child care responsibilities and the need to seek external sources of income. In general, the female participation rate has remained low, despite the availability of employment for women particularly in cleaning and other manual service jobs. In part, this reflects cultural restrictions on female employment, especially amongst the Kurdish group interviewed; however, a number of other factors are also significant. For example, almost all of the women who had been employed in Iran or Iraq were skilled workers or professionals, and they have generally been unwilling to work in unskilled or manual occupations unless absolutely necessary. This is reinforced by the fact that available employment is often at some distance from the home, for example in wealthier districts as domestic workers, with the travelling this involves being incompatible with the demands of child care. In addition, employment in domestic and especially child care roles often carries a requirement for minimal communication skills in Greek, which Kurdish women do not have.

Legal vulnerability

Turning to the question of legal or political vulnerability, a case can be made in Greece in particular that all refugees are vulnerable, in terms of problems over legal status and lack of basic rights such as the right to work and permanent residence. Indeed, the lack of legal protection was the single most important difficulty mentioned by refugees from all ethnic groups during the course of the survey, whether or not they had received recognition by UNHCR. Even those refugees holding a Blue Card have no formal protection in Greek law, and there was at the time of field research a widespread fear of deportation. Previously, this group had been tolerated by the Greek authorities on the basis that they would soon move on to third countries. Residence permits were issued, usually for periods of six months, and were normally renewed until the refugee was granted a visa to travel to another country (or left illegally). However, as part of a tightening of immigration practice early in 1992, the authorities began to routinely refuse the renewal of temporary residence permits, and officially require refugees to leave the country within 15 days, even those holding a 'Blue Card'.

Meanwhile, without a residence permit, it becomes impossible to do some simple things. For example, although it was still possible to be treated in hospital without charge, it was impossible to open a bank account, or to register for baptism at an Orthodox church (several Iranian families interviewed had sought to be baptized since their arrival). Whilst it is certainly possible to continue living in Greece without papers, lack of legal status can nonetheless undermine the social and economic base for survival in Greek society, and certainly makes refugees extremely vulnerable to exploitation in the workplace. Indeed, it was this legal vulnerability that formed the principal motivating factor for those households who reported that they were considering moving (illegally) to other countries in western Europe.

Livelihoods under stress

On any of the indicators touched on in this paper, it is clear that a significant proportion of refugees in Greece are highly vulnerable, whether in socio-economic or especially in legal terms. As such, the need for protection and assistance is also high. However, targeting assistance to the most vulnerable households is problematic, especially in a situation where the budget for assistance is under severe pressure, and where welfare provision for poorer Greek citizens is also often inadequate (see, for example, Triantafillou and Mestheneos 1994, on the case of healthcare for the elderly in Greece). The current policy of targeting financial assistance to those who have been present in Greece for under one year, and then allocating more money to larger households, has some justification, but nonetheless, the data presented in this paper suggests that assistance does not necessarily reach those most in need.

In this context, an important characteristic of households to take into account would appear to be the potential to gain sufficient income from employment, perhaps through the ratio of dependants to potentially active household members. Although such a policy might face the criticism that it would encourage households to have more children, the generally low level of assistance coupled with the initial cost of feeding additional children suggests that this is unlikely to be a serious problem. Meanwhile, although a more sophisticated basis for distinguishing households might be developed, based for example on the range and success of strategies available to households to earn income, in practice, such a system would both be complicated to administer, and suffer from a lack of transparency for the recipients of assistance. Harrell-Bond *et al.* (1992), for example, have challenged the tendency of refugee assistance programmes to seek too precise a measure of the number of needy refugees as 'misguided rigour'.

In a more general sense, however, if the purpose of targeting public action is seen as removing or reducing vulnerability, rather than assisting the most vulnerable, attention needs to be switched away from the internal characteristics of households, and towards what might be described as the underlying causes of vulnerability (Foster 1983). It is argued here that it is the contextual elements of refugee vulnerability that are of most significance for public policy, and on which most leverage to reduce vulnerability can be gained, rather than the inherent characteristics of particular groups within the refugee population.

Once again, this question can be addressed from both a socio-economic and a political standpoint. In terms of socio-economic vulnerability, the policy of focusing on newly arrived refugees underlines the problems felt in particular by new arrivals in participating in informal labour markets, problems which relate largely to the relationship with the wider society rather than a lack of skills or capacities on the part of the refugees. Measures to address vulnerability in labour markets might include the provision of Greek language training, one 'skill' which is lacking amongst most newly arrived refugees, although a number of respondents commented that this would need to be provided during evenings in order to make it possible to work and study at the same time. Alternatively, female participation in paid employment might be increased by provision of day care facilities for children, thus reducing the vulnerability of some households.

Care would need to be taken in such initiatives to avoid the politically divisive feature of providing assistance to refugees that is not available to the local population. For example, over half (56 per cent) of men, and nearly three quarters (73 per cent) of women in employment had found jobs through personal contacts with other Iraqi or Iranian refugees, whilst a further 15 per cent of men and women had found jobs through personal contacts with Greek neighbours or friends. In this sense, vulnerability might be reduced through facilitating the development of the social networks that assist people in entering the labour market. However, such informal networks also form the basis on which Greek workers find employment, such that the creation of a more formal network may be both divisive and unproductive.

Turning to the political side to vulnerability, in the past, UNHCR and the Greek Council for Refugees had encouraged the formation of refugee community associations which partly contributed to the process of building social networks, but support for such groups had been ambiguous due to their tendency to adopt political as well as socio-economic goals. This reflects a characteristic of Greek social organization, namely the lack of 'civil society' structures outside party politics. Thus voluntary associations are generally loci of patronage and clientelism within the Greek party political structure rather than independent units (Mavrogordatos 1993). Meanwhile, Petmesidou (1991: 36) concludes that 'an ever-increasing part of [state] revenue is distributed to households on the basis of their links to the poles of political power . . .' In practice, some of the most successful refugees in the sample interviewed, in terms of those who had found more secure jobs, were those who had found 'sponsorship' by wealthy or influential Greek nationals. However, the best example of this was the case of a small number of politically-active Kurdish refugees who had found secure work in a printing workshop operated by the Greek Communist Party.

Underlying the vulnerability of refugees in Greece is not so much their exclusion from Greek social and economic life: indeed, a separate survey of public attitudes showed a general indifference towards the presence of refugees from the Middle East in Athens (Black 1992b). Rather, it is the context of a lack of secure legal status of refugees in Greece that constitutes the overarching factor in contributing to vulnerability, coupled with their marginalization from the centres of political power. Without secure legal status, any measures to reduce vulnerability are likely to be temporary palliatives, which skirt around the main cause of the problem. Meanwhile, the issue of refugee vulnerability needs to be placed firmly on the political agenda to have a chance of being addressed. What is at stake is not simply an increase in welfare protection for refugees – protection which, as has already been noted, is often unavailable even for the poorer sections of Greek society – but the need for a concerted effort by the Greek authorities to ensure that they are meeting the spirit as well as the letter of the 1951 Convention.

Notes

1 In practice, refugees wishing to work in Greece must apply for a work permit in the same way as other non-EU nationals. Until recently, it was necessary for applicants for work permits to apply from outside the country, clearly an impossibility for refugees. Under pressure from UNHCR, this requirement was relaxed for refugees in 1989, but work permits are still very difficult to obtain.

2 Two exceptions are a group of around 60 Kurdish refugees, granted residency during the 1980s after intense lobbying, and some 200 Vietnamese accepted by Greece as quota refugees.

3 Applications by asylum-seekers from Turkey are treated separately, since the

majority of this group apply directly to the Greek Ministry of Public Order, and not to UNHCR.

4 The figure of 52,000 quoted is based on figures from the Ministry of Public Order, which provides passports to Pontic Greeks once they have been in the country for six months. However, figures from the social welfare office which provides financial assistance to Pontians show a much higher figure of 129,000.

References

Anderson, M. B. and Woodrow, P. J. 1991: 'Reducing Vulnerability to Famine: Developmental Approaches to Relief', *Disasters*, 15(1), 43–54.

Atchley, R. 1990: 'Defining the Vulnerable Older Population'. In Harel, Z., Ehrlich, P. and Hubbard, R. (eds), *The Vulnerable Aged: People, Services and Policies*. New York: Springer, 18–31.

Baldwin-Edwards, M. 1994: 'Immigration and Border Control in Greece'. Paper presented to the workshop 'Police et immigration: vers l'Europe de la Sécurité Intérieure'. Madrid: European Consortium for Political Research.

Black, R. 1992a: *Livelihood and Vulnerability of Foreign Refugees in Greece*, Occasional Paper no. 33, Department of Geography, King's College London.

—— 1992b: 'Political Asylum: is there Racism in Greece?' (in Greek), *Anti*, 551 (25 December 1992), 38–40.

—— 1994: 'Asylum Policy and the Marginalization of Refugees in Greece'. In Gould, W. T. S. and Findlay, A. M. (eds), *Population Migration and the Changing World Order*. Chichester: Wiley, 145–60.

Blaikie, P., Cannon, T., Davis, I. and Wisner, B. 1994: *At Risk: Natural Hazards, People's Vulnerability and Disasters*. London: Routledge.

Daley, P. 1991: 'Gender Displacement and Social Reproduction: Settling Burundian Refugees in Western Tanzania', *Journal of Refugee Studies*, 4(3), 248–66.

Deleeck, H. and Van den Bosch, K. 1991: 'Poverty and the Adequacy of Social Security in Europe: a Comparative Analysis', *Journal of European Social Policy*, 2(2), 107–20.

Dow, K. 1992: 'Exploring Differences in our Common Future(s): the Meaning of Vulnerability to Global Environmental Change', *Geoforum*, 23(3), 417–36.

Earnshaw, J. and Davidson, M. 1991: 'Vulnerable Workers: an Overview of Psychological and Legal Issues'. In Davidson, M. and Earnshaw, J. (eds), *Vulnerable Workers: Psychological and Legal Issues*. Chichester: Wiley, 1–22.

ECRE 1994: *Participants' General Meeting, April 1994: Country Reports*. London: European Consultation on Refugees and Exiles.

EIU 1993: *Greece 1993/94: Country Profile*. London: Economist Intelligence Unit.

ESYE 1993: Monthly Statistical Bulletin. Athens: National Statistical Service of Greece.

Foster, P. 1983: *Access to Welfare: an Introduction to Welfare Rationing*. Basingstoke: Macmillan.

Goodin, R. E. 1985: *Protecting the Vulnerable: a Reanalysis of our Social Responsibilities*. Chicago: University of Chicago Press.

Harrell-Bond, B. E. 1986: *Imposing Aid: Emergency Assistance to Refugees*. Oxford: Oxford University Press.

Harrell-Bond, B. E., Voutira, E. and Leopold, M. 1992: 'Counting the Refugees: Gifts, Givers, Patrons and Clients', *Journal of Refugee Studies*, 5(3/4), 205–25.

Kane, R. 1990: 'Vulnerable and Perhaps Vulnerable: the Nature and Extent of

Vulnerability among the Aged'. In Harel, Z., Ehrlich, P. and Hubbard, R. (eds), *The Vulnerable Aged: People, Services and Policies*. New York: Springer, 4–17.

Kibreab, G. 1993: 'The Myth of Dependency among Camp Refugees in Somalia', *Journal of Refugee Studies*, 6(4), 321–49.

Laizer, S. 1991: *Into Kurdistan: Frontiers under Fire*. London: Zed Books.

Leontidou, L. 1993: 'Informal Strategies of Unemployment Relief in Greek Cities: the Relevance of Family, Locality and Housing', *European Planning Studies*, 1(1), 43–68.

Mavrogordatos, G. 1993: 'Civil Society under Populism'. In Clogg, R. (ed.), *Greece 1981–1989: The Populist Decade*. Basingstoke: Macmillan, 47–64.

Mestheneos, L. 1988: 'The Education, Employment and Living Conditions of Refugees in Greece and the Possibilities for (Self-) Employment', Unpublished report for UNHCR Branch Office, Athens.

—— 1991: 'A Socio-economic Report on the Situation of Foreign Workers in Greece'. Unpublished paper prepared for the Office of the President of the EC, Brussels, and the Marangopoulos Foundation for Human Rights, Athens.

Morris, J., Sherwood, S. and Mor, V. 1984: 'An Assessment Tool for Identifying Functionally Vulnerable Persons in the Community'. *The Gerontologist* 24, 373–9.

Mousourou, L. 1991: *Migration and Migration Policy in Greece and in Europe*. Athens: Gutenberg (in Greek).

O'Higgins, M. and Jenkins, S. 1990: 'Poverty in the EC: Estimates for 1975, 1980 and 1985'. In Teekens, R. and Van Praag, B. (eds), *Analysing Poverty in the European Community*, Eurostat News Special Edition, Office for Official Publications of the European Communities, Luxembourg.

Palmer, J., Smeeding, J. and Torrey, B. (eds) 1988: *The Vulnerable*. Washington, DC: Urban Institute Press.

Petmesidou, M. 1991: 'Statism, Social Policy and the Middle Classes in Greece', *Journal of European Social Policy*, 1(1), 31–48.

Pryer, J. 1990: 'Nutritionally Vulnerable Households in the Urban Slum Economy: a Case Study from Khulna, Bangladesh', Paper presented to the Society for the Study of Human Biology Symposium on Health and Ecology in the Third World, Durham, UK.

Sokou, K. 1987: 'Unemployment in a Developing Country: the Case of Greece'. In Zollner, H. (ed.), *Unemployment, Social Vulnerability and Health in Europe*. Berlin: Springer Verlag, 118–28.

Stevenson, O. 1989: *Age and Vulnerability: a Guide to Better Care*. London: Edward Arnold.

Triantafillou, J. and Mestheneos, L. 1994: 'Pathways to Care for the Elderly in Greece', *Social Science and Medicine*, 38(7), 575–82.

USCR 1993: *World Refugee Survey, 1992*. Washington, DC: United States Committee for Refugees.

Varley, A. (ed.) 1994: *Disasters, Development and Environment*. Chichester: Wiley.

Voutira, E. 1991: 'Pontic Greeks Today: Migrants or Refugees', *Journal of Refugee Studies*, 4(4), 400–20.

Watts, M. J. and Bohle, H. G. 1993: 'The Space of Vulnerability: the Causal Structure of Hunger and Famine', *Progress in Human Geography*, 17(1), 43–67.

Zetter, R. 1988: 'Refugees and Refugee Studies: a Label and an Agenda', *Journal of Refugee Studies*, 1(1), 1–6.

9 Doug Hindson, Mark Byerley and Mike Morris, 'From Violence to Reconstruction: The Making, Disintegration and Remaking of an Apartheid City'

Excerpts from: *Antipode* **26**(4), 323–50 (1994)

The apartheid city was created as a response to the urban crises of the 1940s. In order to protect and enhance the interests of its white constituency, the National Party government drew on past policies of racial segregation and spatial management to restructure and entrench more deeply the racial city form. The processes of urbanization were blocked and racial groups channelled into clearly defined spatial zones within the city or, in the case of illegal entrants, removed from the cities to rural homes or resettlement camps in the homelands.

This created a framework for a period of economic expansion and political stability in which the racial and class antagonisms built into the social structure were held together by a highly repressive regime. The apartheid system accelerated class mobility amongst whites, and to a lesser extent coloured and Indian urban residents, whilst thwarting similar processes amongst Africans. The concentration of socio-economic resources in the white suburbs and extrusion of black city dwellers to the badly resourced urban periphery had produced a particular racial and class map of the city by the early 1970s. However the racial city form became increasingly unworkable from the late 1970s due to its economic inefficiency, and politically unmanageable as a result of growing political mobilization against apartheid.

The accelerating processes of rapid urbanization and class differentiation from the 1980s frayed the tight fabric of the apartheid city and spilled over its constraining boundaries. New social and economic forces in urban life emerged as Africans began to struggle for access to the core areas of the city or for increased resources to make urban life on the metropolitan margins more amenable. This had differential effects on the various urban communities. The white suburbs, by and large, have maintained their social stability and political power. While some black families have been absorbed into the white suburban areas, white urban interests, for the most part, have been able to displace demands for urban resources away from the centers of urban power. This deflection has confined the struggles of the excluded majority to the constricting socio-economic boundaries of the urban periphery. These marginalized areas, with their inadequate resources, have been wholly unable to cater to the urban requirements of their populations. The containment of the struggles within the marginalized peripheries has built up immense political pressure within them, unleashing intermittent violent explosions.

It is becoming increasingly clear that resolution of the conflict within the black residential areas cannot be secured without the developmental needs of those areas being addressed within peace and reconstruction pacts. It is also becoming clear that the problems of the black residential areas, the need for housing, resources, infrastructure, better communications and other resources, cannot be solved by development plans and processes which restrict themselves to the black areas alone. What is needed is a re-examination of the whole city. It is only through a reconceptualization of the apartheid city as a whole that a durable and large-scale solution can be worked out to deal with conflict in the black residential areas.

Out of all the metropolitan areas in South Africa, we feel that many of the connections between the apartheid city form, violence, and reconstruction can best be investigated and illustrated in the context of the Durban Functional Region (DFR). In its racial planning, Durban combined both the policies of residential and territorial apartheid because of its particular geographical location close to the KwaZulu homeland. Furthermore, the disintegration of apartheid controls occurred in the DFR well before the other metropolitan areas. Durban differs from the other cities in that informal urban settlement on a massive scale is a *fait accompli* and the development task is to address the consequences. Other cities have the advantage, on the basis of this experience, of being able to anticipate and plan for settlement and may thus benefit from the store of experience gained in the DFR.

The paper is divided into four sections. Section 1 describes how the city came to be structured racially from the 1940s to the 1980s. Section 2 shows how the nature and spread of the violence in the 1980s related to the racial structuring of the city, rapid urbanization, social differentiation and the emergence of localized power centers in the black residential areas. Section 3 poses the alternatives now facing the city – continuing down the path of exclusion and confrontation with the prospect that the violence will spread from the black residential peripheries to the white core city areas, or peaceful negotiations over a new, more inclusive, city form. Finally, section 4 examines alternative conceptions of the spatial restructuring of the Durban Functional Region. [Editor's note: this section is greatly cut.]

Apartheid spatial planning: 1940s–1980s

The growth of Cato Manor, removals and township construction: 1940s–1960s

Durban in the 1940s exhibited many of the problems faced by the city today, namely, extensive squatting, social upheaval, and violence. With the coming to power of the National Party in 1948 the concerns of many of the white residents and city officials of Durban were to coincide with the aim of the Nationalist Government to rid the inner cities of Africans, Indians and 'Coloureds' (Davies, 1991; McCarthy, 1988).

Industrial expansion and the relaxation of influx control during and immediately after World War II led to rapid urbanization in all of South Africa's major cities. African people came to the cities to find jobs in the context of rapidly expanding employment brought on by wartime demand for industrial products and post-war industrial growth (Maylam, 1982). Influx control throughout South Africa had been relaxed in urban areas proclaimed under the Natives (Urban Areas) Act, although residential controls still prevented Africans from moving into the city itself, except under strict supervision within the hostels or in domestic quarters in white suburbs (Hindson, 1987: 55–56).

Most African migrants who came to Durban ended up settling in Cato Manor, a large squatter area initially outside the municipal boundaries of the city (see Map 9.1). Here they could find relatively cheap accommodation as tenants or sub-tenants on Indian-owned land or on vacant ground. Cato Manor provided an accessible and relatively cheap base for Africans close to employment in the city and grew rapidly from a population of 2,500 in 1936 to reach a peak of approximately 120,000 by 1958 (Townsend, 1991: 21; Davies, 1991: 82).

The area was feared by white residents of Durban as a center of illicit liquor brewing and consumption, vice, disease and violence. The concentration of large numbers of impoverished blacks in close proximity to the suburbs as well as Afro–Asian conflict and clashes between squatters and the police and military aroused white fears about personal safety, crime, health, and economic competition (Maylam, 1982: 10). The concerns of whites fed into government thinking on urban segregation and bantustan policy (Maharaj, 1992; McCarthy, 1988).

Through the various Urban Areas and Group Areas acts the National Party government, in collaboration with the Durban City Council, intensified influx control, embarked on mass forced removals, enforced stricter residential controls on blacks and subsequently enforced residential segregation (McCarthy, 1991). Although for a brief period the possibility of Cato Manor being developed as an African residential area was mooted (Nuttall, 1989: 33), the introduction of Group Areas legislation in Durban in 1958 (Davies, 1991) together with the concerns of city officials regarding the health and safety of white residents, ultimately resulted in the removal of Africans and Indians from Cato Manor. By the mid 1960s virtually all squatters had been removed and this left a large tract of open land to the west of Durban. Only a handful of Indian families were able to cling on to their homes and remain in the area over the coming decades.

Massive housing programs were initiated to the south and north of Durban to house those removed from Cato Manor and other inner city areas. Africans were resettled in the townships of KwaMashu and, to a lesser extent, Umlazi, while Indians were mainly sent to Chatsworth (see Map 9.2). The state provided low cost housing and these areas absorbed large numbers of African and Indian families. In contradistinction to the grievances felt by the richer landowning Indian sector of Cato Manor, many of the poorer Indian tenants responded with mixed emotions; they received their first homes, but as a consequence of forced removal. Only residents of the smaller African townships of Lamontville and Chesterville, which were under municipal control, and a few middle income Indian areas such as Overport and Clare Estate were to escape banishment to the city's far periphery.

By the end of the 1960s, the state-funded housing program in most African areas had essentially drawn to a close. Many Africans who had lived in Cato

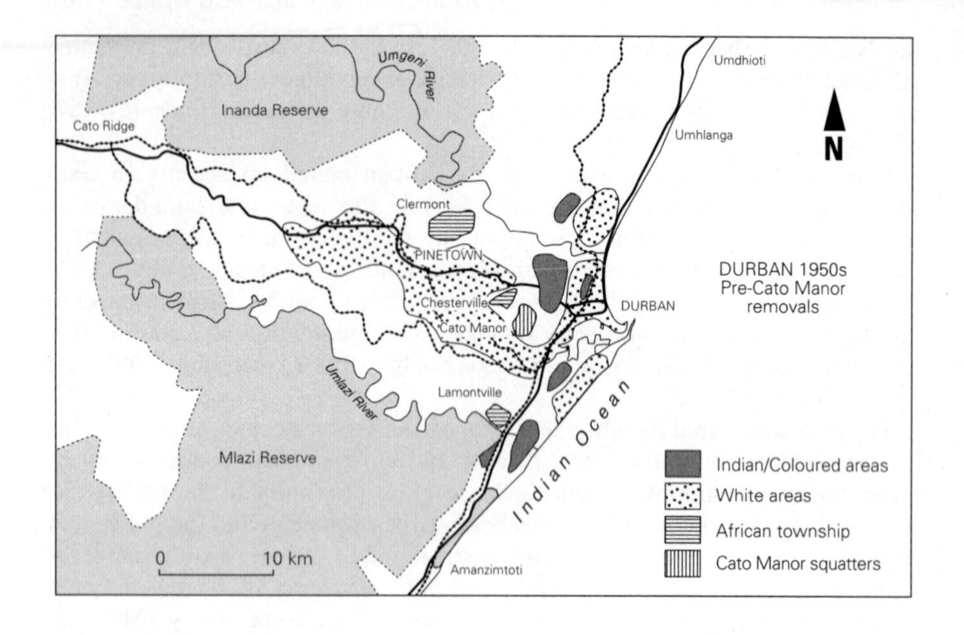

Map 9.1 Sources: Inkatha Institute, 1986 and Davies, 1992.

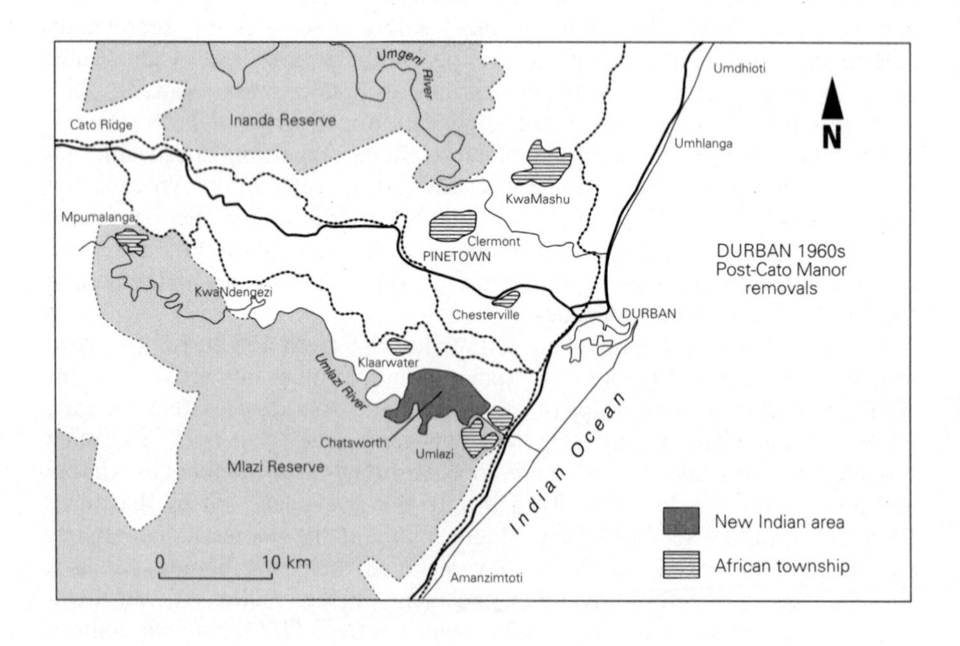

Map 9.2

Manor, however, were unable to acquire housing in the new townships, because they were illegal in the urban areas or unable to pay rents. It is estimated that some 30,000 Africans 'disappeared,' in the sense that they dispersed and resettled outside the process of official removals (Maasdorp and Humphries, 1975). Many of these people took advantage of African and/or Indian freehold land in areas like Inanda and Clermont where they rented rooms from landlords or erected their own dwellings. Others sought the permission of chiefs and indunas in areas abutting the townships.

The removals from Cato Manor and other parts of the city entrenched the racial city form. Whites retained the core city areas, Indians were largely rehoused in Chatsworth which lay between the white and African residential areas in the south of Durban and Africans were relegated to dormitory suburbs on the far peripheries of the city: KwaMashu in the north and Umlazi in the south.

Underlying this racial structure was also a class and income structure. Whites, in the core city areas, comprised a protected skilled working class, middle income professional and managerial classes. Over the 1960s the Indian population in the inner periphery (Chatsworth) occupied an intermediate income group, comprising mainly semi-skilled and skilled workers and lower middle income semi-professionals. Africans, on the outer peripheries (KwaMashu and Umlazi) were at the bottom of the income structure, and comprised mainly the unskilled, semi-skilled and unemployed workers (Hindson and Crankshaw, 1994). The spatial and income profile of the city was thus the reverse of many western cities with their decaying centers and affluent suburban peripheries.

Township construction and the re-emergence of squatting: 1960s–1980s

The short-term viability of urban apartheid was based on the exclusion of Indians and Africans from the centers of economic and political power in the cities, the minimization of social and infrastructural expenditure in the new townships, low wages, and the creation of a differentiated workforce with some urban Africans having minimal access to urban residential rights whilst the majority were prohibited from permanently settling in the urban areas (Hindson, 1987; Morris and Hindson, 1992b).

Although extremely repressive, urban apartheid was a highly effective system in the short term both as a mechanism of urban political control and, for a time, in securing an economically subservient workforce. The 1960s was a decade of political stability and rapid economic expansion. However, during the 1970s the underlying contradictions, costs and inefficiencies of the system began to appear.

This coincides with the beginning of a new stage of urbanization and apartheid spatial planning. The 1970s saw the state grant greater powers to the bantustans either in the form of independence or self-government. This impacted on the process of urbanization in a number of important ways. In the country as a whole, a form of deconcentrated urban settlement occurred on the fringes of the bantustans abutting the metropolitan areas through the settlement of large numbers of workers who commuted daily to work in the core city areas. Deconcentrated urbanization extended urban sprawl, reinforced the racial geography of the city and added to the high costs of infrastructural expenditure and services in black residential areas. Long-distance commuting to the white core city areas added a large additional burden to the costs of reproduction of labor, most of which were misplaced onto individual workers.

The proximity of KwaZulu to Durban made it possible for the state to achieve the objectives of residential and territorial apartheid simultaneously in this city. With the granting of self-governing status to KwaZulu the vast majority of Durban's African people were converted into cross-border commuters. The exceptions were people living in the long-established townships of Chesterville, Lamontville and Clermont, which had been established initially as labor supply outposts beyond the municipal boundaries of Durban and Pinetown and were difficult to move due to the costs involved and the threat of resistance.

The most obvious signs of the failure of territorial apartheid appeared in the early 1970s. By this time most new housing development in the townships had ceased and pressure on existing housing stock increased through natural population growth as well as illegal immigration into the formal townships. During the 1970s township people, particularly the young who were unable to obtain houses, began moving out of the townships onto adjacent land where they joined migrants from the rural areas to form squatter camps (see Map 9.3). The hilly terrain and subtropical vegetation of Durban, as well as the reluctance of tribal authorities to apply influx control in their areas, enabled

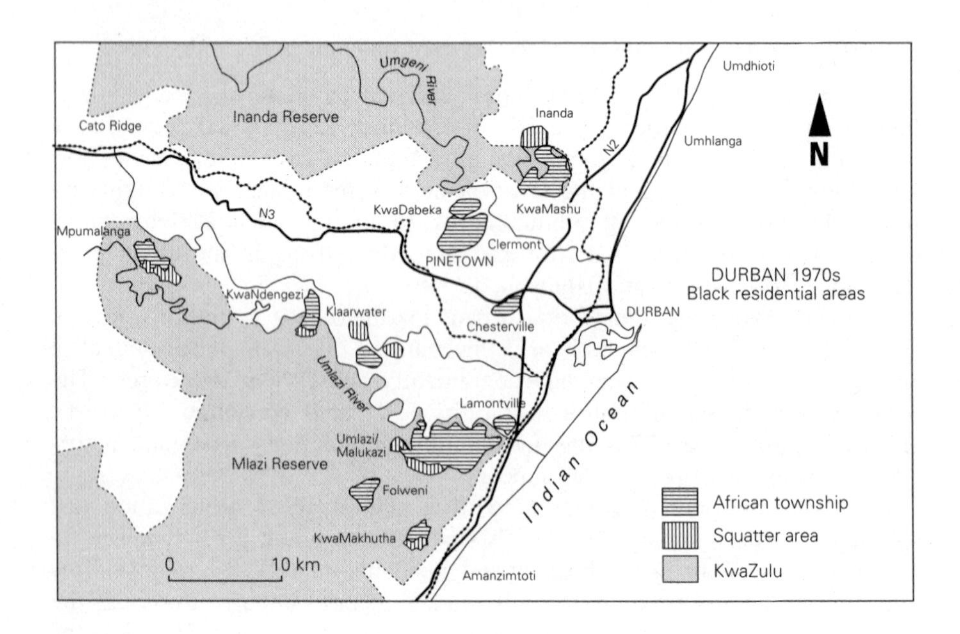

Map 9.3

squatting to go unnoticed by officialdom and the white public for a time. It took the form of slow and sometimes clandestine densification within the freehold and tribal areas as well as on mission and state land near the peripheral black townships, and thus, at first, posed only a remote threat to white residential concerns.

The proximity of tribal and freehold land to the African townships of Durban meant that influx control was far more difficult to apply than in many other metropolitan areas of South Africa. For the Port Natal Bantu Affairs Administration Board (responsible for Durban) the problems of maintaining influx and residential controls increased when KwaZulu gained self-governing status in 1974 (Mare and Hamilton, 1987) because the tribal authorities had neither the means nor the incentives to enforce a system of control devised and administered by the Department of Bantu Administration and Development in the 1960s (Hindson, 1987: 68–74). The tribal labor bureaus, designed as a system of efflux control in the late 1960s, had always been ineffective and were entirely unsuited to the exercise of influx control in rapidly urbanizing tribal areas on the metropolitan peripheries.

Rapid urbanization and residential differentiation: the 1980s and early 1990s

By contrast to the 1970s when squatting was semi-clandestine and relatively slow, the 1980s witnessed the mushrooming of open squatting, at times involving land invasions, in and around the black townships on the metropolitan periphery. Estimates of the size and growth of the African population in the DFR are notoriously unreliable, including official statistics on urbanization. Thus it is necessary to turn to estimates based on the work of local researchers. The best available source on the numbers of Africans living in formal and informal housing is that given by May and Stavrou (1989) for the period 1980 to 1987. According to their calculations, between 1980 and 1985 the total African population increased at an annual average rate of 5.9 percent, while the numbers in informal housing grew at 9 percent per annum. Between 1985 and 1987 the total African population grew at 11.1 percent and the informal at 19.7 percent per annum. By 1987 the informally housed African population comprised 50 percent of the total African population, having increased from an estimated 37 percent in 1980.

Added to the steady weakening of influx controls that had been going on since the early 1970s, a number of factors came together to open the flood gates. Severe drought in the late 1970s drove large numbers of people out of neighboring rural areas towards the cities in search of work. From the early 1980s a series of confrontations occurred between youth and residents' organizations on the one hand and township authorities on the other which led to a weakening of township administration and hence of the capacity to control settlement.

The immediate precipitating factor was the attempt by the authorities to devolve housing and influx control powers to unpopular black local authorities regarded as illegitimate by local residents, while at the same time increasing rents and public transport fares. This abortive application of the recommendations of the Riekert Commission (Riekert Commission, 1979) sparked off major rent and bus boycotts in the early 1980s; actions that marked the beginning of extensive and prolonged political violence in the DFR (Byerley, 1989; Reintges, 1986; Torr, 1985).

The ensuing weakening, and in some cases collapse, of black local authorities meant that control could no longer be exercised over land and housing allocation and hence the pace and form of urbanization. Open land occupations and invasions replaced clandestine squatting, first on vacant land near the townships and then, in the late 1980s, within the townships themselves (Townsend, 1991; see Map 9.4). At a national level, these developments were recognized and given further impetus by the formal abolition of influx control in 1986.

A further stage of squatting began in the late 1980s as squatter settlements sprang up in inner city areas such as Kennedy Road, Clare Estate and, most notably, within Cato Manor itself. Up until the late 1980s Cato Manor had remained empty, apart from a small number of previous residents who had either escaped removal or returned to live hidden in the dense vegetation. The common thread in these developments was the growing incapacity of local authorities to exercise control over African squatting in the face of a mounting challenge to their legitimacy both through struggles mounted in the Durban area over squatter settlement and the continued application of the Group Areas Act, and, at the national level, through the growing opposition to white rule and a state reform strategy which coupled modifications of apartheid with liberal reforms (see Morris, 1991).

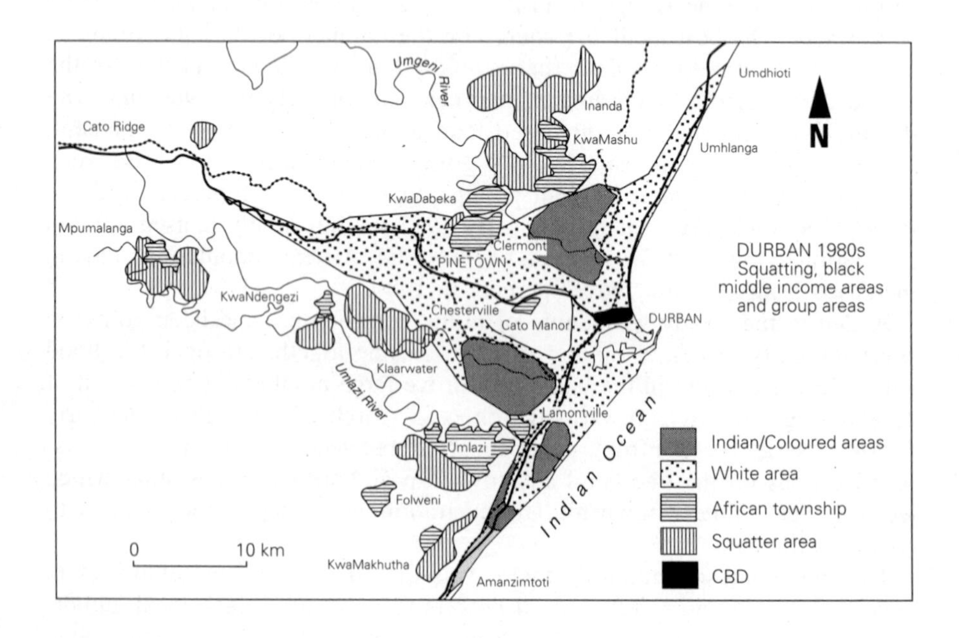

Map 9.4

A second major social development in the 1980s was the growth of middle income suburbs in the black residential areas on the periphery. Whereas classic apartheid attempted to suppress class divisions within the African population, the urban reform strategy pursued by the government in the 1980s sought to foster class divisions within urban black residential areas (Morris and Hindson, 1992a). In the past the primary line of social demarcation within the black urban areas was between temporary migrants in the hostels and permanent residents in formal township houses (Hindson, 1987). During the 1980s, however, two new residential groupings emerged. The established townships were flanked on the one hand by new housing schemes for lower-middle income, semi-professional and better-off working class families and on the other hand by mushrooming squatter settlements dominated by unskilled, marginal and unemployed people. The spatial sorting of classes was reflected in housing type and quality, infrastructure and services and a range of other socio-economic variables.

A survey of social differentiation in three adjacent residential areas on the southern side of Umlazi, undertaken by the authors in 1992 and 1993, suggests that a process of sifting out of income groups has been taking place in the black residential areas (Hindson and Byerley, 1993b). The established townships lost some middle income individuals and families to the new housing estates and many of the youth, poor and destitute to the neighboring squatter areas. Mean monthly household income in Emafezini, a new housing estate which comprises mainly semi-professionals and better paid working class families, was R2,371. This compares with an average household income of R1,697 in U Section of Umlazi, part of an established township with a majority of unskilled and semi-skilled workers. The mean household income of Malukazi, a neighboring squatter area, made up mainly of the unskilled and unemployed, was only R486. When compared to other black residential areas, the average household income of townships and middle income housing areas are similar but the figure for Malukazi is very low (Data Research Africa, 1992; Hindson and Byerley, 1993c, 1993d). The extremely impoverished nature of this particular area is partly due to protracted violence.

The effect of these developments was to produce a three-way class/residential division within the black areas of the DFR, comprising an emerging middle income group with the means to move to the newly constructed private housing estates; the established working class in the old townships, comprising increasingly mixed occupational and income groups, under pressure to move residentially up or down the social ladder; and the new poor crowded into impoverished shack areas in and around the old townships.

The legal and material position of squatters varied considerably from area to area. Within the tribal areas of KwaZulu squatters had rights to sites through allocation by headmen and chiefs. In white areas their position was more ambiguous as the right to reside in such an area depended on the outcome of negotiations and struggles with the local authorities and the strength of those local authorities. A further case was where squatter communities fell under the control of warlords. Here the rights to residence depended on relations of patronage, rent and subservience to the warlord.

It should be noted that not all the residents of squatter areas are equally impoverished. There is differentiation between squatter communities as well as within them in terms of income, education and occupation. The limited provision of African housing from the early 1970s meant that many employed township residents moved into shack areas in order to escape overcrowding in township houses and that a section of these people were employed and earned relatively high incomes.

The appearance of huge squatter areas on the periphery of the Durban metropolitan area marked a clear break with territorial apartheid but it also had the effect of powerfully reinforcing residential apartheid and hence the racial geography of the city. Although the reform strategy of the 1980s accepted the permanence of Africans in the city and, from the mid 1980s, the reality and desirability of African urbanization, the continued application of residential controls under the Group Areas Act as well as private property rights ensured that impoverished blacks were confined to the urban peripheries.

The growth in numbers of people in confined spaces on the black urban peripheries greatly increased the scarcity of residential resources such as land, housing and services and thus heightened competition for these resources. Untrammelled competition for basic resources in the context of the collapse of local government, the reduction of state assistance for housing and transport, and a shrinking resource base for residential life created ideal conditions for communal mobilization around access to and control over these resources. The mobilization of residential communities to defend or extend control over land and other basic resources was a major factor that fed into political mobilization and conflict in the DFR especially from the mid 1980s.

Violence and the breakup of apartheid

Prior to the 1980s political violence exhibited two main characteristics: it tended to be perpetrated by the state against individuals falling foul of the myriad of urban apartheid regulations; and took the form of criminalizing such transgressions (Hindson, 1987: 52–74). Violence in the 1980s emerged in the context of the crumbling of territorial apartheid, rapid unplanned urbanization, the bankruptcy of township administration and abortive attempts to introduce economic liberalization within the framework of residential apartheid.

The inherited racial city structure, coupled with the disintegration of the apartheid controls on the periphery, misplaced and confined urban conflict to the margins of the metropolitan area. A battle by blacks against exclusion from the political and economic power centers of the city – a struggle for greater access to, and control over, the resources of the wider society and economy – was deflected and turned into a fight over the shrinking resources allocated to blacks on the urban perimeters (Morris and Hindson, 1992a).

Both the content and method of implementation of the reforms introduced from the 1970s tended to intensify rather than to reduce social dislocation and the potential for conflict. Piecemeal liberalization opened opportunities for some and gave vent to aspirations long suppressed under apartheid, but this occurred in a context in which not all the major sources of grievance were being dealt with. Economic and political concessions were yielded to the new middle strata of the black population, including a relatively privileged working class, while excluding or marginalizing the impoverished and land-dispossessed. The majority of blacks, even those who benefited from reform, continued to feel the effects of remaining racial controls, an increasingly harsh economic environment and exclusion from national political processes.

During the 1980s the economic costs of residential and territorial apartheid escalated; social divisions sharpened in the black residential areas and were less amenable to simple repressive and undifferentiated mechanisms of control; economic and political repression was increasingly and openly challenged; and competition for access to the urban resources necessary for social and economic life intensified. The apartheid institutions which previously managed conflict through control over the black residential areas began to disintegrate and increasingly lose their effectiveness. This made space for a political struggle for hegemony within the black residential areas; a struggle that was integrally tied up with the capacity of competing and conflicting parties to capture, control, and distribute resources within these areas. The struggle for hegemony was greatly intensified with the rise of Inkatha-linked warlords and African National Congress-linked civic and youth organizations (Morris and Hindson, 1992b; Morris, 1992).

Violence in the early 1980s

Since 1980, township violence in the Durban area has moved through three overlapping stages (Byerley, 1992a). In the first stage, in the early 1980s, it was concentrated in the core area townships, notably Chesterville and Lamontville, that had always been outside the control of KwaZulu. This predominantly took the form of a confrontation between residents on the one hand and community councillors, township administration, police and army on the other. This may be termed 'state–people' violence.

The tension focused on educational, transport, rent and labor issues. It was also exacerbated by the actions of KwaZulu authorities who wanted to continue the 1970s emphasis on territorial apartheid. Ignoring the buildup of pressure for the dismantling of urban apartheid and acceptance of a permanent African presence in the urban areas, the KwaZulu authorities, in collaboration with the central state, sought to incorporate all the Port Natal Administration Board townships (Lamontville, Chesterville and Clermont) into KwaZulu.

This provoked deep antagonism from these communities. Residents first joined together to form the Joint Rent Action Committee (JORAC) which spearheaded the campaign against rent increases and later against incorporation into KwaZulu (Reintges, 1986; Torr, 1985). The formation of JORAC and the subsequent repression levelled against it marks the beginning of a long period of instability and violence in black residential areas.

Violence continued sporadically throughout 1983 and most of 1984 but began to take on a new focus. Clashes between United Democratic Front (UDF) supporting youth and Inkatha supporters became common and unionized workers began to be drawn into the conflict. At its inauguration in late 1985, the Congress of South African Trade Unions (COSATU) openly declared its hostility towards Inkatha, thereby fanning the flames of confrontation. Inkatha supporters reacted towards township youth linked to the UDF and against COSATU affiliated workers.

The KwaZulu Legislative Assembly and Inkatha came increasingly to be seen to be puppets of the apartheid state. At this time the South African Police began to move away from the front lines of the conflict to give way to the KwaZulu Police, a force recently set up under the control of the KwaZulu Legislative Assembly.

The result of these developments was that COSATU lined up against Inkatha, its recently created trade union, the United Workers Union of South Africa (UWUSA), set up with the aim of undermining COSATU in Natal, and the KwaZulu and South African Police (SAP) with a consequent escalation of violence.

Violence from the mid 1980s

In the second stage, from the mid 1980s, the violence shifted towards the urban periphery and became increasingly internecine. The number of people who died from violence in the Durban Functional Region between 1986 and 1992 has been estimated at 3,228. The proportion of deaths due to violence in the core urban areas (Durban Central) remained low compared to the periphery, 9 percent of total deaths in 1986 and 5 percent in 1992. Within the peripheries the locus of violence shifted from the formal townships to the squatter shanty towns, increasing from 30 percent in 1985 to 70 percent of deaths in 1992.

Violence occurred within the black residential areas along different lines in different places. Squatter communities against other squatter communities, township–squatter conflict, hostel–township and hostel–squatter conflicts, and, finally, cases of intra-township conflict. In the established townships on the metropolitan periphery, such as KwaMashu and Umlazi, black opposition, which began as an organized resistance to bus fare and rent increases in the early 1980s, was increasingly propelled into political and internecine violence.

The government's response was to extend the State of Emergency then in force in most of the country to Natal. The draconian measures employed under the emergency resulted in the virtual removal (through detention, death, or going into hiding) of the UDF leadership from the townships.

The removal of leadership promoted the spread of lawlessness within the townships and shantytowns and the youth constituency came to the fore in the ensuing conflict. Clashes between UDF-supporting youth and Inkatha supporters increased. Black councillors and policemen were regularly attacked because they were seen to be instruments of the apartheid state. This period of violence saw UDF-supporting youth – joined, at times, by criminal elements – coming to dominance in most of the formal townships. The result was that Inkatha leaders and their supporters were expelled from large parts of the townships. They were to re-emerge, however, in the squatter settlements on the margins of the townships where they teamed up with local shacklords controlling tenants on a single site and warlords who had effective control over whole squatter communities (Morris, 1992).

From the mid-1980s onwards the violence shifted geographically to the squatter periphery and became increasingly introverted. Political mobilization and conflict frequently coincided with the division between squatters and township dwellers. The UDF and, later, the ANC developed their bases largely in the established townships, especially amongst the youth, while Inkatha retreated into the squatter areas where its political fortunes were tied to the warlords.

Prior to the mid 1980s there tended to be relative harmony between township people and squatters. There was a sharing of resources, such as taps, schools, and other facilities. But with the growing resistance to the state and local councillors, the expulsion of many Inkatha supporters from the formal townships and of ANC youth from some squatter areas, urban turbulence shifted its focus. It turned from a state–people violence into a conflict that was essentially and increasingly intercommunal.

One important line of division was between townships and the new squatter areas, as for example between Ntuzuma and Lindelani near Inanda. However, conflict was also intense within townships and squatter areas and at times between different ethnic groups, notably between Zulus and Pondos in Malukazi and Number 5 near Umlazi (Byerley, 1989, 1992b). A further feature was the superimposition of political allegiance over social and material differences.

Squatters were mobilized by warlords to fight for access to resources in neighboring townships and township youth organized military-style units to defend their areas and counter-attack squatter areas. Local power struggles developed over land, housing, services and business rights. One of the principal reasons why the violence has been so deep rooted is that political allegiance to Inkatha and the UDF/ANC tended at this time to coincide with residential location and differential access to urban resources.

In this period the South African security forces sought to remove themselves from the direct line of fire and present themselves as playing a more impartial policing role. From 1987, with the establishment of the KwaZulu Police force, the South African Police could move a step back from immediate involvement in the violence within KwaZulu. However, they continued to support Inkatha in hidden ways including clandestine funding, training of hit squads and direct covert operations against the UDF/ANC and their supporters. As the violence shifted to the urban peripheries and hence deeper into KwaZulu, the KwaZulu Police came more to the fore as an extension of Inkatha.

From the mid 1980s, then, Inkatha was gradually squeezed out of the formal townships and regrouped and reorganized itself mainly amongst the marginalized squatters on the urban periphery. This resulted in numerous protracted and bitter conflicts between township people and squatters.

Although areas come to be defined as 'ANC' or 'IFP' the reality of political allegiance is more complex. In many instances residents feel compelled to show allegiance to the group that is dominant at any particular point in time.

Violence in the early 1990s

In the third stage (1990s) mass confrontations gave way to more atomized, targeted and clandestine killing, although internecine clashes continued. In this latter stage, two other processes increasingly manifest themselves: a drift into criminality and internecine violence within black residential areas between opposing factions of the same political organization.

The challenge to the racial city

We have seen that internecine conflict in the second half of the 1980s can be explained in part in terms of the racial structuring of the city and the capacity of the white urban constituencies and the state to displace conflict and its consequences onto the black residential areas. The ability of the state to continue in this role began to weaken by the end of the 1980s. Violence became increasingly widespread and in some areas endemic. The economic costs of violence in terms of life and property destroyed, mounting security costs and costs to the economy through disruption of production, absenteeism and worker debilitation continued to increase while the incentive to invest by local and international finance declined. Within South African cities it became increasingly difficult to contain the violence physically, and although white residential areas continued to escape the direct effects of township conflict, robbery and violence became much more frequent and difficult to prevent in the inner city and white suburban areas.

The state was confronted with widening opposition from within key sectors of white South Africa such as business, as well as an increasingly effective international campaign of economic, cultural and other forms of isolation. And, while the extraparliamentary movement was significantly curbed by the State of Emergency, neither the UDF nor COSATU – the two organizations that led open internal opposition during this period – was ever fully silenced, and both continued to place pressure on the state through various forms of protest and other action.

Opposition to the maintenance of racial residential segregation also came from two very different groups seeking to escape the violence on the urban peripheries and to take up opportunities in the core city areas. The late 1980s witnessed the first flow of squatters from periphery to core as pockets of squatters occupied land, in some instances in areas left vacant by apartheid removals. In the climate of growing political instability local authorities became increasingly reluctant to remove these groups despite the continued existence of apartheid and anti-squatter laws.

The second major pressure for racial integration of core city areas was from middle and upper income black groups seeking housing near to places of work and in the more secure white suburban or inner city flatland areas. Well before the abandonment of the Group Areas Act in 1992, incremental deracialization, known as 'greying,' had begun to take place, especially in inner city areas composed mainly of high rise apartment buildings, and also in a number of lower income white suburbs near the central business districts.

By the end of the 1980s, the white state and those interests historically represented by it had to face the stark choice of attempting to maintain racial controls under the Group Areas Act against mounting opposition from within and outside South Africa, or attempting to negotiate a process of deracialization and spatial integration of the city. Following the unbanning of political organizations, and growing acceptance of negotiation as a route to transformation, negotiation forums have been established both nationally and within several cities and towns and it is within these that the possibility of a peaceful negotiated reconstruction of the cities is now being debated.

Peace and reconstruction in the Durban functional region

This paper has argued that the roots of violence lie in deep structural divisions and antagonisms produced by racial policy, spatial fragmentation, and the confinement of the poor to the metropolitan peripheries. It follows that lasting peace of a kind that does not rely on further state repression will require a fundamental racial, spatial and economic integration of the city. This cannot be achieved without threatening interests that are vested in the present spatial geography of the city. It is thus critical that a process is established whereby reconstruction is undertaken with the involvement of all the major interests which can contribute constructively to the reconstruction of the city, or derail that process.

Selected references

Byerley, M. A. 1989: 'Mass violence in Durban's settlements in the 1980s,' unpublished master's dissertation, University of Natal, Durban.

Byerley, M. A. 1992a: 'Conflict in Durban in the early 1980s: From anti-state to internecine violence.' Paper presented to the Critical Studies Group Seminar, University of Natal, Pietermaritzburg, 18 August.

Byerley, M. A. 1992b: Ethnicity, violence and social differentiation: A case study of conflict in Number 5 – a squatter settlement of Durban. Paper delivered to the Conference on Ethnicity, Society and Conflict in Natal, University of Natal, Pietermaritzburg, 14–16 September.

Data Research Africa 1992: *A Study of Income and expenditure and Other Socio-economic Patterns in the Urban and Rural Areas of Kwazulu*, Volume 1, *Overall Results*. Durban: Data Research Africa.

Davies, R. J. 1991: Durban. In A. Lemon (ed.), *Homes Apart: South Africa's Segregated Cities*. Cape Town: David Philip.

Hindson, D. 1987: *Pass Controls and the Urban African Proletariat*. Johannesburg: Ravan.

Hindson, D. and Byerley, M. A. 1993b: A report on class and mobility in the Umlazi/Malukazi/Emaphezini area. Institute for Social and Economic Research, University of Durban-Westville.

Hindson, D. and Byerley, M. A. 1993c: A report on class and mobility in Clermont/KwaDabeka/Fannon Extension. Institute for Social and Economic Research, University of Durban-Westville.

Hindson, D. and Byerley, M. A. 1993d: A report on class, race and movement in Cato Manor. Institute for Social and Economic Research, University of Durban-Westville.

Hindson, D. and Crankshaw, O. 1994 (forthcoming): *South Africa's Changing Class Structure*.

Maasdorp, G. and Humphries, A. S. B. 1975: *From Shantytown to Township*. Cape Town: Juta.

McCarthy, J. J. 1988: Poor urban housing conditions, class community, and the state: Lessons from the Republic of South Africa. In Obduho, R. A. and Mhlanga, C. C. (eds), *Slum and Squatter Settlements in Sub-Saharan Africa: Toward a Planning Strategy*. New York: Praeger.

McCarthy, J. J. 1991: Planning issues and the political economy of the local state: a historical case study of Durban. *Urban Forum* 2.

Maharaj, B. 1992: The spatial impress of the central and local states: The Groups Areas Act in Durban. In Smith, D. M. (ed.), *The Apartheid City and Beyond: Urbanization and Social Change in South Africa*. London: Routledge.

Mare, G. and Hamilton, G. 1987: *An Appetite for Power: Buthelezi's Inkatha and the Politics of* 'Loyal Resistance.' Johannesburg: Ravan.

May, J. D. and Stavrou, S. E. 1989: The informal sector: socioeconomic dynamics and growth in the greater Durban metropolitan region. Rural Urban Studies Working Paper No. 18, Development Studies Unit, University of Natal, Durban.

Maylam, P. 1982: Shackled by the contradictions: The municipal response to African urbanization in Durban, 1920–1950. *African Urban Studies* 14, 1–17.

Morris, M. 1991: State, capital and growth: the political economy of the national question. In S. Gelb (ed.), *South Africa's Economic Crisis*. Cape Town: David Philip.

Morris, M. 1992: Violence in squattercamps and shantytowns: power and the social relations governing everyday life. *Environmental Law Series No. 1*, VerLoren van Thermaat Centre, University of South Africa, Pretoria.

Morris, M. and Hindson, D. 1992a: South Africa: political violence, reform and reconstruction. *Review of African Political Economy*, 53, 43–59.

Morris, M. and Hindson, D. 1992b: The disintegration of apartheid – From violence to reconstruction. In Moss, G. and Obery, I. (eds), *South African Review 6*, 152–70.

Nuttall, T. 1989: The Durban 'Riots' of 1949 and the struggle for the city. Paper delivered to Critical Perspectives on Southern Africa Seminar, University of Natal, Durban, 12 May.

Reintges, C. M. 1986: 'Rents and urban political geography: The case of Lamontville,' unpublished master's thesis, University of Natal, Durban.

Riekert Commission 1979: *Report of the Commission of Enquiry into Legislation Affecting the Utilisation of Manpower*, RP 32–1979.

Torr, L. 1985: 'The social history of an urban African community, 1930–1960,' MA thesis, University of Natal, Durban.

Townsend, M. 1991: 'The rebounding wave – Umlazi land invasions 1987–90,' unpublished MTRP dissertation, University of Natal, Durban.

SECTION THREE

LOCALITY, POLITICS AND CULTURE

Editor's introduction

The 1980s saw the re-emergence of interest in ideas of culture, locality and place in human geography. The three sets of ideas were not initially linked, and Doreen Massey's (1979, 1984, 1985) attempts to show how 'geography matters' in the structuring of economic, social and political life were more concerned to establish a theoretically significant role for geography in the aftermath of the collapse of spatial science as a coherent intellectual project. The solution she and others (Sayer, 1985; Duncan, 1988) arrived at was that there was a reciprocal relationship between society and space, that the social is always and everywhere produced and reproduced in space, and that place and locality make a difference to the nature of social processes (Johnston, 1991). This interest was paralleled by an interest in the development of the local state and variations in local politics (Dickens *et al.*, 1985). Consequently, despite criticism of what has been perceived as a regressive political retreat from the study of general social processes to a concern with the unique (Smith, Chapter 15, this volume), there has been considerable interest in the ways in which social processes are mediated and reproduced in different places and in the role of the local in the constitution of the global (Swyngedouw, 1989). Kay Anderson quotes Geertz as stating that the study of localities affords the opportunity to weave 'the most local detail with the most global structures in such a way as to bring both into view simultaneously' (Geertz, 1983).

The rediscovery of culture in human geography (Cosgrove, 1983; Cosgrove and Jackson, 1987; McDowell, 1994b; Duncan and Duncan, 1988; Duncan and Ley, 1993a) owes a great deal to the work of Clifford Geertz (1973, 1983) who, in a series of superbly written essays, demonstrated the centrality of culture as a set of shared meanings and discourses and, in so doing, helped to create the basis for the so-called 'cultural turn' within human geography. As Geertz (1983) put it:

> Ten years ago, the proposal that cultural phenomena should be treated as significative systems posing expositive questions was a much more alarming one for social scientists – allergic, as they tend to be, to anything literary

> or inexact – than it is now. In part, it is a result of the growing recognition that the established approach to treating such phenomena, laws-and-causes social physics, was not producing the triumphs of prediction, control and testability that had for long been promised in its name (p. 3).

As Geertz points out, there was an aversion in human geography to anything cultural in the 1960s and 1970s, as the notion of culture seemed to embody everything that was literary, descriptive, unique and inexact. It was something that scientific geography needed to leave behind in its search for regularities. But, with the collapse of the spatial science project and the growing doubts about the determinism of structural marxism, there was an intellectual void within social and cultural geography which Geertz, and others such as Said, helped to fill. Consequently, it has come to be recognized that culture is not an awkward residue or 'noise' factor that had to be accounted for or put to one side after the social regularities had been established. On the contrary, culture permeates the whole of economy and society and it has recently been argued that the 'economic', itself, is a social and cultural construct which has to be analysed in these terms. As a result, culture and cultural geography has moved back from the sidelines to intellectual centre stage, as an examination of key journals such as *Society and Space* clearly indicates. How long this will last is difficult to say, as the history of human geography over the last 25 years shows the existence of cycles of intellectual fashion, but it is clear that culture, and with it the role of interpretation, is now firmly re-established.

It is difficult, as ever, to select papers to illustrate some of the important work in contemporary social geography which illustrates the importance of culture and locality, but Kay Anderson's (1988) classic study of cultural hegemony and the race-definition process in Vancouver (Chapter 10, this volume) highlights the importance of culture, power and locality in the process of race definition. As such it holds lessons for a wide variety of different contexts. It would be a mistake to see the lessons of this paper as being specific to a particular place in a particular period. Its implications are far more general.

Kay Anderson points out that the nature of power in Western society has been the concern of many human geographers in recent years. She notes that some have examined the spatial implications of class conflict (Dear and Scott, 1981; Harvey, 1985), and others (Pratt, 1982; Saunders, 1984) have focused on consumption sector cleavages outside the workplace. Yet others such as Pahl (1970b) have examined the institutional control of gatekeepers and urban managers or the influence of coalitions and political movements on the control of urban space (Ley, 1983). But, as she points out, much less attention has been paid to 'the pervasive type of conceptual domination that has been exercised by powerful groups over the *definition* of people and places in Western societies' (pp. 209–10). Such power, Anderson argues, has structured the society and space of Western countries in critical ways and she points to the influence of hegemonic white European racial theorizing.

Her paper traces the 'cultural domination by one white European community over a set of racially defined outsiders known as the "Chinese" in Vancouver, British Colombia'. She argues that it is important to examine the process by which 'racial categories are . . . constructed and transmitted', and suggests that the force and resilience of racial representation is such that its contribution to systems of inequality should be 'examined on its *own* terms rather than simply "read off" from putatively deeper causes' such as capitalism (p. 211). Her paper attempts to situate the concept of race in historical and political context, to demonstrate the role of government agents in the construction of the 'Chinese' as a racial category, and to trace the articulation of the category at the local level where 'socially created difference is actually organized and reconstituted in space' (p. 211). She points to the importance of Darwinian evolutionary metaphor in the late nineteenth century in reinforcing prejudice, and draws on Gramsci to argue that racial ideology has long been a 'critical unifying principle' in justifying the rise of the white European historical bloc to its position of hegemony.

Kay Anderson argues that the sphere of politics and the state is intimately linked to the cultural realm and shows how the establishment of white settler power in British Columbia in the nineteenth century led to the development of *conceptual control* over the definitions and status of settlers and residents in the colony. But she notes that the race-definition process has not operated simply in the minds of historical actors, and argues that relationship between consciousness and structure is crucial and that they can have an important geographical dimension: 'the geographic sphere provides a critical nexus in which racial ideology is ontologically realized and reconstituted'. She points to the importance of taking a historical perspective to the development of social processes, arguing that:

> Within the social geography of the postpositivist era, there has been a tendency to treat history as past time, as a backdrop to the present, as if the present were almost an autonomous creation. The advantage of an analysis that is situated in both space and time is that the mutually confirming relationship between the 'ideal' and the 'material' can be demonstrated and demystified (pp. 214–15).

Anderson concludes by asserting that:

> Racial ideology has been a significant cultural and political force in structuring the social order of Western countries such as Canada. During the evolution of white European domination in Vancouver . . . conceptions of a 'Chinese race' became inscribed in institutional practice and reconstituted through the locality known, and produced, as 'Chinatown' (p. 230).

With Gillian Rose's paper (Chapter 11, this volume), we move from Vancouver to a study of the inter-relationships of locality, politics and culture in the East London borough of Poplar, in the 1920s. Their concerns, however, have a certain amount in common. Rose begins her paper by suggesting that: '"Culture" is rapidly becoming a buzz-word for social theorists, especially for locality-study theorists and others struggling to explain local differences in social and political behaviour; it is now often evoked as the cause of otherwise inexplicable geographic variations in social activity or political allegiance'.

Looking at the rise of local political studies in the 1980s, Rose argues that:

> What is missing from all these structural accounts of local politics is contextualised analysis. The expression and formation of local opinions and attitudes and understandings within civil society are ignored, and so the contingent nature of the politics which develop from them is never acknowledged. What is lacking is recognition of the sheer complexity of the politicisation process (p. 237).

Rose draws on Geertz's (1973) notion of 'socially established structures of meaning', and his notion of 'culture' as 'local knowledge', 'local frames of awareness' and 'communal sensibility', to argue that this is what is needed in studies of local politics, and she argues that it was 'the communal sensibility of Poplar in the 1920s which . . . was formative of its political behaviour' (p. 237).

She shows how Poplar, the most easterly borough of London's East End, developed industrially in the nineteenth century, and how cheap, speculative housing rapidly turned into overcrowded slums, with a fifth of the population living more than two to a room. In the early 1930s, it was a wholly working-class area, and the poorest borough in London. As Rose points out, 'Poplar was obviously a proletarian locality, and on class-theoretical logic it was ideal Labour Party territory', and the Labour Party swept into power in the borough after the 1918 Representation of the People Act (which reduced the property qualification for a vote in local elections and enfranchised most of Poplar's adult population). But, as Rose points out, Poplar's allegiance to the Labour Party cannot be understood as a direct reflection of working-class interests, partly because the Poplar Labour Party was a maverick both within London and nationally. She argues that there is 'no one natural and inherent working-class politics' and suggests that it is necessary to look at the contingent nature of local politics.

In the body of her paper she attempts to link, 'somewhat overschematically', four aspects of Poplar's cultural life to its Labour Party's politics and beliefs. These are its class structure, its neighbourliness, religious faith and its love of performance and melodrama. Whether these are the four most appropriate or relevant aspects of Poplar's cultural life can be debated, and Rose gives no rationale for the choice of these areas. She suggests, however, that:

> In the 1920s, all these aspects of Poplar's political culture – its class loyalty, its community morality, its religiosity, its riotousness – all these coalesced to encourage a participatory form of politics, with . . . many of Poplar's ordinary people becoming involved in the council's battle against the Poor Law (p. 245).

Rose concludes by suggesting that although local politics are shaped by local cultural values, the political implications of this local culture are by no means straightforward, and she argues that in Poplar the same cultural context sustained two types of relations between the local state and local people: participatory and autocratic. Thus, she states that:

> 'Culture' . . . is not a simple solution to the problem of geographic variation in social and political action; its social implications are not straightforward or consistent over time or space, and this means that locality-study theorists should be much more wary than they have been hitherto in evoking 'culture' as some sort of ultimate explanation of social and political behaviour (p. 251).

Finally, Rose concludes by arguing that there is another reason to be wary of the way in which the concept of culture is used. She suggests that Geertz can be criticized for his lack of interest in the social sources of power structures, and that until these are more fully investigated, the idea of culture can only be a partial solution to the problems which analyses of local politics have encountered to date. This is an important point, and the sources of social power clearly need further investigation (Saunders, 1979)

Cindy Katz's paper (Chapter 12, this volume) is important for a number of reasons. First, it looks at the issue of social reproduction in a non-western, agriculturally based economy, which helps overcome some of the strong western, urban bias in much of the social geographic literature. Second, it employs an ethnographic, participant-observation methodology in a powerful way, and thirdly, it addresses the vital question of the relationship between structure and agency, focusing on the role of agency, and the ways in which knowledge is transmitted and learnt. As Katz points out in her introduction:

> It is difficult . . . to move analytically between structure and substance, and most studies end up concentrating on socioeconomic and political structures without any real analysis or understanding of the practical activities of the people who create and are constituted by these structures (p. 257).

She adds

> All too often, people are recognized as 'making their own history', but this history is rather thin and pale, a narrative of socioeconomic change that points to but glosses over the material social practices of everyday life to focus only on their temporal and spatial outcomes (p. 257).

Part of the problem, says Katz, is that few analyses that claim to address the empirical relationship between structure and agency are grounded in theory that can analyse the mutual relations between the two and their outcomes in time and space. She suggests that serious consideration of feminist theory and research would help as it was among the first work to examine the links between agency and structure. Her fieldwork was conducted in an area of Sudan where most of the population were subsistence cultivators until 1971, when the area was included in a state-sponsored agricultural development project which introduced irrigated cultivation of cotton and groundnuts and undermined local control over production and reproduction by specifying which crops and agricultural practices were acceptable. This disrupted the rhythm of daily and annual work and altered the relationship between the local population and the land. Thus, although local farmers welcomed some of the changes they actively resisted the loss of local agricultural land and the disruption to local agricultural practices and ways of life. What is novel about Katz's study, however, is that she focused her work on children (52 per cent of the population was under 15 years old) and studied a sample of children in depth using participant observation, asking children to produce a model of their village and the production of oral geographies, to:

> discover, document, and describe the range of children's environmental interactions and the forms and content of their environmental knowledge in all their complexity; and to analyze these in relation to the larger context in which they occurred' (p. 261).

Rituals of the streets: the geography of everyday life

The late 1980s and the 1990s have seen the emergence of social geography as an exciting area of research once again, largely as a result of its close link with the new cultural geography. The narrowly circumscribed boundaries of traditional urban social geography have been broken, and social geography has emerged as a brilliantly coloured butterfly from a rather dull pupa. It has been realized that the traditional focus of social geography on residential patterns, migration, social areas, community and the rest are but a small part of a much wider sphere of spatially grounded social behaviour. One area of work which has rapidly come to prominence is that of street life, culture and public rituals and what can be termed the 'geography of everyday life'. Since the late 1980s, there has been a sudden explosion of research in this area. In retrospect, the closest parallel is with the work of Park and other members of the Chicago school of urban sociology in the early decades of this century. In fact Park (1904) produced his doctoral thesis 'The crowd and the public' on the subject. Intriguingly, however, it was not Park's work which was taken up by social geographers in the 1960s, but Ernest Burgess's, with its emphasis on urban spatial patterns. Indeed, there is a strong argument for saying that, until recently, much social geography has concentrated on the relationships between home and work at the expense of studying the geography of everyday life in a wider sense (Eyles, 1989). The 1970s French graffito, 'Home, Metro, Work, Metro, Home, Sleep' can be seen as an accurate picture of much urban social geography until the 1980s.

As Smith (1993) points out, the study of crowd behaviour has interested social researchers for over a hundred years, and one of social geography's foci in studying collective action has been the analysis of riots (Peach, 1985; Keith, 1987; Charlesworth, 1983). But riots and street disturbances are only one form of collective public action, and Smith suggests that the current interest in collective action is in the localized character of struggle, and the relationship between 'claiming space and making place'. It is an example of the renewed interest in locality and place and rituals of resistance (Hall and Jefferson, 1976) and urban social movements. Smith suggests that this focus is not unexpected given that most research to date has focused on attempts made by marginalized minorities to win space within the metropolitan cores. She argues that:

> *Urban* space has effectively become the terrain on which analysts have anchored the restructurings of late modernism; *world cities* are the symbols of postmodern culture . . . and it is on *city sites* that collective behaviour (most notably in the form of urban social movements) has been portrayed as a means to claim space, assert identity and resist oppression (pp. 291–2).

There are many examples of such work in recent years and research has been carried out on contemporary carnivals (Cohen, 1980, 1982; Jackson, 1988, 1992), and on processions, demonstrations and street life in the nineteenth century. Goheen (1993), for instance, states that in many Canadian cities during the middle decades of the nineteenth century, 'The central streets were magnets attracting demonstrators and spectators who by their presence gave support to many causes through performances that ranged from meticulously arranged civic festival to unrestrained riot' (p. 127). Goheen suggests that the variety of collective public performance attests to a diversity of outlook among a population largely composed of immigrants. It also 'reminds us of the strength of a tradition, carried to America as cultural baggage from Europe, of largely unhindered access to the streets'. The streets were the principal area for both consensual and contentious demonstrations. Goheen suggests that this was particularly true of the central streets, which had the highest symbolic value and where the economic, social and political life of the city was focused and where the inhabitants most frequently interacted. This suggests, says Goheen, that these streets were the most important public space in the city, and their significance was reinforced by the concentration of important institutions which located on them. Similar arguments have been made by Marston (1989) regarding the relationships between public rituals, public space and community power in St Patrick's Day parades in Lowell MA. There are a variety of historical and anthropological studies regarding 'the language of sites and the politics of space' (Kuper, 1972; see Cotterell, 1992; MacMaster, 1990). But it is important not to fall into the trap of seeing these simply as a phenomenon of purely historical interest. Streets, parks and other public spaces (such as shopping malls) remain the focus of a great deal of public life in the current era, as Jackson (1988, 1992) has shown in his studies of Caribbean carnivals. Smith (1993) examines the role of rituals in rural Scotland, looking instead at the festival celebrations of a rural majority. She notes that in early summer each year, 'the streets of many rural towns in the Scottish borders provide sites for the enactment of public rituals – galas, ridings and other anniversary celebrations – which celebrate and affirm the character, value, history and continuity of local life' (pp. 292–3).

More generally, Berman (1984) views 'the street and the demonstration as primary symbols of modern life' and for Smith 'The street is an important cultural site', prone, at certain times and for certain groups, to 'become associated with particular values, historical events and feelings' and likely to be adopted variously as a symbol of good, evil, normal, deviant, right or wrong (Shields, 1991, p. 29).

Similarly, Nuala Johnson (1995) points out in her study of monuments, geography and nationalism, that since the nineteenth century, if not before, public monuments have been the foci for collective participation in the politics and public life of villages, towns and cities. She explores the ways in which 'the sociology, inconography, spatialisation and gendering of statues reveal ways in which national "imagined" communities are constructed' (p. 51). She demonstrates the importance of this for contemporary events in her discussion of the removal of many monuments to Marx and Engels and leaders of communist regimes in Eastern Europe since 1989, and the conflict over the erection of a statue to Sir Arthur 'Bomber' Harris outside the air force church in London in 1992. Harris was the architect of the saturation bombing policy which destroyed Dresden and Köln at the end of the Second World War.

One of the most fascinating of the new types of work being done today is Charlesworth's (1994) 'Contesting places of memory: the case of Auschwitz' (Chapter 13, this volume). The paper examines the contest over symbolic space at the Auschwitz death camp, the postwar symbol of the Holocaust. He examines both communist and Catholic attempts to de-Judaize the place and hence turn an 'icon of remembrance' into an 'idol of remembrance'. It is commonly estimated that some six million Jews were murdered in the Holocaust, along with Russians, Poles, Gypsies and others. Auschwitz alone is estimated to have accounted for over a million deaths and, as such, it constitutes a horrific reminder of 'man's inhumanity to man', and it constitutes a focus of forced migration and death for millions of people during the Second World War. To me, its geographic importance is only too evident and its significance greatly outweighs many current social geographic concerns which, not to put too fine a point on it, seem positively trivial by comparison. It is impossible to estimate the number of deaths associated with colonialism, from the Spaniards onward, but along with slavery, the Holocaust must rank as one of the most appalling examples of the treatment of the 'Other' ever to have taken place. For this reason alone it is important to include Charlesworth's paper.

But, as Charlesworth points out, it is also important to consider the choice of Auschwitz as a site of commemoration of the Nazi terror. At least six camps whose primary function was the extermination of European Jewry were all possible commemorative sites, but only Auschwitz and Majdanek remained intact. Charlesworth argues that though Auschwitz has the strongest claim in terms of the numbers who died there, the choice also reflected the way 'the Polish–Soviet relationship could be refracted through the memorialisation of each camp'. In particular, Charlesworth notes that although Majdanek had a much wider 'victim profile', only 30 per cent of the victims being Jews, its choice could have undermined Polish–Soviet solidarity and fuelled Polish nationalist sentiment in that it was taken over by the NKVD (the Soviet secret police) at liberation and had held thousands of members of the Polish Home Army, the noncommunist resistance movement. Second, it was adjacent to the city of Lublin, which was the seat of the Soviet–Polish puppet government at liberation in 1944. Thirdly, it was too close to the eastern border of pre-war Poland and could remind Poles of their 'subordination and victimisation by the Soviet Union both between 1939 and 1941 and after 1944 . . . Two million Poles had been taken to the Gulags, of whom half were dead within a year of arrest. At Katyn 4,500 Polish officers had been murdered by the NKVD' (p. 274).

Finally, Charlesworth points out that, after 1945, Majdanek was less than 90 km from the new Polish–Soviet border, and would thus have been able to 'act as a place of remembrance for Poles of their latest traumatic loss of territory'. Auschwitz 'with its more western location was much less problematical and could be portrayed as symbol solely of fascist aggression' (p. 274). But, as Charlesworth points out, the memorialization of Auschwitz in terms of Soviet–Communist ideology dominated until the early 1970s when a Polish Catholic Auschwitz began to emerge. He notes that the process of Catholicizing Auschwitz began at a crucial point in Church–State relations, when the Poles' sense of national identity and unity was strengthened. Auschwitz began to be seen by Poles as a site of Polish Catholic martyrdom which brought it into strong conflict with the Jewish world, particularly over the presence of the Carmelite nuns in a building adjacent to Auschwitz. As Charlesworth notes, the process was assisted by Cardinal Karol Wojtyla, the present Pope John Paul II, who was Archbishop of Krakow, and in whose diocese Auschwitz is located. Wojtyla understood the significance of Auschwitz in terms of Catholic martyrdom and he successfully campaigned to have Father Maximilian Kolbe, who died at Auschwitz, beatified. He also proposed that a church should be erected as Christian tradition dictated for examples of sainthood and martyrdom. The first Catholic mass and sermons were given at Birkenau, the adjacent camp, and the altar was sited on the very ramp where from spring 1944 Jews were unloaded from the trains and where they were selected for life or gassing. The offence to Jewish sensibilities is clear, and it reinforced a long tradition of Polish anti-Semitism. The point of Charlesworth's paper, however, is to argue that although Auschwitz is *the* symbol of the Holocaust to many people, it has for long taken on different meanings, and is a socially contested space. As Anderson (Chapter 10, this volume) points out, although many human geographers have examined the nature of power in western society, much less attention has been paid to the pervasive type of conceptual domination that has been exercised by powerful groups over the *definition* of people and places in western societies. Yet such a 'power of definition' (Western, 1981 p. 8) has structured the society and space of western countries in critical ways (pp. 209–10). The struggle over the definition of Auschwitz is one such example.

10 K. J. Anderson,

'Cultural Hegemony and the Race-Definition Process in Chinatown, Vancouver: 1880–1980'

Excerpts from: *Society and Space* **6**, 127–49 (1988)

Introduction

On 9 February 1985 the Toronto *Globe and Mail* reported that almost 800 people in South Africa had had their racial classifications changed in 1984 under apartheid policies. South African Home Affairs Minister Frederick De Clerk said in Parliament that:

> 518 coloureds became whites, 14 whites became coloureds and 17 Indians became Malay. There were also 89 blacks who became coloured and five coloureds who became black, three blacks who became Indian, one who became an Asian and a Malay who became a Chinese (page 8).

Although few countries have traded so transparently in the currency of race as has South Africa, there are many parallels throughout the Western world to the process by which powerful institutions, such as the state, confer arbitrary racial identities. Classifications – whether of 'West Indians' in Britain, 'Aboriginals' in Australia, or 'Blacks' in the United States of America – differ from the South African example in the degree of force with which they have been wielded, but in kind they bear the same stamp of a majority society conferring identity. In this paper, I will argue that the social definition of the 'Chinese' in Canada as a racial group has been a comparable cultural abstraction, and Chinatowns stand to this day in North America and Australasia in large part as physical manifestations of that European abstraction.

The nature of power in Western society has been the concern of many human geographers in recent years. Some have examined the spatial configuration of a system of economic production which is predicated on endemic conflict between social classes. Power is linked to economic control which is said to permeate, and be reproduced in, the superstructural domains of capitalist society (Dear and Scott, 1981; Harvey, 1985). Elsewhere, the cleavages created by 'consumption sectors' have been identified in order to demonstrate that independent sources of economic and political power reside outside the workplace (Pratt, 1986; Saunders, 1981). Others, also relying on Weberian stratification and political theory, have studied the institutional control of 'gatekeepers' (Williams, 1976), and the influence of coalitions and political movements which attempt to mould urban form in the image of their own interests (for example, Cox and McCarthy, 1980; Hasson and Ley, forthcoming; Ley, 1983).

Much less attention, however, has been paid to the pervasive type of conceptual domination that has been exercised by powerful groups over the *definition* of people and places in Western societies. Yet such a 'power of definition' (Western, 1981, page 8) has structured the society and space of Western countries in critical ways. It has, as one instance, been the property of a hegemonic white European historical bloc that rose to world domination from the sixteenth to the late nineteenth century. By that time, racial theorizing pervaded the social consciousness of Europeans and a rigid epistemology of separation had been constructed between those classified as 'us' and those deemed to be outsiders. It was only the culmination, however, of Europe's long cultural history of perceiving and acting toward the foreign in ways that polarized, rather than blunted, the distinctions.

In this paper, I propose to trace the century-long exercise of cultural domination by one white European community over a set of racially defined outsiders known as the 'Chinese' in Vancouver, British Columbia. Since race is something which must be taken as problematic rather than as axiomatic (Anderson, 1986, chapter 1; Fenton, 1980; Gates, 1985; Husband, 1984; Miles, 1982), it is important to examine the process by which racial categories are themselves constructed and transmitted. Such a concern with the etiology of systems of racial classification has not been widely developed in the social sciences. 'Race' and race 'prejudice' have more often been taken for granted than made objects of explanation. Indeed the long liberal tradition of segregation studies in the geography, sociology, and history of race relations has been much more concerned with the consequences or 'evils associated with racial classification' (Banton, 1977, page 19) than with the race-definition process itself. Elsewhere in urban studies, the assumption of class primacy (Saunders and Williams, 1986, page 398) has produced an oversight or dismissal of such 'non-class-based social identities' as those constructed around the idiom of race.

The cognitive leap that has been made in many Western societies from physical differences to something more fundamental called 'race' has been sufficiently enduring and politically significant to warrant more rigorous attention from social scientists. In the case of the 'Chinese' in Vancouver, the racial category has persisted in different guises in white European culture for over a century from the late nineteenth century right up to the present. That history is replete with evidence of the depth and tenacity of forms of consciousness and social organization which have a 'racial' character. Precisely how this 'collective representation' (Prager, 1982; 1987) has been fashioned and recast in Vancouver society and space is the concern of this paper.

Indeed, there has been nothing inevitable about the pressures that brought the notions of 'Chinese' and 'Chinatown' into European cognition and sustained them in Vancouver culture and territory. The social tensions between people of Chinese and European origin did not 'inhere in the racially plural situation', as Ward (1978, page 169) has argued for the British Columbia context. Race is no once-and-for-all happening that, of its own accord, inspired the 'Sinophobia' about which so much has been written since Allport's (1954) influential work (for example, Lyman, 1974; Palmer, 1982; Price, 1974). The only reality that race has possessed is that which has been socially assigned to colour distinctions. Nor can 'race' be traced 'without remainder' to the production process under capitalism, with its need for a cheap and dispensable labour force (for example, Gabriel and Ben-Tovim, 1978; Greenberg, 1980; Warburton, 1981). Racial consciousness was undoubtedly systematized during the development of capitalist labour markets, and a more ambitious study would examine the *mutual* structuring of economic pressures and the cultural conceptions emphasized here. But the force and resilience of racial representation suggest that its contribution to systems of inequality should be examined on its *own* terms rather than simply 'read off' from putatively deeper causes.

Last, although sensitive to the problematic nature of our cognitive categories, the tradition of symbolic interactionism has, as a whole, been impoverished by its 'eclipse of history' (Zaret, 1980), and its silence on issues of power (Giddens, 1976). With regard to the race question, these are serious omissions indeed. The existence of socially based insider and outsider tendencies (Suttles, 1968) does not alone explain why race became an idiom of exclusion and inclusion in Vancouver society. Nor does it account for the relative power of the reference groups so defined or the configuration of social relations in that setting.

The objectives of this paper, therefore, are threefold: (1) to situate the concept of race in historical and political context; (2) to demonstrate the role of a key set of government agents in the social construction of a racial category; and (3) to trace the articulation of the category at the microlevel or local level, where socially created difference is actually organized and reconstituted in space. First, each of these concerns will be addressed in general terms, after which empirical material from the Vancouver case study is presented in outline form.

The race idea in historical context

Europeans have always interpreted the world with the use of concepts that have dramatized the distinctions between themselves and others. Just as generations of Chinese have long held myths of a civilized in-group (see Tuan, 1974, page 38), so Europeans since Aristotle have regionalized the globe against a romanticized idealized version of 'Europe' itself (Hay, 1966). Already in the fifth century BC, the term 'Asia' connoted 'despotic authority and barbaric splendour' (Dawson, 1967, page 91), whereas Europe was rational, free, civilized, complex, and later on Christian. Such was the system of knowledge developed by European poets, travellers, and intellectuals that served to filter the 'Orient' *for* Europe from antiquity (Said, 1978).

The early antinomies of 'East' and 'West', master and slave, civilized and uncivilized, and Christian and heathen, provided the root for the development of deterministic thinking in Western culture and science. Increasingly, with the extension of Portuguese and British military, industrial, and missionary influence overseas (see Boxer, 1963; Brantlinger, 1985; Ross, 1982), the ethnocentrism of the past gave way to the modern doctrine of racism. Phenotype was gradually added to the European cognitive package, and by the late seventeenth century, skin colour was an independent justification for the enslavement of American 'Negroes' (Jordan, 1968, page 96). At the time, however, 'Negro' connoted heathenism, more than a biological status (Montagu, 1965). Not until the mid-nineteenth century were Western scientists arguing that a *naturally* rooted relationship existed between phenotype, culture, and civilizing capacity. Not only that but a chain of 'races' existed, at the apex of which stood the white race as the supreme measure of biological and cultural worth. It was a dramatic shift, Stepan (1982, page 4) argues, 'from a sense of man as primarily a social being, governed by social laws and standing apart from nature, to a sense of man primarily as a biological being, embedded in nature and governed by biological laws'. By the late nineteenth century, intellectuals and others were so persuaded by the notion of a discrete and immutable racial type that they could see none of the egalitarian conclusions in Darwin's theory of the common origin of the human species. On the contrary, Darwin's evolutionary metaphors seemed to offer an irresistible idiom for old prejudices, so strong was the European will to identify a natural underpinning to the social order (Jones, 1980).

Whatever the precursor, the twin ideas that, first, every human belonged to an essence or type whose physical expression was an index of innate biological and cultural organization, and, second, that the types were ranked by nature in a struggle for survival, were axioms of late nineteenth century science and culture. It was a deterministic and pessimistic turn indeed. But biological determinism tapped a cultural tradition long disposed to viewing the world's populations in typifications. Fredrickson (1981, page 199) has taken this tradition as evidence for his claim that 'Industrial capitalism may be the major cause of social and economic inequality in the modern world, but it makes little historical sense to view it as the source of ideologies directly sanctioning racial discrimination.'

The state and racial ideology

Racial classifications have been cultural and historical ascriptions, but they have assuredly also been political phenomena. Race has not been just another social construction. To use Gramsci's terms (1971), racial ideology has been a 'critical unifying principle' in consolidating and justifying the rise to hegemony of a white European 'historical bloc'. In using the term 'hegemonic', one goes beyond culture defined as a way of life, to insist 'on relating the "whole social process" to specific distributions of power and influence' (Williams, 1977, page 108). A hegemonic culture, therefore, is not simply a configuration of manifestly cultural values or activities, but a system of ideas, practices, and social relations that permeate the institutional and private domains of society. Just 100 years ago, the colony of British Columbia was over 50 per cent 'Indian' (Canada *Census 1880–81*, page 299). But through initial strategies of force to routine forms of immigration regulation and control over the division of labour and access to power, British Columbia imperceptibly but surely *became* the society of white European 'institutional completeness' (Breton, 1964) that it is today.

In such a transformation, the Legislative Assembly established by British officials on Vancouver Island in 1849 and the Provincial Parliament of British Columbia after 1871, assumed ownership of the means of *conceptual control* over the definition and status of settlers and residents in the colony. Clearly, the state must be conceptualized as more than mere government. In Canada, all three levels of government have participated in building a 'white' European society in all its symbolic and institutional dimensions. Recurring rounds of politicians and officials did not simply react to the currents of popular or economic pressures, but actively sponsored and enforced the global extension of European hegemony in Canada.

The relationship between the political and cultural spheres of Western societies has received relatively little theoretical and empirical attention in the recent writings on the state in capitalist countries. Yet, as Breton (1984, page 127) argues,

> The state is involved not only in the economy's management, the pursuit of growth, and the initiation or support of changes found necessary in that sphere. It is also engaged in managing the symbolic system, the protection of its integrity and its adaptation to new circumstances.

The link between culture and power was one of Gramsci's primary concerns, and his work offers some fertile concepts for further theoretically informed work in the field of race relations. Though faithful to the Marxist point of departure that capitalism is a contradictory and historically limited system of production based on the exploitation of wage labour, Gramsci distanced himself from the determinism of the base–superstructure model of explanation. No simple deduction of ideology from deeper production forces was appropriate, he held, because, in organizing action and practice, ideology assumes a dense material basis. 'Mental life is more than a pale reflection of more basic developments in material life', one interpreter of Gramsci has said (Lears, 1985, page 570); 'The link between the two realms is not linear causality but circular interaction within an organic whole.'

Gramsci was absorbed by the problem of the superstructure and, in particular, the manner in which dominant 'historical blocs' articulate the interests of other social groups to their own. To the extent they are successful, there comes a historical 'moment' when the philosophy and practice of a society are fused; when a concept of reality is diffused through a society. The bloc achieves such 'hegemony' not through indoctrination, but as more and more people come to interpret their *own* interests and consciousness of themselves in the 'unifying discourse'. In this process, political office provides the dominant sector with the critical means for 'nationalizing' and legitimizing its intellectual control. The sphere of politics is therefore organically linked with the cultural realm in Gramsci's scheme. The state is not a separate 'level' from civil society, but part of it, evolving and legitimizing the rules and framework within which social life is structured.

Although many interpreters of Gramsci, and indeed Gramsci himself, see the ideological struggle as one ultimately between Marx's two 'fundamental classes', others have found it useful to extend the notion of the 'historical bloc' to other categories that have amassed cultural and economic solidarity (for example, Lears, 1985; MacLaughlin and Agnew, 1986). Potentially hegemonic groups may be bound by religious or other ideological ties, as well as by economic interest. Here it seems helpful to include one of the most influential of socially based hegemonies, that of the white European historical bloc with which this paper is concerned. As I have said, the race idea has been a unifying concept in the evolution of white European global hegemony. With more or less force in different colonial settings, racial ideology was adopted by white communities, whose members (from all classes) indulged it for the definition and privilege it afforded them as insiders. And to the present day in Canada, racial ideology has commanded a following, not as some colonial legacy that has resisted the passage of time, but through daily regeneration in different guises over time.

Indeed, without the moral and legal authority granted racial ideology by successive administrations of the Canadian state, it might not have been so resilient or effective. As Sibley (1981) has argued for all 'outsiders' in Western societies, the state has been widely implicated in their constitution and preservation because it has officially specified the characteristics that socially differentiate them (see Foucault, 1977). In doing so for the 'Chinese' in Canada, the Canadian state has dignified not just a set of benign cultural images, but a politically divisive system of racial discourse. Within the limits of their judicial framework, the levels of state have been the 'private apparatus' of a socially based hegemony, to use Gramsci's words, centrally implicated in the structuring of Canada's social order.

The geographical articulation of racial ideology

Needless to say, the race-definition process has not operated simply in the minds of historical actors. Indeed it will be clear that the study of racial classification is not only about the forms and roots of cultural conceptions, but about the dialectical relationship between consciousness and structure. In such a process, the geographic sphere provides a critical nexus in which racial ideology is ontologically realized and reconstituted. The sphere's importance for shaping social relations is perhaps at its most transparent in South Africa (Western, 1981), while elsewhere there have been influential systems of imagery attached to racially defined places such as 'Chinatown' (Anderson, 1987). Indeed, neither space nor place has been incidental to the structuring of the race-definition process in British settler societies. On the contrary, they have been integrally part of its making and unfolding. As Cosgrove and Jackson (1987, page 99) have written more generally: 'The geography of cultural forms is much more than a passive spatial reflection of the historical forces that moulded them; their spatial structure is an active part of their historical constitution.' This is not because of any *intrinsic* properties of space and place. Their structuring role cannot be asserted on logical grounds, but is contextual and must be empirically demonstrated.

Within the social geography of the postpositivist era, there has been a tendency to treat history as past time, as a backdrop to the present, as if the present were almost an autonomous creation. The advantage of an analysis that is situated in both space and time is that the mutually confirming relationship between the 'ideal' and the 'material' can be demonstrated and demystified. In the Vancouver example discussed below, I seek to show that behind what is often taken to be the objective category of race has been an historically evolving dynamic between racial discourse, Chinatown, institutional practice, and power. To that end, I begin my interpretation in the late 1870s when 'race' was a key organizing principle in Vancouver society and space, and conclude in 1980 when a new form of government targeting of Chinatown carried the racial category forward.

The 'Chinese race' and the Canadian state: legislating a category of outsiders, 1875–1903

The presence of people of Chinese origin in British Columbia was associated with the main pioneering industries of the colony. Gold discoveries in the middle course of the Fraser River in 1858 attracted the first substantial contingent of approximately 2,000, and when the boom subsided, wage labour was plentiful in the sawmills of Burrard Inlet, the canneries on the Fraser and Skeena rivers, the coal mines at Nanaimo and Wellington, and in other unskilled occupations elsewhere. In each of these capacities, Chinese workers were paid one half to two thirds the wage rates of other workers. In addition, hours of work were long and much of it seasonal. Not surprisingly, therefore, the predominantly male settlers were roundly praised from some quarters of British Columbian society in the early years of the colony's development. Entrepreneurs coveted what was held to be the uniquely 'docile and industrious' Chinese labour, and, after hearing their evidence in 1885, Commissioner J. Gray, a Supreme Court judge, went so far as to say, 'The Chinese in British Columbia, as affecting the rapid development of the country, are living machines' (RCCI, 1885, page LXX).

But the praise heaped upon residents from China must have afforded them little comfort. Vancouver entrepreneur Henry Bell-Irving of the Anglo-British Packing Company expressed a typically paternalistic view when he said in 1901, 'It is the destiny of white men to be worked for by the inferior races' (RCCJI, 1902, page 144). Only in those industries where Chinese secured an early monopoly, such as laundering and vegetable gardening, did they escape almost uniform relegation to the lowest occupational tasks. 'It was the natural outcome', Judge Gray said, 'of the dispensations of Providence by which the highest good can be obtained for mankind' (RCCI, 1885, page LXXIII).

The alien status of settlers from China was still more bluntly apparent in the conduct of white labourers. Without regard for the demonstrated willingness of early Chinese workers in British Columbia to strike, white workers shared their employers' belief that the Chinese were endowed *as a category* with a natural capability to undersell their own more deserving labour. 'The degraded Asiatics . . . live crowded together in such numbers as must utterly preclude all ideas of comfort, morality and decency', said a representative of the Nanaimo Knights of Labour in 1885 in what was to become a familiar charge over the decades (RCCI, 1885, page 156). For the white workers of British Columbia, then, just as for the colony's entrepreneurs, Chinese pioneers were somehow beyond the body of eligible citizenry to Canada. It was no blind 'prejudice' on their part, but rather a matter of their *respect* for the ideological formation of the day which deemed people from China unalterable outsiders.

In the hands of British Columbia's early politicians, the race idea was a persuasive symbol around which to rally communal consciousness after the colony joined Confederation in 1871. Mr J. Robson, Provincial Secretary, declared that the anti-Chinese agitation 'commenced as a political question in 1872' (RCCI, 1885, page 72), as did Surveyor-General B. Pearse who claimed it 'was begun and carried forward chiefly by politicians' (RCCI, 1885, page 95). Theirs was the expedient political ambition to win electoral support and the moral mandate to build a British outpost on the Pacific Coast. And both missions seemed to require and justify targeting of the 'Chinese' who signified in British Columbian culture – above everything that might have been said about them – *non*-white, *non*-Christian, 'them' as opposed to 'us'.

In 1875, the Provincial Government recognized the exclusive claims of 'whites' to insider status when it removed the voting rights of 'Chinese' (and native 'Indians') (BCS, 1875). In addition to the symbolic implications of disenfranchisement, the Act precluded Chinese from entering most professions. Other forms of employment were restricted from 1878, when a clause was inserted in provincial public works contracts prohibiting the employment of Chinese labour in government-assisted projects (JLABC, 1871, volume VII, page 82). Nor could Chinese buy or lease Crown land (BCS, 1884). Further to these forms of legal targeting, the provincial administrations of the late nineteenth century lent moral force to an undifferentiated conception of the immigrants from China. Committees were appointed to report on the 'baneful presence of the Chinese in our midst' (JLABC, 1879, volume VIII, page xxix), and numerous resolutions were drafted for transmission to the Federal Government in Ottawa. Their primary objective was to protest 'the hordes of Chinese [who] . . . carry with them the elements of disease, pestilence and degradation over the face of the fair land' (British Columbia, 1885, page 2). The intention was also to urge Ottawa to 'prevent the province being over-run with a Chinese population to the injury of the settled population of the country' (JLABC, 1876, volume V, page 46).

There were limits on the political will of the provincial authorities, however. Following the precedent set by the Supreme Court of British Columbia over the Chinese Tax Act of 1878, at least three attempts by different administrations to impose immigration restrictions on the 'destructive incursion of Asiatics' (JLABC, 1889, volume XXVIII, page 10) were disallowed by Ottawa's Minister of Justice (La Forest, 1955). As Fredrickson (1981, page viii) has written of another context, 'White supremacy was no seed planted by the first settlers that was destined to grow at a steady rate into a particular kind of tree.' The judicial constraints on British Columbia's anti-Chinese policy-making, for example, were more intrusive than were those on the separately governed colonies of Australia (see Markus, 1987).

It was the Parliament in Ottawa, however, that had ultimate power over the definition and status of people from China in Canada. Few politicians were more sympathetic than Prime Minister Sir J. A. MacDonald, but his first loyalty was to the Canadian Pacific Railway Company whose project to build a national railway had much to gain, he believed, from the uniformly cheap 'Chinese' labour. Besides, Chinese were not 'permanent settlers' (Canada Parliament, 4 May 1885, page 1589), MacDonald assured the Pacific Coast lobby during a session in which he introduced an amendment to the Franchise Act excluding any 'person of Chinese or Mongolian race' from the federal vote (CS, 1885a).

Only when the railway was completed did the MacDonald government see fit to wield the racial category in a manner more acceptable to the British Columbian members. Their concession finally came in July 1885, when the Federal Government decided to pass the Act to Regulate the Immigration of Chinese to Canada (CS, 1885b) which effectively divided immigrants to Canada into two classes – one comprising 'Chinese' who were obliged to pay a $50 entry tax, and another composed of non-Chinese who could expect to become 'Canadian'. By 1903, Prime Minister Wilfrid Laurier was so convinced of the 'invincibility' of 'race antagonism' that he agreed, despite the protest of the president of the Canadian Pacific Railway, to raise the entry tax on Chinese to a massive $500 (RCCJI, 1902, page 210; see Canada Parliament, 5 May 1903, page 2399). Henceforth, Chinese family settlement in Canada became the preserve of an elite few.

The classification 'Chinese' was based on an appeal to origins that justified a cultural–political order created by and for Europeans. Parkin (1979, page 85) has emphasized the importance of the state in such category legislation. 'In all known instances where racial, religious, linguistic, or sex characteristics have been seized upon for closure purposes', he argues, 'the group in question has already at some time been defined as legally inferior by the state'. Federal officials, like their provincial counterparts, saw it as both their moral calling and as a means of political legitimacy to secure a white European order against those who would pass their deficiency on. By the close of the nineteenth century, the morality of separation had all the force of national official status, and as we shall see at the local level, all the confirmation of territorial boundaries.

Amid the growth and change wrought by the completion of the trans-Canada railway at its western terminus in 1885, the once peaceful relations between the small settlements of whites, Chinese, and Indians on Burrard Inlet grew competitive and violent. The first Mayor of Vancouver, Mr A. Maclean, lost little time when a 'small horde of unemployed Chinamen' tried to locate in the emerging business section on Water Street. It was the 'thin end of the wedge', a news report stated (*Vancouver News* 2 June 1886, page 2), and, in the same month, he and Alderman Hamilton presided over three street meetings convened to find means of preventing Chinese settlement (Roy, 1976, page 45). By November, with Chinese scattered in various locations,

Hamilton pressed in Council for further 'action in trying as far as possible to prevent Chinese from locating within the city limits' (CVA *Council Minutes* volume 1, 8 November 1886, page 164). To aid the cause, Council prohibited the employment of 'any person of Chinese race' on municipal contracts or City-assisted projects (see CVA *Incoming Correspondence* volume 1, 8 August 1887, page 82).

The City was helpless, however, to constrain people like Mr J. McDougall who hired 'batches of Mongolians' in 1886 for his clearing contract in Vancouver's West End (*Vancouver News* 9 January 1887, page 4). Tension built among the city's white population, and by late February, after a crammed meeting at City Hall, some 300 of them tried their own hand at ousting the Chinese. While the authorities looked on, they attacked the West End camp, after which they burned the homes of a settlement of Chinese on Dupont Street (*Vancouver News* 25 February 1887, page 1). Ironically, it was the Provincial Government in Victoria, not known for its sympathy to the Chinese, that intervened on behalf of the victims. The Attorney-General looked unfavourably on Vancouver's decline into mob rule, and, within days of the riot, he suspended the Vancouver City Charter and sent constables to take charge of law and order (BCS, 1887).

With this protection, the dispersed Chinese returned to Vancouver and established a concentrated pattern of settlement in the swampland around Dupont Street, later Pender Street (Figure 10.1). Mr Justice Crease explained their 'tendency to congregate' as 'directly owing to the fact that as foreigners, held in dangerous disesteem by an active portion of whites, they naturally cling together for protection and support' (RCCI, 1885, page 143). Chinese

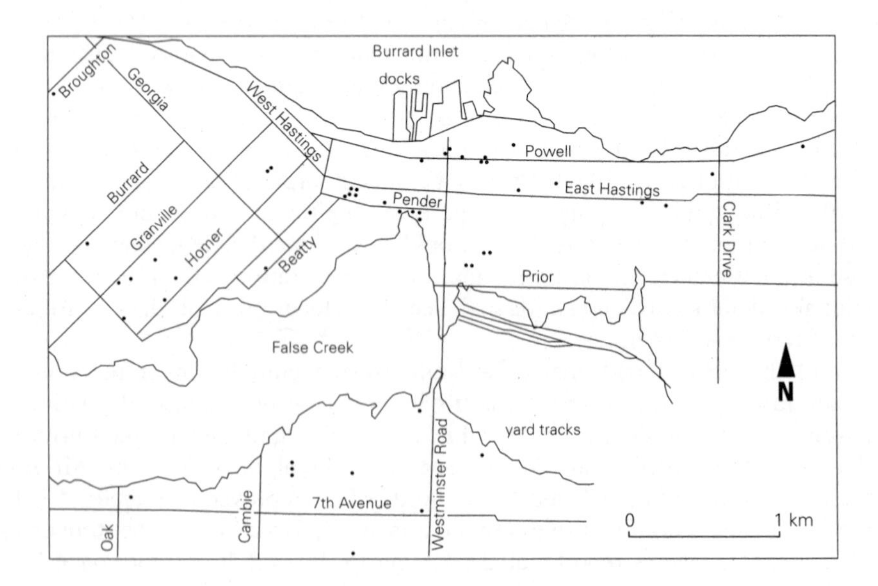

Figure 10.1 Distribution of Chinese in Vancouver, 1910 (dots indicate Chinese residential, commercial, or associational use; based on *Henderson's Directory* 1910)

spatial concentration was also a product of the host of *secondary* effects that their racial definition supported. The shanties around Dupont (later Pender) Street were home to a pool of politically vulnerable, single, and poorly paid men who were often dependent on Chinese bosses for employment, lodgings, and in some cases, their head-tax payment (Chan, 1983). Merchants lived more comfortably, often in elegantly decorated homes (see *Province* 5 February 1910, page 4), but they too were constrained in their residential choices and forms of livelihood by the prevailing culture of race.

Dupont Street as 'Chinatown': the ideologies of race and place in Vancouver, 1886–1920

The spatial manifestation of the racial category in late nineteenth century Vancouver was only part of the Chinatown story. Equally important in the construction of the racial category were the early images that gave the district a discreteness in the social consciousness of its representers – regardless of how Chinatown's residents defined themselves and their new home (Anderson, 1987). Well before any substantial settlement of Chinese was identified as such in Vancouver, a 'place' for them already had a distinct reality in local vocabulary and culture. 'Had Dante been able to visit Chinatown, San Francisco', said Secretary of State J. Chapleau in 1885, 'he would have added yet darker strokes of horror to his inferno' (RCCI, 1885, page 369). Chinatown was a peculiarly Oriental phenomenon, Chapleau insisted, because unlike the familiar situation where 'those who inhabit Whitechapel are the dregs of a population, thousands of whom live surrounded by the most refined civilization . . . Chinese immigrants will herd together in a quarter of their own' (RCCI, 1885, page LXXX).

Such official views attuned white citizens of Vancouver to the 'evil [that] is only beginning to take shape' (*Vancouver News* 7 December 1886, page 2) on Dupont Street in the period leading up to the 1887 riot. And within months of the violence, the *News* was calling for 'strict surveillance by the City . . . to prevent the spread of this curse' (1 May 1887, page 4). For at least thirty years, Vancouver's elected officials, medical health officers, license and building inspectors, and policemen adopted this mandate. European conceptions of civilization and propriety were their guide, and it was against these benchmarks that 'Chinatown' received its cultural profile and public status.

The Chinatown idea was made up of European assumptions about the inherent dirtiness, amorality, criminality, cunning, and unassimilability of 'John Chinaman'. Health Inspector Marrion was especially well versed in the language of race, and from the time of his appointment in 1893, many shacks in the area were destroyed and owners served notices under the health, boarding house, and laundry bylaws (see CVA *Incoming Correspondence* volume 17, 26 November 1900, page 13298). Chinatown was also listed as a separate category in the medical health officer rounds (CVA *Health Committee Minutes* 1899–1906) – a designation which facilitated and justified neighbourhood targeting well into the twentieth century, despite the fact that no evidence for diseases originating in Chinatown was ever provided in the civic records or the vigilant local press. By 1915, the guardian of Chinese residents in Canada, Ambassador Lim, had taken to publishing warnings of civic visits in the Chinatown press, so strict had the City's surveillance of bylaw standards become (for example, see *The Chinese Times* 25 February, 4 March 1915).

'Race' and 'place' were also conflated in a set of moral equations. In the eyes of many white Vancouver residents, including aldermen, there was an irresistible relationship between Chinatown and vice. A 1912 profile of 'Vancouver's plague spot', for example, described men stupefied by opium, engrossed in gambling, surrounded by 'stagnant air', eating 'dead birds', and worshipping strange idols, while 'two white women repos[ed] on couches'. It was a 'most repellent sight', wrote the reporter, informing Chinatown's first tourists that the most 'loathsome spectacle' was to be had in the 'small hours of the morning' when the majority were at home 'passed out' (*World* 10 February 1912, page 6). It was only for a *few* qualities, then, that Chinatown was known – the ones that in their selectivity revealed more the desire to characterize Chinese residents as alien than some blind impulse toward 'prejudice'.

The cultural leadership provided by City Hall lent considerable influence to understandings about the Chinese and their place in Vancouver society. In the case of gambling, for example, raids on Chinatown were routine for the five decades from 1890 to 1940 (see Won Alexander Cumyow Papers, Court Notebook, Box 2). Such attention legitimized the belief that Chinatown was united by a habit 'ingrained in the Celestial nature' (*Province* 5 February 1906, page 15), and served to obscure the conflict over gambling that actually existed in Chinatown between Chinese Christian missions, resort operators, and their customers, some of whom became police informants and almost all of whom showed themselves quite capable of dispensing with the 'habit' come the Sino-Japanese War in the late 1930s, when they were urged by neighbourhood organizations to give generously to the war effort of their homeland (*The Chinese Times* 22 December 1917, 16 January 1939).

In so defining and targeting Chinatown, the European authorities of early Vancouver ensured that the racial category 'Chinese' would be carried forward in the society and space of their city. 'Chinatown' was *their* unit of knowledge and its 'Chineseness' belonged to them. Both notions continued to be reinvented at the moral and material levels in the years to come, and, while various governments were mapping the career of the Chinese, they would at once be defining the insider community whose boundaries and privilege it was their ambition to protect.

Chinatown as a maximum entitlement: the consolidation of the racial category, 1920–35

Samuels (1979, pages 71–72) has claimed that 'landscape impressions although subjective in origin . . . acquire an objective content in so far as they have a history of authorship, diffusion and impact'. In Vancouver, 'Chinatown' was one such abstraction to shape practices that in turn further objectified the idea and its material referent. As we have seen, Chinatown was something of a 'counter-idea' (Voegelin, 1940) into which were herded things held to be in opposition to mainstream society. It was 'their' home away from home, 'their' doing, 'their' evil. So in the late 1910s, when some ambitious Chinese merchants attempted to breach the moral order of race and place with moves to the suburbs, the disturbance was a particularly jarring one. In the context of heightened racial self-consciousness after World War 1, Chinese mobility represented a 'yellow peril' of the most undignified order, and white citizens soon looked to their governing bodies for legislative solutions.

Between 1890 and 1920, the race idea received fresh sanction in North America, Britain, and Germany from a branch of science and a social programme called 'eugenics' (Haller, 1963; Livingstone, 1984). Essentially, the science argued that the selection in nature of adaptive physical characteristics applied equally to behavioural, mental, and moral conditions. Of course, biologists had long believed that innate differences in ability and behaviour separated the world's populations, but eugenics gave the idea of the pure racial type more precision by explicitly linking it to Darwinian concepts of fitness and heredity. Humankind could even improve its quality, scientists argued, by promoting the breeding of the 'fit', discouraging the reproduction of the 'unfit', and preventing deterioration of the organism through 'intermarriage' with distant 'strains'. In practice, much dispute about the mechanisms of genetic transmission diluted the force of the science as a *science* (Stepan, 1982, chapter 5), but methodological disputes did not challenge the scientific and lay faith in the priority of heredity over environment.

By the late 1910s, the retail and property-owning sectors of British Columbia's European community had taken on an advocacy role in the anti-Chinese agitation. For them, the wartime 'infiltration' of Chinese challenged the sociospatial order that was their hold on power and privilege. The Vancouver Board of Trade, for example, told Ottawa's 1921 Special Oriental Immigration Committee,

> During late years the matter has become serious in that the Oriental is no longer content to seclude himself in the Chinese or Japanese quarters of the towns of the Province, but is either occupying land in advantageous localities or is branching out in the retail or wholesale trades in the best districts in the cities. . . . We strongly feel that we should retain British Columbia for our own people (CVA *Additional Manuscript 300* volume 1, 21 July 1921, page 106).

The Minister of Lands, Hon. T. Patullo, was quick to support the Board on the grounds that 'biologically there is a great difference between white and Oriental races. We must trade together', he said, 'but I think it is sufficient to suggest that we should occupy our own spheres' (*Colonist* 2 November 1921, page 7). Attorney-General A. Mason was also concerned for 'race hygiene', and used this theme effectively between 1921 and 1925 to persuade the province's major railway companies and some manufacturing industrialists to employ only white labour (for example, *Colonist* 5 August 1922, page 21). Premier J. Oliver assisted too, by reenacting a 1902 order-in-council that excluded Chinese (and Japanese) from employment on Provincial Government contracts (JLABC, 1920, volume XLIX, page 246).

In Vancouver, the worst casualties of European territoriality during the postwar years were itinerant Chinese peddlers, whose 'tremendous energy and frugal habits of life' (Johnston, 1921, page 316) were thought to present 'unfair competition' to white grocers. For years, Vancouver's shoppers had happily avoided long trips to the City market by buying produce from the peddlers. But the Vancouver Retail Merchants' Association was an influential lobby, and in 1915 Council appeased it by imposing a hefty annual licensing fee on peddling (*The Chinese Times* 26 May 1915). Three years later, neither legal action on the part of the peddlers nor appeals from the Chinese Consul-General dissuaded Council from raising the fee to $100 (CVA *Incoming Correspondence* volume 69, Yih to Finance Committee, 13 December 1918).

The civic persecution of peddling had an unintended consequence of graver proportions for the retail lobby. By the late 1910s, with an assured suburban market, Chinatown merchants began to open small stores with attached residences in locations more convenient to their clients. In 1920, thirty of the eighty-eight Chinese grocers and as many as twenty-five of the thirty-one Chinese greengrocers in Vancouver were located outside Chinatown (CVA *Incoming Correspondence* volume 126, L File, 1929). 'John Chinaman's' entrepreneurial spirit had been unleashed, Alderman J. Hoskins warned at a meeting of the Grandview Chamber of Commerce in February 1919, and he 'promised his every endeavour to stamp out the evil' (cited in CVA *Incoming Correspondence* volume 74, C File, Yih to Mayor, 28 February 1919). The Central Ratepayers' Association and the merchants of Upper Granville, Davie, and Robson Streets also made a case to Council (*Province* 20 February 1919, page 12) which responded with a resolution calling for immigration exclusion of Chinese and an investigation into the possibility of confining 'Asiatic retail business' to 'some well defined area' (CVA *Council Minutes* volume 22, 8 March 1919, pages 413–414). Such official support did not, however, guarantee the municipalities their legislation. Legal limits on the City's will denied them formal means, but, in select suburbs, the City turned a blind eye to restrictive covenants until the late 1940s when the legal validity of such gentlemen's agreements was finally challenged.

By the early 1920s, Vancouver displayed all the excesses of a solidly rooted cultural hegemony fuelled by eugenicist prophecies about the dilution of the 'white' race. Civic harassment of Chinatown continued apace, and from 1920 the image of Chinatown as an opium den became assimilated into a more fearful perception of the territory as a dangerous narcotics base. It was a charge British Columbia members of Parliament, Mr H. Stevens and Mr L. Ladner, exploited to advantage in 1922 when they described Chinatown's 'snow parties' to the House of Commons (Canada Parliament, 8 May 1922, page 1529). In the same year, the Asiatic Exclusion League was revived in Vancouver and every household was canvassed for support, 'with considerable success' (PAC, 1922, Jolliffe to Scott, 7 February).

The 1920s represented the height of biological determinism in Canada. In the House of Commons, more consensus on the wisdom of anti-Chinese policies existed during that period than at any other in its history. The challenge in the eyes of almost all members was a kind of negative eugenics: to keep the national organism free of those whose 'stock' threatened to undermine the whole. The agenda was presented as a patriotic one, and no agent was more keenly committed to it than Prime Minister W. K. Mackenzie King. When 'two kinds of metals are in circulation as coinage', he said in 1922, 'if one [is] of finer quality than the other, the baser metal tend[s] to drive the finer metal out of circulation' (Canada Parliament, 8 May 1922, page 1559). By November of that year, with concern mounting among administrators over loopholes in the Chinese Immigration Act, his government saw fit to close Canada's doors to Chinese immigration altogether (CS, 1923).

The exclusion legislation was an extreme form of official sanction of the racial category, 'Chinese'. Although at the civic level it had not been possible to administer a direct remedy to Chinese mobility, the power to further engrave the social and spatial contours of a hegemonic European order was supremely exercised at the federal level. If leakage from the ghetto could not be directly plugged, at least flow into Chinatown could be stemmed at the source, and for the following twenty-five years, only a handful of Chinese entered Canada. Meanwhile in Vancouver, people of Chinese origin were left to decline into the depths of the Depression when seventy-five Chinatown residents died of starvation from inadequate relief provisions (CVA *Incoming Correspondence* volume 198, 1935, Unemployed Organizations File).

'The Little Corner of the Far East': an expanding interpretation of the racial category, 1935–49

The study of race relations has typically been concerned with situations of negative ascription and conflict. Rex's views can be taken as representative when he states that one of the 'necessary elements' of a 'race relations situation' is 'abnormally harsh exploitation, coercion and competition' (1982, page 198). Such a narrow conceptualization of racism has for many decades limited the types of questions asked in the field of race relations. In particular, it has foreclosed the possibility of seeing connections between racially defined places such as Chinatown and deeper ideological processes: of seeing the continuity of the European racial frame of reference beneath changes in its expression over time.

During the Depression, Chinatown found its first allies. Abdication by the state of responsibility for the Chinese became, in the eyes of some nascent political reform groups, an obstacle to the ideological struggle against a stratified society. Such lobbies as the Trades and Labour Congress and the Vancouver Trades and Labour Council began to argue that legalized discrimination ultimately militated against the long-run class solidarity of all workers (Ireland, 1965). In other quarters, concerned Canadians began to speak out against the Chinese exclusion legislation and the denial of political rights to minorities and their local-born children (for example, *Province* 13 November 1934, page 8; Angus, 1933). The period was one of critique and unrest, and out of it came an expansion of the historically established body of theory and practice that delivered 'Chinatown' to white Vancouver.

By the mid-1930s, a somewhat innocuous vision of 'Little China' began to be articulated in Vancouver society that fed on age-old fantasies about China's ancient and venerated civilization. 'In our little city-within-a-city', said a *Province* feature in May 1936, 'yellow gods rule and the Occident fades into the background' (21 May 1936, page 43). At a time when Chinatown's facade was in fact becoming more 'westernized' and 'modernized' (*Province* 11 October 1929, page 6), Europe's once romantic conception of the East filtered back into Western consciousness (Dawson, 1967, chapters 1 and 6). The new mood in Vancouver society was signalled by Council's approval in May 1936 of an application from the Chinese Benevolent Association to 'erect Chinese buildings adjoining the south-east corner of Pender and Carrall during the Golden Jubilee Celebrations' (CVA *Council Minutes* volume 37, 11 May 1936, page 265). The City was captivated by the appeal to opulent Old Cathay, and Chinatown merchants, eager to find ways of alleviating the rigours of Depression, emulated the route to profit and status that had proved effective for their counterparts in San Francisco and New York City (see Light, 1974; Salter, 1984). Their guide to Chinatown thus promised 'hundreds of Oriental splendours' and '100 per cent Celestial atmosphere' (Yip, 1936).

Vancouver's 'Chinese village' was officially opened on 17 July 1936 by Mayor M. McGeer who spoke of the City's pride at 'the work our local Chinese have done toward the establishment of a Chinese village for the Jubilee'. The moment was captured in a front-page photograph of the Mayor on the arm of 'Chinese queen', Grace Kwan (*Province* 18 July 1936, page 1), and for two weeks, the Vancouver press ran praiseworthy reports on the quaint village where 'East and West are meeting each other' (*Sun* 2 August 1936, page 7). So popular was the display that the City's Jubilee Management Board decided to extend it for two weeks, after which the Board paid tribute to the 'contribution of our Oriental friends' (CVA, 1936).

The Jubilee event was a form of incorporation of the Chinese that relied on historically established conceptions of this undifferentiated category of people. Although more benign than the earlier vice definition, the new imagery expressed and confirmed the same epistemological separation between 'us' and 'them', and the definition of Canada as a white European society to which the Chinese and Chinatown were exotic contributions. With the assistance of the Chinese themselves – who chose to act in the dominant categories when it served their own interests – Vancouver's Chinatown was becoming a European commodity. Indeed within two years of the Jubilee celebrations, Vancouver's Chinatown was officially opened up to tourism (*Sun* 16 April 1938, page 14). 'No other city of Canada', boasted a news feature in 1940, 'possesses a Chinatown which has retained the glamour, fascination and customs of the parent country to the same extent as this Little Corner of the Far East' (*Province* 6 July 1940, page 8).

The European definition of Chinatown was not a simple linear process in which one image was born to be replaced by a second which was in turn transformed by a third. Rather, 'Chinatown' became a store of competing conceptions, where residues of the vice definition informed, and became sedimented within, the later tourist definition. In the same year as the Jubilee celebration, for example, a group of hard-line Vancouver officials launched a crusade in Chinatown that resembled far more closely the civic posture of the past. In their opinion, it was not just the Chinese, but some whites who needed to be kept in their properly appointed place.

For the purpose of attracting non-Chinese customers and maintaining a cheap supply of employees, Chinese restaurateurs, like others in Vancouver in the 1930s, saw the advantage of employing young white women. Only in Chinatown, however, did Alderman H. Wilson and Chief Constable W. Foster see a sinister motive afoot, and in 1935, Foster invoked the Act for the Protection of Women and Girls against three restaurants in Chinatown which employed such women (CVA *Council Minutes* volume 36, 6 November 1935, page 682). Action was needed on 'moral grounds', the retired Colonel said, because waitresses were being 'induced to prostitute themselves with the Chinese' (CVA *Mayor's Correspondence* volume 37, Foster to King, 18 July 1938). By the end of November 1936, eight Pender Street restaurants (employing a total of twenty-nine women) had been served notices under the Act to dismiss their white labour (Foster to King, 18 July 1938). Council also rallied to the cause, and on 14 September 1937 – just one year after Mayor McGeer had publicly praised 'our local Chinese' – it gave authority to cancel the licenses of three Chinatown restaurants (*Sun* 16 September 1937, page 1). The new mayor, G. Miller, announced he was 'out to clean up Chinatown' and dismissed a delegation of waitresses who, in the most revealing protest of all, marched to City Hall to condemn the actions of the 'self-appointed directors of the morals of the girls in Chinatown' (*Sun* 25 September 1937, page 1).

For the following ten years, Chinese occupational and residential mobility in Vancouver became Alderman Wilson's major political cause (see Halford Wilson Papers). Increasingly, however, his fundamentalist invocations of the race idea lost favour; for example, in 1940 the chairman of the Private Bills committee of the Provincial Government rejected Council's request for greater power to control licensing in Vancouver as a request to 'persecute Orientals. Substitute the Jews for Chinamen', he remarked, 'and you are copying Hitler' (*Province* 7 November 1940, page 24). Similar charges were soon heard in Council and the House of Commons, and, within a relatively brief time span, most forms of legally enforced discrimination were dismantled. In 1947, the conspicuous Chinese exclusion legislation was repealed (if not every discriminatory clause) (McEnvoy, 1982), and two years later, voting rights were extended to Canadians of Chinese origin (Lee, 1976).

In the wake of such reforms, the European perception of the 'Chinese' as an exotic people congealed. By the late 1940s, Foster's vice characterization had been submerged (but not dissolved) by the new romantic connotation. Chinatown was 'a glint of the Orient in an Occidental setting' (*Sun* 1 May 1943, page 9), and Pender Street and its environs were speedily adapted to match their representation. The tourist and restaurant industry blossomed during the 1940s, encouraging a vertical expansion of grocery stores, butcher shops, fish markets, import outlets, and curio stores (Gilmour, 1949). Merchants added neon exteriors to their stores as part of what Light (1974) has called the 'purposeful Orientalizing' of North America's Chinatown. Notwithstanding the rise of liberalism in the immediate postwar era, 'Chinatown' still embodied, and was being constructed to project, the irreducible essence of a foreign race and culture.

The legacy of a defenceless past: rebuilding the 'slum', 1958–70

As long as Chinatown's 'Chineseness' was its defining characteristic, the neighbourhood of Chinese-origin settlement in Vancouver remained vulnerable to external agendas. In 1954, the National Housing Act was amended to allow a federal contribution of 50 per cent to the cost of housing acquisition, slum clearance, and public-housing projects in Canada (Rose, 1971). Such enabling legislation paved the way for local bureaucracies to identify areas in need of 'rationalization', and in Vancouver, the first area designated was Strathcona, east of Main Street, and home to the majority of the city's Chinese (Marsh, 1950). 'Area A', as it was classified, was a 'revenue sink' of 'critical town planning importance', whose housing was in a state of 'chronic deterioration', the 1950 report stated. Moreover, while the area west of Main Street 'can be said to be a tourist attraction, the remainder of the Chinese quarter to the east of Main Street is at present of significance only to the people who live there' (CVTPB, 1956, page 26).

In February 1958, the Vancouver City Council approved in principle a strategy for bringing the area east of Main Street into 'more productive land use' (CVTPB, 1957, page 111). The residents initially welcomed the prospect of improved public services, but in time it became clear that improvement would entail demolition of their homes and their relocation to high-density public housing. Confiscation of property would leave homeowners 'only tenants', while discrimination restricted homeownership possibilities elsewhere in the city, charged the spokesman for the newly formed Chinatown Property Owners Association (CTPOA) (*Province* 6 April 1959, page 21). Alderman Wilson gave his blithe assurance the plans were 'not complete', but by then, Council had rezoned part of Area A to residential use as its first step toward implementing the project (CVTPB, 1959, page 9).

The postwar era was an optimistic age of modernization in the 'free world' when it was believed the ghettos of the inflexible past would be 'assimilated', and the benefits of unprecedented economic growth enjoyed perhaps by all. Progress was as inevitable as 'prejudice' had been in the eyes of Prime Minister Wilfrid Laurier, and throughout North America racially defined and low-income neighbourhoods were renewed in its name (Mollenkopf, 1975). In Vancouver, Council no sooner heard a protest from a contingent of fifty representatives for the CTPOA in October 1960, than it decided to hold a special meeting on Stage Two of the Strathcona project (CVA *Council Minutes* volume 74, 4 October 1960, page 675). 'We're sure the Chinese will be very surprised when they find they can have modern Western accommodation at prices they can afford', Director of Planning, Mr G. Fountain reassured the protesters when expropriation began in 1961 (*Chinatown News* 18 February 1961, page 6).

The project for Stage Two earmarked for demolition blocks that were in intensive residential use by Chinese. Not surprisingly therefore, there were protests from over twenty Chinatown organizations (CVTPB, 1963). In response, Mayor W. Rathie conceded one block of Area A to private development designed by Chinese architects, after which Council approved the second clearance scheme (*Sun* 23 January 1963, page 23). But the federal housing authority with ultimate control over all renewal projects never had any intention of allowing private developments in Strathcona. Not even the Chinese proposal to build 'an Oriental City' of private family units persuaded the Federal Government against the original concept (CVTPB, 1964). Significantly, such an attempt by the Chinese to manipulate the Chinatown landscape according to European images – as leverage for their *own* interests – proved ineffective as long as it conflicted with the agenda of more powerful architects. In 1964, the City began to acquire property under Stage Two of redevelopment, while west of Main Street, in a crude juggling of images, the City flirted with a short-lived plan to strengthen Chinatown's tourist potential (CVPD, 1964).

The decision of the City of Vancouver to implement a massive and untried public venture in Chinatown in defiance of the residents' appeals highlighted the long-standing tendency of the dominant society to perceive *the* Chinese as typifications – as objects for the projects of others. Council might have rested easy in the knowledge that the City was providing housing. But in June 1967, with the residential base of Chinatown fast diminishing. Council's first commitment was bared, when it elected to build an eight-lane freeway down Chinatown's west wing (Pendakur, 1972, page 60). In so doing, it galvanized a 'long-divided, fragmented and quarrelsome community' (*Chinatown News* 3 November 1967, page 3) with new heights of fear and indignation.

At that time, however, Vancouver's Chinese community was joined by an unexpected set of allies with a fresh interest in Chinatown. 'If the freeway comes about, it will destroy Vancouver's largest tourist attraction and ancient history landmarks', said an architect from the influential firm of Birmingham and Wood in June 1967 (*Sun* 10 June 1967, page 7). Sensing a new wave of external interest in Chinatown, a delegation from the Chinese Benevolent Association submitted a plea to Council that demonstrated again how the Chinese themselves participated in the culture of racial representation: 'Chinatown is a tradition, a landmark and a major tourist attraction. . . . Vancouver's Chinatown has potentials of being the largest Chinatown in North America' (CVA, 1967). Although rattled by the delegation's popularity, Council was unconvinced, and after instructing the freeway consultants to consider alternative routes, reaffirmed its preference for Carrall Street (*Sun* 18 October 1967, page 17).

With that decision, the City of Vancouver became the target of an urgent round of protest. This hailed not only from Chinatown, whose spokesman, Mr Foon Sien, saw the freeway decision as 'the latest abuse in 81 years of discrimination' (*Province* 2 December 1967, page 9), but also from academics, lawyers, and community workers. Within a short space of time, Chinatown became a liberal cause célèbre through which to challenge the prevailing logic of progress. By late 1967, reformist agitation had made the freeway politically suicidal for the incumbent administration, and early in the new year, Council voted to rescind the project (Pendakur, 1972, page 73). Even Alderman Wilson conceded: 'I can see their complaints now, it would wipe out Chinatown' (*Province* 12 December 1967, page 8). A year later, in a remarkable turn in Chinatown's fortunes, the Federal Government imposed a freeze on spending for redevelopment in Strathcona and accepted the proposal of a local organization to 'rehabilitate' the area back to private residential use (see CVA *Additional Manuscript 734*).

Chinatown as an 'ethnic neighbourhood': invoking the benign myth, 1970–80

During the 1950s and 1960s, there was a gradual shift in emphasis in Canada's language of race from the presumed biological differences between the world's populations, to the 'ethnic' or cultural distinctions that were also assumed to separate them. That which was once the fearful embodiment of an alien and inferior 'stock' had by 1969 become – in the words of a government report – 'an inestimable enrichment that Canadians cannot afford to lose' (RCBB, 1969, volume 4, page 3). In 1971 a Federal Government policy of multiculturalism gave official recognition and financial support to Canada's 'ethnic groups', the 'other' (non-British and non-French) communities 'that give structure and vitality to society' (Canada Parliament, 8 October 1971, page 8545).

'Multiculturalism' was a harmonious metaphor for fashioning a nationalist ideology in Canada (Moodley, 1983), but what went unacknowledged was its subtle assimilation of the long-standing European tendency to impute qualities to national types. In 1970, Vancouver Citizenship Court Judge Oreck praised the Chinese as 'industrious, fastidious and courteous' (*Chinatown News* 18 January 1970, page 51). The point is not so much the 'masked negativity' of such 'positive stereotypes' (Wong, 1985), but the manner in which cultural conceptions about race were being recreated by those who controlled the knowledge about, and practice toward, people of Chinese origin in Canada. Liberalism had long erased Darwinist notions of a race hierarchy, but faith in the 'Chinese' category itself – with its own sui generis culture – had found a new footing.

At the local level, the tourist definition of commercial Chinatown was recycled and refined in the 1970s. The functional relationship between the residential and retail sectors of Chinatown became recognized, and in combination they were said to constitute an 'ethnic neighbourhood' whose uniqueness the state should help preserve and enhance. Neighbourhood targeting of Vancouver's Chinatown had also come to assume a radically new meaning.

In 1969, a City-commissioned *Restoration Report* pointed out the contribution of the 'Pender Street community' to the 'quality and richness of city life'

(CVPD, 1969, page 16). Soon after, Council requested the Provincial Legislature to designate Chinatown a 'historic site' under the Provincial Archaeological and Historic Sites Protection Act. Such a classification would change Chinatown 'without endangering the area's essential character', Director of Planning, Mr W. Graham advised (CVA, 1970, File 17, Director's Report, 12 June). In February 1971 Cabinet unanimously approved the initiative, and the Province became vested with control over all significant changes, demolitions, and renovations to the buildings of Chinatown (CVA, 1970, File 18). It was a form of control not lost on the *Chinatown News* which noted 'Our preservation and restoration regulations require the Planning Department approval for all facelifting work done to buildings in Chinatown before a single penny is spent' (18 November 1972, page 3).

Despite the lack of interest of Chinatown's merchants in any improvements to the area (save for increased parking space), City planners were committed to their new vision of 'Chinatown' (CVPD, 1972–74). In 1974, for example, a set of sign guidelines was devised to ensure that Chinatown's 'unique ethnic and visual character' was preserved (CVA, 1974b, page 2). Equally important in further realizing the enduring idea of the 'East' in the 'West' was a Historic Area zoning classification approved by Council in 1974. This schedule designated new physical perimeters of the district and gave the City its own power to control certain building changes as well as new construction in the area (CVA, 1974a).

The new definition of Chinatown as an 'ethnic' asset was not unilaterally imposed upon a passive Chinese community, however. The processes of hegemonic control have been infinitely more subtle than that; the imprint of the past always contained within the present. With time, some Chinatown residents came to identify interests of their own in the implementation of the dominant view. Some of their receptivity was economically based, geared to the tourist trade; for others, the courting of Chinatown offered a form of power and status to community advocates, and a pride in the Chinese experience in Canada. Whatever the motivation, their acceptance was part of a significant secondary process that contributed to the perpetuation and toleration of European appellations of identity and place.

After a great deal of effort, the Chinatown Historic Area Planning Committee (CHAPC) rallied the necessary grass-roots support for the City's Chinatown beautification project. In 1977, the Pender Street merchants saw their way to contributing $300,000 toward a $700,000 project that would upgrade Chinatown's 'ethnic character' (CVPD, 1977). Chinese merchants and organizations also found compatible the proposal to build a Chinese Cultural Centre and Garden complex in the late 1970s. The initiation came again from the Canadian government which was keen to enhance the legibility of Chinatown's Oriental motif. The City of Vancouver agreed to lease the site at the corner of Carrall and Pender Streets to the Chinese Cultural Centre organization (*Chinatown News* 3 September 1976, page 2), and the following year, the Federal Government granted $1.5 million toward a garden that would 'conform [to] and enhance that community's architecture and character . . . arrest the dispersal of this unique community . . . and provide a major tourist attraction' (CVPD, 1978–79, Ouellet to Volrich, 20 April 1978).

However well intentioned the Canadian state's mosaic ideology, it represented at root a reaffirmation of the century-old moral order of 'us' and 'them'. 'Chinese' still signified nonwhite. And, as long as the classification was given new forms of currency within the symbolic system of the dominant society, cultural relativism could as easily give way to classical forms of targeting (Prager, 1982). Indeed it did in Vancouver in the late 1970s when, for example, the City Health Department ruled that Chinatown's barbecued meats industry was not an appropriate 'cultural contribution' (see *Chinatown News* editorials, 1978 and 1979), and the Federal Government committed $100,000 of multiculturalism funds toward a film on Vancouver's Chinatown that the Chinese Benevolent Association described as 'a way to perpetuate the worst stereotypes' (*Chinatown News* 3 October 1979, page 10). Concern in the House of Commons over immigration from nontraditional sources during a downturn in the Canadian economy also revealed the flip side of the distinctiveness officials were keen to court at the local level. One British Columbia member captured the sentiments of many politicians in March 1977, when he said 'Multiculturalism is all right in its place' (Canada Parliament, 11 March 1977, page 3904).

Conclusion

Racial ideology has been a significant cultural and political force in structuring the social order of Western countries such as Canada. During the evolution of white European domination in Vancouver, British Columbia, conceptions of a 'Chinese race' became inscribed in institutional practice and reconstituted through the locality known, and produced, as 'Chinatown'. In this paper, I have attempted to show that cultural hegemony at work; to demonstrate how the micro-order of cognitive representation, as mediated through Chinatown, informed the macrostructure of European hegemony in a thoroughly reciprocal dynamic which must be situated in history.

The study of localities affords the opportunity to weave the 'most local detail with the most global structures in such a way as to bring both into view simultaneously' (Geertz, 1983, page 69). The challenge, however, is not only to read off the larger processes which in part construct, and are themselves constructed through, the locality. It lies also in bringing a *critical* perspective to the hermeneutic circle; in examining the functions that landscape representation performs for us and the political pressures governing it (Cosgrove, 1985; Fabian, 1983). Said (1978, page 23) has argued there has been nothing so innocent as the single European idea of the 'Orient'. Similarly, the space of knowledge called 'Chinatown' grew out of, and came to structure, a politically divisive system of racial discourse that justified domination over people of Chinese origin in late nineteenth century Vancouver. Henceforth, for 100 years, the filter of knowledge through which the 'Chinese' and 'Chinatown' were screened tended not to be radically edited, but rather reworked and refined. It could have been otherwise. Indeed the pressures which have made the Other into our designated image are never complete and always reversible. Our racial categories are not inherently enslaving but open to challenge and scrutiny. As Olsson (1979, page 304) has said: 'In order to live and create, we must destroy the tradition from which we stem'.

References

Allport, G. 1954: *The Nature of Prejudice*. Reading, MA: Addison-Wesley.

Anderson, K. 1986: '"East" as "West": Place, State and the Institutionalization of Myth in Vancouver's Chinatown, 1880–1980'. PhD dissertation, Department of Geography, University of British Columbia, Vancouver.

Anderson, K. 1987: 'Chinatown as an idea: the power of place and institutional practice in the making of a racial category'. *Annals of the Association of American Geographers* 77, 580–98.

Angus, H. 1933: 'A contribution to international ill-will'. *The Dalhousie Review* 13, 23–33.

Banton, M. 1977: *The Idea of Race*. Andover, Hants: Tavistock Publications.

BCS, 1875: *British Columbia Statutes* 35 Victoria, chapter 26, section 22.

BCS, 1884: *British Columbia Statutes* 47 Victoria, chapter 16.

BCS, 1887: *British Columbia Statutes* 50 Victoria, chapter 33.

Boxer, C. 1963: *Race Relations in the Portuguese Colonial Empire, 1415–1825*. Oxford: Clarendon Press.

Brantlinger, P. 1985: 'Victorians and Africans: the genealogy of the myth of the Dark Continent'. *Critical Inquiry* 12, 166–203.

Breton, R. 1964: 'Institutional completeness of ethnic communities and the personal relations of immigrants'. *American Journal of Sociology* 70, 193–205.

Breton, R. 1984: 'The production and allocation of symbolic resources: an analysis of linguistic and ethno-cultural fields in Canada'. *Canadian Review of Sociology and Anthropology* 21, 123–44.

British Columbia, 1885: *Sessional Paper* Correspondence in connection with the mission of Hon. W. Smithe to Ottawa. Victoria: Richard Wolfenden, Government Printer.

Canada *Census 1880–81*, Volume 1.

Canada Parliament, *Debates of the House of Commons*, miscellaneous years.

Chan, A. 1983: 'Chinese batchelor workers in nineteenth century Canada'. *Ethnic and Racial Studies* 5, 513–34.

Chinatown News 1961–79: miscellaneous issues, Special Collections Division, Main Library, University of British Columbia, Vancouver.

Chinese Tax Act, 1878: *British Columbia Statutes* 42 Victoria, chapter 35.

Chinese Times (The) 1915–39: miscellaneous issues, Chinese–Canadian Project, Special Collections Division, Main Library, University of British Columbia, Vancouver.

Colonist 1921–22: miscellaneous years, Main Library, University of British Columbia, Vancouver.

Cosgrove, D. 1985: 'Prospect, perspective and the evolution of the landscape idea'. *Transactions of the Institute of British Geographers New Series* 10, 45–62.

Cosgrove, D. and Jackson, P. 1987: 'New directions in cultural geography'. *Area* 19, 95–101.

Cox, K. and McCarthy, J. 1980: 'Neighbourhood activism in the American city'. *Urban Geography* 1, 22–38.
CS, 1885a: *Canada Statutes* 48–49 Victoria, chapter 40.
CS, 1885b: Act to Regulate the Immigration of Chinese to Canada. *Canada Statutes* 48–49 Victoria, chapter 71.
CS, 1923: *Canada Statutes* 13–14 George, chapter 38.
CVA, City of Vancouver Archives, 1150 Chestnut St, Vancouver, BC.
CVA *Additional Manuscript 300*, Vancouver Board of Trade papers.
CVA *Additional Manuscript 734*, Strathcona Property Owners and Tenants Association Papers.
CVA *Council Minutes*, miscellaneous years, Office of the City Clerk.
CVA *Health Committee Minutes*, 1899–1906, Office of the City Clerk.
CVA *Incoming Correspondence*, miscellaneous years, Office of the City Clerk.
CVA, 1936: 'Report of the Managing Board', *Special Committee File Number 1*, Vancouver Golden Jubilee, Office of the City Clerk.
CVA, 1937–38: *Mayor's Correspondence*, volumes 26 and 27, Office of the City Clerk.
CVA, 1967: 'Brief re proposed freeway through Chinatown', CVA public document 788, Chinese Benevolent Association.
CVA, 1970: Files 17 and 18, City of Vancouver Social Planning Department.
CVA, 1974a: *Chinatown Sign Guidelines*, CVA public document 163, City of Vancouver Planning Department.
CVA, 1974b: *Historic Area Zoning Schedules*, CVA public document 165, Schedule F, Appendix B.
CVPD Publications from the City of Vancouver Planning Department may be consulted at the Planning Department Library, City Hall, West 12th Ave, Vancouver.
CVPD, 1964: *Chinatown Vancouver: Design Proposal for Improvement.*
CVPD, 1969: *Restoration Report: A Case for Renewed Life in the Old City.*
CVPD, 1972–74: Chinatown Beautification Files.
CVPD, 1977: *Chinatown Streetscape*, Chinatown Beautification Files 1976–77.
CVPD, 1978–79: Chinese Cultural Centre Files.
CVTPB Publications from the City of Vancouver Technical Board may be consulted at Fine Arts Division, Main Library, University of British Columbia, Vancouver.
CVTPB, 1956: *Downtown Vancouver, 1955–76.*
CVTPB, 1957: *Vancouver Redevelopment Study.*
CVTPB, 1959: *Redevelopment: Acquisition and Clearance.*
CVTPB, 1963: *Redevelopment Project 2.*
CVTPB, 1964: *Urban Renewal Program, Project 1, Area A-3.*
Dawson, R. 1967: *The Chinese Chameleon: An Analysis of European Conceptions of Chinese Civilization.* Oxford: Oxford University Press.
Dear, M. and Scott, A. (eds) 1981: *Urbanization and Urban Planning in Capitalist Society.* New York: Methuen.
Fabian, J. 1983: *Time and the Other: How Anthropology Makes its Object.* New York: Columbia University Press.
Fenton, S. 1980: ' ''Race relations'' in the sociological enterprise'. *New Community* 8, 162–8.
Foucault, M. 1977: *Discipline and Punish: The Birth of the Prison.* New York: Pantheon Books.
Fredrickson, G. 1981: *White Supremacy.* New York: Oxford University Press.
Gabriel, J. and Ben-Tovim, G. 1978: 'Marxism and the concept of racism'. *Economy and Society* 7, 118–54.

Gates, H. 1985: 'Writing ''race'' and the difference it makes'. *Critical Inquiry* 12, 1–20.

Geertz, C. 1983: *Local Knowledge*. New York: Basic Books.

Giddens, A. 1976: *New Rules of Sociological Method*. London: Hutchinson.

Gilmour, C. 1949: 'What, no opium dens?' *Maclean's* 15 January, 27–30.

Globe and Mail 1985: 9 February (Toronto), page 8.

Gramsci, A. 1971: *Selections from the Prison Notebooks*. London: Lawrence and Wishart.

Greenberg, A. 1980: *Race and State in Capitalist Development: Comparative Perspectives*. New Haven, CT: Yale University Press.

Halford Wilson Papers, Catalogue Number E/D W69, Provincial Archives of British Columbia, Government St, Victoria, BC.

Haller, M. 1963: *Eugenics: Hereditarian Attitudes in American Thought*. New Brunswick, NJ: Rutgers University Press.

Harvey, D. 1985: *The Urbanization of Capital*. Oxford: Basil Blackwell.

Hasson, S. and Ley, D. forthcoming; *Neighbourhood Organzations and the Welfare State*. Cambridge: Cambridge University Press.

Hay, D. 1966: *Europe: The Emergence of an Idea*. Edinburgh: Edinburgh University Press.

Husband, C. (ed.) 1984: *'Race' in Britain: Continuity and Change*. London: Hutchinson.

Ireland, R. 1965: 'Canadian trade unionism and Oriental immigration: a study of ideological change'. *Indian Journal of Economics* 46, 1–31.

JLABC *Journals of the Legislative Assembly*, British Columbia, miscellaneous years. Victoria: Richard Wolfenden, Government Printer.

Johnston, L. 1921: 'The case of the Oriental in British Columbia'. *Canadian Magazine* 57, 315–18.

Jones, G. 1980: *Social Darwinism and English Thought*. Brighton, Sussex: Harvester Press.

Jordan, W. 1968: *White over Black: American Attitudes toward the Negro, 1550–1812*. New York: Penguin Books.

La Forest, G. 1955: *Disallowance and Reservation of the Provincial Legislation*. Department of Justice, Justice Building, 239 Wellington, Ottawa.

Lears, T. 1985: 'The concept of cultural hegemony: problems and possibilities'. *American Historical Review* 90, 567–93.

Lee, C. 1976: 'The road to enfranchisement: Chinese and Japanese in British Columbia'. *BC Studies* 30, 44–76.

Ley, D. 1983: *A Social Geography of the City*. New York: Harper and Row.

Light, I. 1974: 'From vice-town to tourist attraction: the moral career of American Chinatowns'. *Pacific Historical Review* 43, 367–94.

Livingstone, D. 1984: 'Science and society: Nathaniel S. Shaler and racial ideology'. *Transactions of the Institute of British Geographers New Series* 9, 181–210.

Lyman, S. 1974: *Chinese Americans*. New York: Random House.

McEnvoy, F. 1982: 'A symbol of racial discrimination: the Chinese Immigration Act and Canada's relations with the Chinese, 1924–47'. *Canadian Ethnic Studies* 14, 24–42.

MacLaughlin, J. and Agnew, J. 1986: 'Hegemony and the regional question: the political geography of regional industrial policy in Northern Ireland, 1945–1972'. *Annals of the Association of American Geographers* 76, 247–61.

Markus, A. 1987: 'Australian governments and the concept of race: an historical perspective', unpublished manuscript courtesy of author, History Department, Monash University, Clayton, Victoria.

Marsh, L. 1950: 'Rebuilding a neighbourhood', Research Publications 1, Department of Sociology and Social Work, University of British Columbia, Vancouver.

Miles, R. 1982: *Racism and Migrant Labour*. Andover, Hants: Routledge and Kegan Paul.

Mollenkopf, J. 1975: 'The post-war politics of urban development'. *Politics and Society* 5, 247–95.

Montagu, A. 1965: *The Idea of Race*. Lincoln, NE: University of Nebraska Press.

Moodley, K. 1983: 'Canadian multiculturalism as ideology'. *Ethnic and Racial Studies* 6, 320–31.

Olsson, G. 1979: 'Social science and human action or on hitting yourself against the ceiling of language', in Gale, S. and Olsson, G. (eds), *Philosophy in Geography*. Dordrecht: D. Reidel, 287–308. .

PAC, 1922: Public Archives of Canada Immigration Branch, Volume 121, File 23635, Part 5 in Chinese–Canadian Project, Box 28, Special Collections Division, Main Library, University of British Columbia, Vancouver.

Palmer, H. 1982: *Patterns of Prejudice: A History of Nativism in Alberta*. Toronto: McClelland and Stewart.

Parkin, F. 1979: *Marxism and Class Theory: A Bourgeois Critique*. New York: Columbia University Press.

Pendakur, S. 1972: *Cities, Citizens and Freeways*. Vancouver: V. S. Pendakur.

Prager, J. 1982: 'American racial ideology as collective representation'. *Ethnic and Racial Studies* 5, 99–119.

Prager, J. 1987: 'American political culture and the shifting meaning of race'. *Ethnic and Racial Studies* 10, 62–81.

Pratt, G. 1986: 'Housing tenure and social cleavages in urban Canada'. *Annals of the Association of American Geographers* 76, 366–80.

Price, C. 1974: *The Great White Walls are Built*. Canberra: Australian National University Press.

Province 1906–67: miscellaneous issues, Main Library, University of British Columbia, Vancouver.

RCBB, 1969: *Report of the Royal Commission on Bilingualism and Biculturalism*, volume 4. Ottawa: Queen's Printer.

RCCI, 1885: 'Sessional paper', 54a, Report of the Royal Commission on Chinese Immigration (printed by order of the Commission, Ottawa).

RCCJI, 1902: 'Sessional paper', 54, Report of the Royal Commission on Chinese and Japanese Immigration (printed by S. E. Dawson, Ottawa).

Rex, J. 1982: 'Racism and the structure of colonial societies'. In Ross, R. (ed.), *Racism and Colonialism: Essays on Ideology and Social Structure*. The Hague: Martinus Nijhoff.

Rose, A. 1971: *Citizen Participation in Urban Renewal*. Centre for Urban and Community Studies, University of Toronto.

Ross, R. (ed.) 1982: *Racism and Colonialism: Essays on Ideology and Social Structure*. The Hague: Martinus Nijhoff.

Roy, P. 1976: 'The preservation of peace in Vancouver: the aftermath of the anti-Chinese riots of 1887'. *BC Studies* 31, 44–59.

Said, E. 1978: *Orientalism*. New York: Random House.

Salter, C. 1984: 'Urban imagery and the Chinese of Los Angeles'. *Urban Review* 1, 15–20, 28.

Samuels, M. 1979: 'The biography of landscape'. In Meinig, D. (ed.), *The Interpretation of Ordinary Landscapes*. New York: Oxford University Press, 51–88.

Saunders, P. 1981: *Social Theory and the Urban Question*. London: Hutchinson.

Saunders, P. and Williams, P. 1986: 'Guest editorial'. *Environment and Planning D: Society and Space* 4, 393–9.
Sibley, D. 1981: *Outsiders in Urban Society*. Oxford: Basil Blackwell.
Stepan, N. 1982: *The Idea of Race in Science: Great Britain, 1800–1960*. London: Macmillan.
Sun 1936–67: miscellaneous issues, Main Library, University of British Columbia, Vancouver.
Suttles, G. 1968: *The Social Order of the Slum*. Chicago, IL: University of Chicago Press.
Tuan, Y.-F. 1974: *Topophilia*. Englewood Cliffs, NJ: Prentice-Hall.
Vancouver News 1886–87: miscellaneous issues, Main Library, University of British Columbia, Vancouver.
Voegelin, E. 1940: 'The growth of the race idea'. *Review of Politics* 2, 283–317.
Warburton, R. 1981: 'Race and class in British Columbia: a comment'. *BC Studies* 49, 79–85.
Ward, P. 1978: *White Canada Forever: Popular Attitudes and Public Policy toward Orientals in British Columbia*. Montreal: McGill-Queen's University Press.
Western, J. 1981: *Outcast Capetown*. Minneapolis, MN: University of Minnesota Press.
Williams, P. 1976: 'The role of institutions in the inner London housing market'. *Transactions of the Institute of British Geographers New Series* 1, 72–82.
Williams, R. 1977: *Marxism and Literature*. Oxford: Oxford University Press.
Won Alexander Cumyow Papers, Special Collections, Main Library, University of British Columbia, Vancouver.
Wong, E. 1985: 'Asian American middleman theory: the framework of an American myth'. *Journal of Ethnic Studies* 13, 31–88.
World 1912: 10 February, Main Library, University of British Columbia, Vancouver.
Yip, Q. 1936: *Vancouver's Chinatown: Vancouver Golden Jubilee, 1885–1936*. Vancouver: Pacific Press.
Zaret, D. 1980: 'From Weber to Parsons and Schutz: the eclipse of history in modern social theory'. *American Journal of Sociology* 85, 1180–1201.

11 G. Rose, 'Locality, Politics, and Culture: Poplar in the 1920s'

Excerpts from: *Society and Space* **6**, 151–68 (1988)

Introduction

'Culture' is rapidly becoming a buzzword for social theorists, especially for locality-study theorists and others struggling to explain local differences in social and political behaviour; it is now often evoked as the cause of otherwise inexplicable geographic variation in social activity or political allegiance (for example, Agnew *et al.*, 1984; Johnston, 1986; Rees *et al.*, 1987; Savage, 1987). Attempts to define exactly what is meant by culture are much rarer, yet ever since Williams's seminal work (1963) we have known that culture is far from being an unproblematic term. I ought to begin this paper then by immediately declaring which concept of culture I will be utilising. But the reasons for my use of a particular notion of culture will be much clearer if I start by outlining not the literature of cultural theory, but the literature on local politics and the local state. Analyses of the local state have encountered great difficulties in satisfactorily explaining local politics (hence the recent appeal to vague notions of local culture) and I will conclude this first section by offering a specific interpretation of culture as a solution to some, and only some, of these problems.

Writers on the local state, somewhat paradoxically, have always tended to ignore the local. They almost always look at the relationship between the central state and the local in order to characterise the nature of local government, and this is as true of Marxists like Cockburn (1977) as it is of neo-Weberians like Saunders (1984). Johnston (1985, page 172) has recently summarised this orthodoxy by saying that

> ... local government is part of the apparatus of the central state. It has no independent existence, and the functions it is called upon to perform represent – in both quantity and quality – the perceptions of the role of local government held by those who control the central state apparatus.

This theorisation of the actions of the local state in terms of central–local relations alone is by no means new. Political theorists of the interwar years such as Finer (1933) and Robson (1931) also saw local authorities as no more than the administrative arms of the central state, and, as a general verdict on the balance of power between the central and the local states over the past 150 years or so, the analysis is probably justified; local authorities which consistently flout the wishes of central government have been legislated out of existence. But that is not an insight which is very helpful in understanding the politics of particular local authorities at specific times.

In the present context of an increasingly centralist Thatcherite state, however, interest in the 'local' of the local state is growing because local political relations are coming more and more to be seen as crucial to the general policies adopted by local authorities, and as especially crucial for their willingness to defy the dictates of central government. Activists of the 'new urban left' (Gyford, 1985) such as David Blunkett (1984) and Ken Livingstone (1984) see local popular involvement in local government as an essential precondition for assertions of autonomy from the central state by the local. And academics, too, like Byrne (1982) and Duncan and Goodwin (1982) have begun to focus on the importance of the locality to local politics by assessing local social relations and the political implications of the connections between a local society and the local state. They argue that a study of these connections is fundamental to an understanding of local political behaviour.

Yet both Byrne, and Duncan and Goodwin construct the link between a local society and its polity from class and class alone. Duncan and Goodwin use the histories of Oldham in the 1840s, Poplar in the 1920s, and Clay Cross in the

1970s to argue that socialist councils are the natural result of proletarian political activity; the Poplar council of the 1920s is often used as evidence of a class-conscious rebellion against the bourgeois state (for example, see Branson, 1979; Stone, 1985). Byrne studies Newcastle in a similar manner, seeing the local Labour Party's policies as inevitable given the working-class character of the locality. These writers assume that the real interests of the working class lie in voting Labour and thus that the support of the working class for the Labour Party is unproblematic. However, this assumption creates a distorted perspective on local politics in two ways. First, any social force other than that of class is marginalised as irrelevant to political activity, and, second, detailed examination of the process of politicisation and of the specific reasons for party loyalty are made unnecessary. The questions of exactly how, why, and what party politics become the fundamental politics of the working class – questions which seem to me to be crucial to a full analysis and made even more so by recent electoral history – are ignored in such a theoretical strategy because socialism is somehow seen as natural to working-class voters.

Locality-study theorists, dissatisfied with this reductionist view of political behaviour, have suggested that the notion of civil society might be of some help in problematising political behaviour (Urry, 1981). Civil society allows theoretical space for a geographically variable arena of social life outside politics, which nonetheless shapes political expression in locally unique ways. Rees (1985) has recently argued this case, and civil society certainly offers an opportunity to explore the politicisation process, as the work of Mark-Lawson *et al.* (1985) and Warde (1986) demonstrates. But many locality-study theorists and others still utilise the concept in a structural manner, trying to read off the politics of a locality from its labour-market structure (Cooke, 1986), or from its degree of formal organisation (Cronin, 1984).

What is missing from all these structural accounts of local politics is contetxualised analysis. The expression and formation of local opinions and attitudes and understandings within civil society are ignored, and so the contingent nature of the politics which develop from them is never acknowledged. What is lacking is recognition of the sheer complexity of the politicisation process, recognition of the importance of local socially established structures of meaning. The phrase 'socially established structures of meaning' is taken from the writings of the US anthropologist Geertz (1973, page 12); it summarises his formulation of culture and in this paper I will use the term in his specific sense. For his notion of culture best describes what studies of the local state and local politics have so far omitted: local ideas and attitudes. Geertz elsewhere (1983) describes culture as 'local knowledge', as 'local frames of awareness' (page 6), and as 'communal sensibility' (page 12), and it is the communal sensibility of Poplar in the 1920s which I now want to argue was formative of its political behaviour.

Poplar's politics in the 1920s

Poplar was the easternmost borough of London's East End. It was incorporated in 1894, but in the beginning of the nineteenth century it was a small and quiet hamlet of skilled shipbuilding craftsmen. The major period of growth of the locality was the third quarter of the nineteenth century, prompted by the building of the large import and export docks needed to cope with the growth in Britain's colonial trade (Rose, 1957). This comparatively late industrialisation meant that most of the industry in Poplar was not carried out in the small workshop typical of the East End, but in large factories like those of Spratt's, and Bryant and May's which at peak times each employed nearly two thousand people (Bedarida, 1975).

Jerry-built housing soon covered the fields north and south of Poplar village. These terraces were built as quickly and as cheaply as possible by speculative builders, and they soon turned into slums, usually sublet and very overcrowded. Conditions did not change until the London County Council began a major slum-clearance programme in the 1930s; the Census in 1921 and again in 1931 found about a fifth of Poplar's population living more than two to a room. Conditions were appalling, with rats and vermin everyday hazards to be faced, but these houses were cheap and this attracted the poor to Poplar. Many men came initially to excavate the docks and then stayed on to work as dock labourers. By the 1920s the docks were still the main employers of men, but the engineering works and the railways, and general unskilled labouring, also engaged large numbers. Women – and just over half of Poplar's female population over the age of fourteen were in paid employment – were most likely to be tailoresses, homeworkers, or to work in the large food and drink processing factories in the district.

So Poplar was a heavily industrialised area, and when the London School of Economics published an updated version of Booth's late nineteenth-century survey of London's social structure in the early 1930s, the extent to which it was a wholly working-class area became very clear. Despite its very demanding definition of poverty, it was found in the new survey that no less than one quarter of Poplar people fell below it. The next poorest borough was Shoreditch, also in the East End, with a mere 18 per cent of its population in poverty (*New Survey* 1932, page 365).

Poplar was obviously a proletarian locality, and on class-theoretical logic it was ideal Labour Party territory. Sure enough, after the 1918 Representation of the People Act (which dramatically reduced the property qualification for a vote in local elections and thus enfranchised most of Poplar's adult population), the Labour Party swept into power in the borough. In the 1919 local elections it won thirty-nine of the forty-two seats on the council and a majority of nineteen on the twenty-four-member Board of Guardians, and it remained in office in Poplar until the 1985 local elections. Poplar's allegiance to the Labour Party cannot be understood as a straightforward reflection of working-class interests, however, for a reason quite apart from the theoretical objections to such a claim outlined in the introduction. The Poplar Labour Party was a maverick, often in conflict both with the national Labour Party and with the regional London Labour Party. Its frequent contrariness suggests that there is no one natural and inherent working-class politics and this points, from another direction to the introductory arguments, to the contingent nature of local politics. I will now look at that contingency by linking, admittedly somewhat overschematically, four aspects of Poplar's cultural life to its Labour Party's policies and beliefs.

Class

The first element of Poplar's culture to be focused on here is its class politics. The socialism of Poplar Labour Party was more genuine – or at least more extreme – than that of other sections of the Labour Party. The event which brought Poplar's Labour council to the attention of the nation (and of local-state theorists like Duncan and Goodwin) was the 'Poplarism' episode of 1921. Poplarism was the name given to Poplar Council's decision in 1921 not to remit its rates precept to various London-wide bodies like the London County Council and the Metropolitan Police so that it could spend all its local rates, quite illegally, on its own poor; this action resulted in all but six of Poplar's councillors going to gaol in order to defend their high-spending relief policy (Branson, 1979; Keith-Lucas, 1962). The leader of the Parliamentary Labour Party, Ramsay MacDonald, and the secretary of the London Labour Party, Herbert Morrison, both condemned 'Poplarism' (Jones, 1973; MacDonald, 1924). Poplar's unusual radicalism, its willingness to treat the poor decently and to ignore the fiscal logic of the rates structure, its readiness to break the law in defence of a principle it believed to be right, stemmed in large part from the existence of several left-wing groups in Poplar and from the cooperation between them and Poplar Labour Party.

The organisations of civil society which are most often seen as the catalysts of a locality's socialism are the trade unions. Unions are supposed to demonstrate the strength of organised labour to workers and to educate them in socialist theory, and Bush (1984) has argued that this occurred in east London during the First World War. However, it seems that in the 1920s unions were not especially important to Poplar's politics. They were weakly organised in the borough, despite the large factories and docks in the district, and this was a result of the prominence in Poplar's economy of industries notoriously resistant to unionism, like tailoring, and to the persistence of homeworking and of the casual labour system (Gillespie, 1984). The practice of hiring workers casually, that is for short periods of time, was common well into the interwar years even in Poplar's large factories (*New Survey*, 1931; 1933; 1934), and the consequent high level of labour turnover made trade union organisation very difficult. The lack of influence from the trade unions on Poplar Labour Party was starkly demonstrated in 1920 when the council unilaterally imposed a weekly minimum wage of £4 on its employees without bothering to consult their unions (PMBC meeting minutes, 27 May 1920). So the unions had little political muscle in Poplar, and in any case their politics were not necessarily the same as those of the local Labour Party; in 1927 the Transport and General Workers Union official who was the Member of Parliament for South Poplar was heavily criticised by the local Party for being too right-wing (ELA, 10 December 1927). But if the trade unions had no great impact on Poplar's politics, what did?

First, there was Poplar's branch of the Independent Labour Party (ILP). The

ILP saw itself as a revolutionary organisation and it kept its affiliation to the Labour Party in the 1920s only in order to act as a left-wing ginger group. The Poplar branch was especially militant. The Revolutionary Policy Committee of the ILP was a communistic clique which agitated against the ILP's advocacy of a peaceful and bloodless revolution; its bulletin was published from the offices of Poplar's ILP branch in the East India Dock Road and was largely written by the branch's secretary, C. K. Cullen (Cullen collection). One sixth of the people who served Poplar for more than three years between 1919 and 1929 as councillors or guardians were members of the ILP, so the ideas of the ILP clearly had a hold on Poplar Labour Party.

Many of Poplar's Labour councillors and guardians were also active in another left-wing organisation in the borough: Sylvia Pankhurst's Workers' Socialist Federation (WSF). The WSF began in 1913 when Sylvia arrived in Poplar to bring the battle for the vote to working women (Pankhurst, 1931), and she received much help in her struggle from Poplar Labour Party activists. By the end of the First World War, however, her politics had changed, shifting from women's suffrage to adult suffrage to socialism to communism, and finally to a left-wing communism (*Workers' Dreadnought* passim). Despite its dislike of Poplar Labour Party's involvement in the democratic system (a system which it saw as a bourgeois sham meant to fool the workers into thinking they had some power under capitalism), the WSF worked closely with the Party in the early 1920s. A Labour Party alderman (*sic*) was the secretary of the WSF until her death in 1922. There was also what became known as the '*Jolly George* incident' in 1920. The *Jolly George* was a ship berthed in Poplar's East India Dock, which was being loaded with guns for the Polish armies to use against the Bolsheviks in Russia. Poplar Labour Party members worked hard to stop this, cooperating with the WSF rather than with the committee set up by the national Labour Party and the Trades Union Congress to support the Bolsheviks (Council of Action archive); members of the WSF and Poplar Labour Party plastered the walls of Poplar with stickers demanding 'HANDS OFF RUSSIA' and gave copies of Lenin's leaflet 'An Appeal to the Toiling Masses' to dockers as they left work each day (Pollitt, 1940). The toiling masses responded in May and refused to load the arms meant to kill their Russian brothers – a fine example of international working-class solidarity, said the WSF, and also an example of Poplar Labour Party's radicalism in conjunction with a communist group in its civil society.

The third left-wing influence on Poplar Labour Party was the organised unemployed. After a very brief post-war boom, an economic slump set in and unemployment in Britain began to rise, peaking in the summer of 1922. In Poplar in June of that year nearly 16,000 people were receiving unemployment benefit, which was over 10 per cent of the total population (PRO, LAB85/92). It seems that Poplar Labour Party initially helped to politicise the local unemployed by introducing them to the National Unemployed Workers' Movement (NUWM), an organisation run by the Communist Wal Hannington (PBMA archive letter, 25 July 1923). George Lansbury, Poplar's leading politician for fifty years, knew Hannington well and fully supported his campaign (Hannington, 1936, pages 16 and 53). The NUWM demanded that the central government provide 'WORK OR FULL MAINTENANCE' for the unemployed, and in January 1922 the enthusiasm of Poplar's unemployed for this slogan took even the Poplar guardians by surprise; the Poplar branch of the NUWM barricaded the guardians' meeting and refused to let the guardians leave the building until they agreed to a higher rate of relief: full maintenance (Hannington, 1936, pages 53–54; PRO, MH68/214). The guardians did agree, and this pressure from the unemployed kept Poplar's level of poor relief one of the highest in the country, both in terms of how much was paid and in terms of how many applicants were helped (PRO, HLG68/14).

These three left-wing groupings kept a critical eye on Poplar Labour Party; the WSF, for example, published a free newspaper, the *Workers' Dreadnought*, which was always quick to point out any deviations from the socialist path. Poplar Labour Party's involvement with these left-wing organisations increased the conflict between Poplar and the Labour Party nationally and in London. The national Labour Party was actively opposed to any sort of cooperation with communists, its conference in 1924 voting first to refuse to allow the Communist Party to affiliate to it, and then to ban individual Communist Party members from becoming Labour Party members. Herbert Morrison of the London Labour Party was a renowned anti-Communist polemicist. So, although both the Labour Party in Poplar and nationally described themselves as 'the organisation of workers . . . against the landlords and capitalists' (Poplar Labour Party election leaflet, George Lansbury collection), Poplar's actions demonstrated that class allegiances were strongly influenced by local left-wing groups.

Neighbourliness

That self-description of Poplar Labour Party as the organisation of all workers leads to the second aspect of Poplar's culture which was central to the politics of the locality, and that was the morality of its neighbourhood communities. The Party saw itself as the party of the workers, but the source of the workers' virtue was not their industrial strength or labour value, but the ethical quality of their home life. John Scurr, a councillor and guardian in Poplar for many years, claimed that he would find more loving kindness in a Poplar slum than he would ever discover in a Hampstead drawing-room (Scurr, 1924, page 85), and George Lansbury's son said that his father found his inspiration for politics in the generosity and kindness of working people (E. Lansbury, 1934). In the interviews I have had with people who lived in Poplar in the 1920s, the goodness, kindness, and friendliness of people then is a constantly recurring theme, and one which is invariably discussed in the context of Poplar's neighbourhood life.

These neighbourhoods revolved around their corner shops, and one woman I spoke to described their sociality like this:

> . . . there was a sort of atmosphere then, you knew if you were in need there was somebody. And of course there was nearly always family near, and families weren't scattered.

Family bonds were particularly important in maintaining neighbourhood bonds. Here is a woman from the Isle of Dogs, a district in the south of the borough:

> . . . my husband lived next door to me from a child, and of course then he moved across the road and of course when my dad married again his wife was my husband's father's sister. She was his sister, see, it made my stepmother his aunt.

Another man described his father to me:

> . . . nearly everyone he passed 'e knew 'em by their christian name, 'cause 'e either worked with 'em, 'cause there was about 750 men at this one factory and they all lived thereabouts, within walking distance see, and 'e'd been to school at the school opposite and 'e knew their christian names and in many cases the christian name of their wives.

That man was in fact Charlie Sumner, a long-serving Poplar councillor and guardian. Like the vast majority of Poplar's elected representatives, he was working class himself and he lived among Poplar's poor all his life; Poplar councillors almost always lived in the ward they represented in the council chamber.

The structure of this neighbourhood life so well known to Poplar's Labour men and women affected their politics profoundly. The importance of the family in the community fuelled their hatred of the Poor Law, for one of the most loathed aspects of the workhouse was the way in which families were split up when they entered. Several of Poplar's guardians were poor enough to have had personal experience of this trauma and were determined never to let the same thing happen to anyone else – hence their preference for outdoor relief, which was relief given in cash or kind and which did not involve the recipient entering the workhouse.

But the major political consequence of Poplar's neighbourhoods was not the reality of their social structure; rather it was the morality which underpinned the practices of the neighbourhood. Although in Poplar neighbours did help one another, lending and borrowing and sharing among themselves, I have also been told tales of domestic violence, family feuds, and street fights. Nonetheless, the *idea* of community, of cooperation and mutual aid, was a strongly held one in Poplar, one which is presented to outsiders not as a factual description of what went on (which is how these accounts were taken in the 'community studies' of the late 1950s and 1960s), but as an expression of what people want to believe went on, as an ideal. A woman describing the Isle of Dogs of sixty years ago told me

> . . . everybody was friendly, it was – well, everybody said it was like a little village and everybody was friendly with one another.

'*People said* it was like a little village': it was this self-image, this shared sensibility of what Poplar ought to be like, which influenced the politicians of

Poplar. Poplar Labour Party believed, like any good neighbour, that those in need should be helped by adequate levels of relief (for example, see Key, 1925); it believed, like any good neighbour, that poverty was not a crime and that the poor therefore should not be treated like criminals and incarcerated in the workhouse (G. Lansbury, 1907; 1928).

Religion

Another communal sensibility which affected Poplar Labour Party deeply was the church, and that is the third aspect of Poplar's culture to be discussed here. The strength of the church in 1920s Poplar was unusual, for it does not seem to have been especially important in any other radical locality of the period. In his study of the 'little Moscows' of the interwar years Macintyre dismisses religious faith as irrelevant to their politics (1980). Although the proportion of Poplar residents who went to church regularly is quite low – it is impossible to calculate reliable statistics since small chapels which kept no records were particularly popular (Grant, 1930, page 62), but the percentage was certainly never more than 10 per cent – the absolute numbers who attended were fairly impressive. About a thousand women turned up each week at the women's meeting held at the Wesleyan Chapel on the East India Dock Road for example, and when its minister died in 1937 the streets of Poplar were packed with people watching his funeral cortege pass (ELA, 13 February 1937). However, churchgoing among Poplar's Labour councillors was much more popular than among the mass of the population and thus the influence of the church on Poplar politics is quite marked.

Poplar Labour councillors were churchwardens, they helped to organise church fetes, they opened Baptist chapels' temperance pubs, they founded men's chapel meetings, and several were Catholic (ELA, passim). George Lansbury was emphatic that his socialism sprang from his Christian faith (G. Lansbury, 1924; 1928), and many members of the ILP shared his logic (Hardie, 1905). Christ had condemned poverty, exploitation, and greed, and, since capitalism was the prime cause of these evils, all true Christians should also be socialists, he argued. This was the standard logic of Christian Socialism, which had begun as an organised movement in the 1870s. By the 1920s, most clergy of all denominations accepted the idea of the 'social gospel' (Binyon, 1931) – that is, the idea that the Christian faith had social implications for this world as well as the next – but not all saw political agitation as its logical result.

Poplar clergy had no such qualms. Often radicalised by the sheer poverty of Poplar, they became involved in the politics of the place. All Hallows' church by the West India Docks was known locally as the 'red church' and the pastor at Bruce Road Congregational Church was elected a Labour guardian in 1928. William Dick of Trinity Congregational on the East India Dock Road opened his church in 1925 to seamen for a strike meeting and to the No More War Movement for a peace demonstration (Dick, 1924–26, passim). John Groser, a socialist curate at St Michael and All Angels, was among the injured when police attacked a crowd outside Poplar Town Hall at the end of the General

Strike (Groser, 1949, page 48; and see Brill, 1971). The Wesleyan Simeon Cole organised discussions of 'Socialism, Fabianism, and Bolshevism' for his congregation, and in a debate on 'Is Labour fit to govern?', he took the affirmative and won (ELA, 20 October 1928).

This interweaving of politics and religion in Poplar had one specific policy result, and that was a strong commitment to pacifism in Poplar Labour Party. Several of Poplar's councillors, including George Lansbury, were associated with Kingsley Hall, a Christian settlement in the north of the borough run by an ardent pacifist, Muriel Lester (Lester, 1937; G. Lansbury, 1937). This pacifism was yet another source of antagonism between Poplar and other sections of the Labour Party, and Lansbury lost the leadership of the Parliamentary Labour Party in 1935 as a result of it [for contrasting accounts of his resignation, see Bullock (1960) and Postgate (1951)].

Poplar Labour Party's religiosity also gave it a crusading air and a faith in apocalyptic social upheaval, with the Apocalypse acting almost as a metaphor for the revolution. The language of the news-sheet which Lansbury published during the General Strike conveys the Nine Days almost as the millennium (*Lansbury's Bulletin*), and this ardent ethical sensibility was a further cause of Poplar's loathing of the Poor Law's inhumane treatment of the poor.

Melodrama

This fervour and passionate faith leads to the fourth and final aspect of Poplar's culture I will stress in this paper: its love of performance and melodrama. Poplar people relished drama and sentiment. Poplar's music-hall, the Queen's in Poplar High Street, was still extremely popular in the 1920s and apparently its audiences were so involved in the performances there that the actors playing villains dared not leave the theatre for fear of being recognised and abused (Blake, 1977, page 32). The ability to 'do a turn', to sing a sentimental song, or, for the better-off, to play a maudlin tune on the piano, was an obligatory social skill in Poplar. No pub closed without an impromptu street concert from its patrons (interviews) and no public meeting was complete until entertainment had been provided (ELA, passim), and this was as true of Poplar Labour Party as of any other organisation.

The Poplar Labour Women's Guild met every Wednesday evening in the Town Hall to hear a Labour speaker, drink a cup of tea, and watch a few turns on the stage. About three hundred people came to these meetings which were attractive as much for the entertainment they provided as for the political enlightenment they offered, if the account of the organiser's son is to be believed. The Poplar councillors in gaol in 1921 were serenaded by bands and choirs of supporters, George Lansbury's magazine *Lansbury's Labour Weekly* released a gramophone record in 1926 of English socialist songs with an introduction by Lansbury (who loved singing), and the 'Red Flag' was always sung at moments of high political tension.

And Poplar loved oratory. The women who flocked to the Wesleyan women's meeting on Monday afternoons went to hear the Reverend William Lax give one of his rousing sermons which moved his listeners from laughter to tears and back again with consummate ease (Walters, 1927, page 10). The only man to rival Lax was Lansbury, whose emotionalism and idealism and passionate rhetoric always gave him an attentive audience and in Poplar guaranteed him an adoring one; for George, a Poplar man himself, knew exactly how to talk to these rowdy and emotional people and how to move them to righteous anger against the injustices of the world (Postgate, 1951). This extravagant speechmaking and emotionalism fed Poplar's defiance; its reply to a government inquiry of 1922 which was highly critical of its Poor Law policy was a pamphlet called 'Guilty and proud of it', hardly a conciliatory gesture but one which was absolutely typical of Poplar's flamboyance, and one which rubbed salt into wounds already made by the Party's extremism on the national and London Labour Parties.

Poplar's political relations in the early 1920s

In the 1920s, all these aspects of Poplar's political culture – its class loyalty, its community morality, its religiosity, its riotousness – all these coalesced to encourage a participatory form of politics, with a great many of Poplar's ordinary people becoming involved in the council's battle against the Poor Law.

The left-wing commitments of Poplar Labour Party certainly encouraged the council to seek the active involvement of its electorate in its policies, and this was largely because of its radical belief in the paramount importance of local democracy. This was a belief at the heart of the ILP's policy on local government; and, as I have argued, the ILP was an influential organisation in Poplar. An ILP motion at the 1918 Labour Party conference opined that

> . . . liberty of action should be given to municipal bodies; Government interference in local affairs should be confined to giving information, to giving grants for work done, and not in fettering the local authorities as hitherto (Labour Party, 1918).

This notion of the local state being accountable to local needs and not to the central government wishes was crucial to the ideology of Poplar Labour Party. When George's son Edgar Lansbury was chair of Poplar's Board of Guardians in 1922 he told the ministry responsible for local government that it should 'assist and advise' Poplar, not command and compel it (PRO, MH68/215); and Poplar Labour Party constantly defended its policies by pointing to its massive local electoral mandate, for no more than one or two ratepayer candidates were ever even remotely successful in Poplar's local elections. And if Poplar people voted for them, Poplar Labour Party felt it in turn had an obligation to serve them well. Edgar Lansbury saw this obligation, an obligation *to* the locality imposed *by* the locality, as inescapable. 'We have no option,' he told the ministry, 'we were elected to carry out a certain policy, we honestly believe in that policy and will use every endeavour to carry it out' (PRO, MH68/215). The Party's belief in its duty to its locality encouraged its defiance of the wishes of central government; it also created an enthusiasm for participatory politics.

Poplar Labour Party's commitment to democracy also involved a strong dislike of strict party organisation and hierarchy. This was yet another source of conflict between Poplar and the London Labour Party, for Herbert Morrison was an administrator extraordinaire who at times gave the impression of believing that the route to revolution lay via the card indexes in his office (Donoughue and Jones, 1973). A strict party structure steamrollered individuals into submission, argued George Lansbury, and he for his part was not interested in encouraging submissiveness. He wanted to 'teach the masses to think and act for themselves' (G. Lansbury, 1928, page 2). This faith in mass political education and mobilisation (similar in fact to the strategies of the new urban left today) were shared by the WSF, which wanted society after the revolution to be organised in the form of soviets precisely because soviets were truly democratic and gave everyone a say in communal decisions (*Workers' Dreadnought* 19 June 1920).

As a result of these left-wing commitments to the locality and against outside interference, Poplar Labour Party did all it could in the early 1920s to persuade its electorate to become engaged in the politics of their local state. The Labour council constantly held public meetings. In 1921 it sent letters to every household in the borough explaining its actions and to every schoolchild encouraging them to break the law too if they felt it was an evil one. They organised processions and demonstrations and deputations, they organised outdoor speakers and ran social events, all to publicise their ideas. The effect was described in a report on the unemployed young men of Poplar in 1922:

> They are less submissive and more defiant [than before the war]. The present period of unemployment is making them rebellious rather than submissive and there is a disposition to attribute it to what is called, without economic or political analysis, 'the breakdown of the present system' (Toynbee Hall, 1922, page 31).

The general outlines of Labour's arguments were common currency in the locality by the 1920s.

The sociability of neighbourhood life also encouraged mass involvement in local politics, by making each councillor personally known to their constituents, to be addressed on first-name terms, accessible, familiar, their door always open. They listened to individuals and deputations and they rejected the alienating pomp and ceremony of municipal tradition; Labour mayors refused to wear their regalia of office in the council chamber, for example (ELA, 5 February 1921). All this culminated in the scenes outside the Town Hall in 1921 when the councillors were taken to gaol. Twenty thousand people turned out to watch the most unusual melodrama Poplar had ever seen, a street theatre scene which the councillors carefully directed to create the most dramatic event possible, assembling together outside the Town Hall, photographers and newsreel cameras in place: the arrest of thirty-six councillors for a cause they felt to be morally unassailable. For these Christian men and women, their imprisonment was their martyrdom, and they had the support of their people who turned out en masse to participate in the drama themselves.

Poplar's political relations in the late 1920s

By the late 1920s, this participatory politics had disappeared from Poplar. The Labour council had become more elitist and more right-wing (and, as will be argued, these two developments are not unrelated); in other words, the council was becoming increasingly separate from the institutions of its civil society. Yet it retained local loyalty, continuing to achieve very large majorities on the council and Board of Guardians and very high election turnouts (London County Council, 1915–30), and this was because it never became alienated from Poplar's local culture. This apparent paradox of a continuous cultural milieu coexisting with changing social relationships will be discussed in terms of the four aspects of Poplar's culture already outlined, plus a new fifth one which helped to keep Labour in power.

Class

The class awareness of Poplar Labour Party depended upon a sense of the injustice of working-class poverty and hardship. That awareness motivated Poplar men and women into politics, into socialist politics, but once they were elected into office it seemed to make them only too willing to grab the fruits of their success with both hands, almost as if they wanted to compensate personally for their own underprivilege which their socialism made them see so clearly (Gyford, 1984, page 63).

This emerging gulf in social status between the councillors and their constituents was noted by a disapproving Ministry of Health inspector in 1929 who remarked that 'Poplar councillors are far too apt to "hob-nob" with their [middle-class, professional] officers, to drink with them, and to be on terms of close personal friendship' (PRO, MH68/213, minute, 10 May 1929). Poplar's councillors were also remarkably successful in getting themselves council housing, when one considers that the waiting list for council accommodation in Poplar was closed in July 1924 when 3,500 families were on it and the council built a mere seven hundred dwellings during the 1920s (ELA, 25 April 1931). Poplar people were very sensitive to the status differences based on residence – all the men and women I have interviewed readily and clearly distinguish between the 'nice' and 'not-so-nice' streets of their districts – and these perks of office would therefore have been seen as creating a distinction between the councillors and their voters. That distinction was also felt by the Poplar Labour Women's Guild, a popular institution in both senses of the word; it was increasingly ostracised by what the son of its secretary described to me as 'an elite of snobby councillors' by the end of the 1920s. The Labour Party in Poplar was breaking off its populist social relations.

Neighbourliness

The social bases of Poplar's neighbourhoods, particularly the family, were the roots of a growing corruption in Poplar council by the late 1920s. Poplar people were motivated into office by the desire to help others; that was the foundation of their socialism, it was built on their love of community. But as well as leading to socialism it also encouraged jobbery and nepotism. Friends and relations were found jobs by councillors, especially in the council's Electricity Department, and from 1926 until the mid-1930s the Irish of Poplar were especially involved. It was virtually impossible to find employment in the council in those years unless you were a member of the Knights of St Columbus, a Catholic version of the Freemasons, and the council was known locally as Hubbart's band after the councillor who organised the Knights in Poplar (interviews).

Religion

That brings us to the church, the third aspect of Poplar's culture. Poplar Labour Party's religiosity was fine and fervent, but it did not contain a clear political strategy. Its strident moralism gave the Party a very broad set of priorities, some 'simple Christian moorings' (Groser, 1949, page 27), but it was unmarked by any detailed socialist theory. This allowed Poplar Labour Party to commit itself to some decidedly nonsocialist policies. In 1924, for example, the Housing Committee of the council decided not to rent council houses to any Asians (there were some Chinese living in the borough) (PMBC Housing Committee minutes, 7 December 1925), and the WSF had earlier accused George Lansbury of criticising a political opponent for being foreign (*Workers' Dreadnought* 6 October 1923). Such racism would never have been countenanced by a scientific socialist; to them, all workers were identical because all suffered under the capitalist yoke. But, as was argued above, Poplar's socialism was not based on Marx but on the moral virtue of Christ and the working class; such moral criteria made racist decisions by councillors possible because foreigners could be characterised as immoral. This was certainly true in the case of the Chinese, who were perceived as gambling opium dealers in the 1920s, in the East End novels of Thomas Burke, for example.

Melodrama

Poplar's love of performance and oratory, the great power which an eloquent individual could wield over Poplar people, was also implicated in Poplar's changed relationship to the institutions of its civil society, because it created a political loyalty not to party ideology but to individual Labour activists. George Lansbury in particular inspired a remarkable affection from nearly everyone who knew him, but all the councillors and guardians relied on personal familiarity with their constituents for their popularity. George Lansbury's biographer, who knew both George and Poplar very well, claimed that

George could do anything at all in the 1920s and Poplar would still vote for him (Postgate, 1951, page 220), and Lansbury's successor as Member of Parliament for North Poplar was recommended to voters not on the basis of his political beliefs but because George would have approved (Key collection).

This absence of electoral vulnerability in Poplar enabled Poplar Labour Party's politicians to do whatever they liked and still be returned to office, and that made Poplar's councillors, on their own admission, lazy, uncritical of themselves, and unwilling to listen to others (Overland, 1975, page 92). More and more often as the 1920s progressed they refused to see deputations from local groups since they had nothing to lose by so doing; the Works Committee of the council in 1927, for example, refused to see deputations from the Transport and General Workers Union and the local costermongers whose activities they had begun to police (PMBC Works Committee, 16 May 1927 and 11 October 1927). By the end of the decade this insularity led to candidates from the Communist Party and the ILP opposing the Labour Party in Poplar's local elections, accusing the Labour Party of selling out and of class collaboration (Tower Hamlets Local History Library local election pamphlet collection).

Labour

By the end of the 1920s it is possible to identify a new fifth element in Poplar's local culture: an unwavering support for the Labour Party in the borough. Poplar Labour Party had its own impact on the locality. It had imposed its own frame of awareness on the local civil society, making voting Labour a more or less inevitable part of living in Poplar. It had achieved this not only by over thirty years of political education in the area (G. Lansbury, 1928; Postgate, 1951), but also by its high levels of relief and its very good provision of services when it gained office: housing, public wash-houses and laundrettes, maternity and child welfare clinics, electricity, street cleaning, public conveniences, a recreation ground (PMBC, 1925). By the late 1920s, voting Labour was one of Poplar's habits, and the Labour Party no longer had to work to gain the allegiance of the electorate (interviews).

Nor did it have to justify its actions or defend its corruption to local people, because the ordinary voter in the borough never challenged them. My interviewees saw Poplar's corruption more as a joke than a problem and they responded to it by trying to find their own strings to pull on the council. This response suggests that they accepted the relevance of family ties and so on to social and political life and that the council's corrupt practices never involved a cultural alienation from the Poplar electorate. Thus Poplar Labour Party remained in power despite its increasing remoteness from the formal organisations of its civil society.

The causes of this shift in the power relationship between the local state in Polar and its local civil society are complex. Most important was the decline in the militancy of Poplar Labour Party's socialism. It has been argued that much of the Party's commitment to participatory politics sprang from its radicalism and, more specifically, from the ideas of the ILP and WSF about local democracy and mass involvement in the political process. Thus when the revolutionary ardour of the Party began to fail, so too did its attempts at participatory politics. The radicalism of the labour movement in general declined during the 1920s, and especially after the failure of the General Strike to help the miners win their battle against wage reductions, but in Poplar the process was exacerbated by the disappearance of several of its most left-wing activists. Some, like the communists Minnie Lansbury and Alf Watts, died, some like Sylvia Pankhurst, left the area, and George Lansbury was increasingly busy in the House of Commons. Added to this, Charlie Key emerged as Poplar Labour Party's leader. Key was an expert on local government law, but he was an expert in exploiting existing legislation to Poplar's best advantage; he was a reformist, not a revolutionary. The drift to the right in Poplar's politics which resulted from these changes in leadership was not checked by the warnings which socialist theory would have given because, as has been pointed out, no such benchmark of radicalism existed in Poplar Labour Party.

Equally relevant as an explanation for the growing autocracy of the local state in Poplar was the power of the central state over the local, which increased significantly over the decade. The government passed legislation to give itself greater control over the actions of local authorities, and this curtailed Poplar's ability to act as it chose and be as radical as it liked. The Guardians (Default) Act of 1926 was especially significant because it allowed the central state to replace locally elected guardians with civil servants if a Board of Guardians refused to follow the ministry's wishes; it was seen by the government specifically as an act to coerce Poplar into submission (PRO, MH79/305). Poplar's Board, with its great belief in local democracy, refused to be replaced by officials of central government and in order to avoid this fate it began to alter its poor relief policies. The scale of relief was cut and so too were the numbers of people in receipt of relief, and the guardians began to prosecute anyone who claimed relief under false pretences, something it had never done before (PRO, MH68/218). The government also expanded its powers over the local state by increasing its control over their expenditure. Since Labour councils were committed to providing good services, the limits set by central government on their spending in effect restricted their political radicalism. This control grew in two ways in the 1920s. First, the powers of the district auditors to decree what was necessary expenditure and what was not were increased (Robson, 1925), and, second, there was the secret 'Q'-list (Ward, 1986). This was a list of local authorities whose finances were felt by the Ministry of Health to be insecure because of high relief spending and large debt burdens; localities on the list had their applications for loan sanctions carefully scrutinised and revised by the Ministry, and Poplar's more ambitious projects were reduced in scale by civil servants because of Poplar's 'Q'-listing (PRO, MH52/191).

The control of central government over Poplar was tightened too by the arrival in 1926 of a new Clerk to the Guardians more convinced than his predecessor of the need for 'sound' orthodox finance, who assured the Ministry in early February 1928 that he could and would persuade Poplar's

Board to cut their spending and reduce their loan debt, and by the end of the month had succeeded in doing so (PRO, MH68/218). As a result of these changes, as the 1920s progressed, Poplar's Labour council became less and less guided by local needs. It was turning into what orthodox political theory suggests it should have been all along: an administrative arm of central government.

Conclusions

I hope it is obvious that local politics *are* shaped by local cultural values, by the local configuration of 'frames of awareness' and 'communal sensibilities' which Geertz has talked about. Certain formal institutions in Poplar's civil society – the left-wing political groups and the churches – and certain informal understandings which were held more widely in Poplar – the morality of the neighbourhood and the love of performance and melodrama – both shaped Poplar's politics throughout the decade. And the Labour Party forged a mass communal sensibility of its own too: the naturalness of voting Labour.

But the political implications of this local culture – by which I mean the social power relationships developing from these awarenesses and sensibilities – are by no means straightforward. In Poplar, the same cultural context sustained two types of relations between the local state and its local people: participatory and autocratic. 'Culture' then is not a simple solution to the problem of geographic variation in social and political action; its social implications are not straightforward or consistent over time or space, and this means that locality-study theorists should be much more wary than they have been hitherto in evoking 'culture' as some sort of ultimate explanation of social and political behaviour.

There is a further reason for caution in the use of the concept of culture. Geertz, whose conceptualisation of culture I have used in this paper, is very keen to stress the social reproduction of cultural symbols, but he can be criticised for his lack of interest in the social sources of power structures. Geertz is little interested in the origins of the power relationships which are fused into cultural practices, and it is for this reason that the idea of culture can only be a partial solution to the problems which analyses of local politics have so far encountered. As I have shown in this paper, local politics cannot be fully understood without the social relationships involved within them. And to grasp these social relations, we need detailed historical analysis, as well as geographical analysis. Investigation must be historical because, as I have shown, the transformation of the power structures of a political culture can be explained by looking at the changing circumstances of the locality. For, as the local and the national contexts of Poplar changed, so too did the political expression of its culture.

Acknowledgements

A version of this paper was presented at the 'New Directions in Cultural Geography' conference held at University College, London, in September

1987, and I would like to thank the participants there for their comments on it. I am also grateful to Roger Lee and David Ley for their help, and to the people who have talked to me about their memories of life in Poplar in the 1920s.

References

Agnew, J. A., Mercer, J. and Sopher, D. E. 1984: *The City in Cultural Context*. Hemel Hempstead, Herts: Allen and Unwin.

Bedarida, F. 1975: 'Urban growth and social structure in nineteenth-century Poplar'. *London Journal* 1, 159–88.

Binyon, G. C. 1931: *The Christian Socialist Movement in England*. London: Society for Promoting Christian Knowledge.

Blake, J. 1977: *Memories of Old Poplar*. London: Stepney Books.

Blunkett, D. 1984: interview. In Boddy, M. and Fudge, C. (eds), *Local Socialism? Labour Councils and New Left Alternatives*. London: Macmillan, 242–60.

Branson, N. 1979: *Poplarism 1919–1925: George Lansbury and the Councillors' Revolt*. London: Lawrence and Wishart.

Brill, K. 1971: *John Groser, East London Priest*. Oxford: A. R. Mowbray.

Bullock, A. 1960: *Ernest Bevin:* Volume One, *Trade Union Leader 1881–1940*. London: Heinemann.

Bush, J. 1984: *Behind the Lines: East London Labour 1914–1919*. London: Merlin Press.

Byrne, D. 1982: 'Class and the local state'. *International Journal of Urban and Regional Research* 6, 61–82.

Cockburn, C. 1977: *The Local State: Management of Cities and People*. London: Pluto Press.

Cooke, P. 1986: 'The changing urban and regional system in the United Kingdom'. *Regional Studies* 20, 243–51.

Council of Action archive, Labour Party Library, 150 Walworth Rd, London SE17.

Cronin, J. 1984: *Labour and Society in Britain 1918–79*. London: Batsford.

Cullen, C. K., collection of papers, Marx Memorial Library, 37a Clerkenwell Gn, London EC1.

Dick, W., manuscript diary June 1924–December 1926, Tower Hamlets Local History Library, 277 Bancroft Rd, London E1.

Donoughue, B. and Jones, G. W. 1973: *Herbert Morrison*. London: Weidenfeld and Nicolson.

Duncan, S. S. and Goodwin, M. 1982: 'The local state and restructuring social relations: theory and practice'. *International Journal of Urban and Regional Research* 6, 157–86.

ELA, 1919–29: *East London Advertiser*. Tower Hamlets Local History Library, 277 Bancroft Rd, London E1.

Finer, H. 1933: *English Local Government*. Andover, Hants: Methuen.

Geertz, C. 1973: *The Interpretation of Cultures*. New York: Basic Books.

Geertz, C. 1983: *Local Knowledge: Further Essays in Interpretative Anthropology*. New York: Basic Books.

Gillespie, J. A. 1984: 'Economic and political change in the East End of London during the 1920s'. PhD dissertation, Department of History, Cambridge University, Cambridge.

Grant, C. 1930: *Farthing Bundles*. Fern St Settlement, Leadenham Ct, Spanby Rd, London E3.

Groser, J. B. 1949: *Politics and Persons*. London: SCM Press.
Guardians (Default) Act, 1926: *Public General Acts – George V* 16 and 17, chapter 20. London: HMSO.
Gyford, J. 1984: *Local Politics in Britain*. Beckenham, Kent: Croom Helm.
Gyford, J. 1985: *The Politics of Local Socialism*. Hemel Hempstead, Herts: Allen and Unwin.
Hannington, W. 1936: *Unemployed Struggles 1919–1936*. London: Lawrence and Wishart.
Hardie, J. Keir 1905: *Can a Man be a Christian on a Pound a Week?* London: Independent Labour Party.
Johnston, R. J. 1985: 'Local government and the state'. In Pacione, M. (ed.), *Progress in Political Geography*. Beckenham, Kent: Croom Helm, 152–76.
Johnston, R. J. 1986: 'The neighbourhood effect revisited: a spatial science or political regionalism?' *Environment and Planning D: Society and Space* 4, 41–55.
Jones, G. W. 1973: 'Herbert Morrison and Poplarism'. *Public Law* 18, 11–31.
Keith-Lucas, B. 1962: 'Poplarism'. *Public Law* 7, 52–80.
Key, C., collection of papers, Bodleian Library, Oxford.
Key, C. 1925: *Red Poplar: Six Years of Socialist Rule*. London: Labour Publishing.
Labour Party, 1918: *Report of the Eighteenth Annual Conference of the Labour Party*. London: Labour Party.
Lansbury, E. 1934: *My Father*. London: Sampson, Law and Marston.
Lansbury, G., collection of papers, London School of Economics, London.
Lansbury, G. 1907: *The Development of the Humane Administration of the Poor Law under the Poplar Board of Guardians*. Woolwich: Labour Representation Company; copy can be consulted at Tower Hamlets Local History Library, 277 Bancroft Rd, London E1.
Lansbury, G. 1924: *Jesus and Labour*. London: Independent Labour Party.
Lansbury, G. 1928: *My Life*. London: Constable.
Lansbury, G. 1937: 'Why pacifists should be socialists'. *Fact* 7, 5–74.
Lansbury's Bulletin May 1926: Tower Hamlets Local History Library, 277 Bancroft Rd, London E1.
Lax, W. 1927: *Lax of Poplar*. London: Epworth Press.
Lester, M. 1937: *It Occurred to Me*. London: Harper.
Livingstone, K. 1984: interview. In Boddy, M. and Fudge, C. (eds), *Local Socialism? Labour Councils and New Left Alternatives*. London: Macmillan, 260–383.
London County Council, 1915–30: *London Statistics* volumes 11–19, London County Council, London.
MacDonald, J. Ramsey 1924: *Socialism: Critical and Constructive*. London: Cassell.
Macintyre, S. 1980: *Little Moscows: Communism and Working-class Militancy in Inter-war Britain*. Beckenham, Kent: Croom Helm.
Mark-Lawson, J., Savage, M. and Warde, A. 1985: 'Gender and local politics'. In Lancaster Regionalism Group (ed.), *Localities, Class and Gender*. London: Pion, 195–215.
New Survey of London Life and Labour 1931: Volume II. *London Industries*. London: P. S. King.
New Survey of London Life and Labour 1932: Volume III. *Survey of Social Conditions I: Eastern Area*. London: P. S. King.
New Survey of London Life and Labour 1933: Volume V. *London Industries II*. London: P. S. King.
New Survey of London Life and Labour 1934: Volume VIII. *London Industries III*. London: P. S. King.

Overland, A. 1975: interview. In Richman, G. (ed.), *Fly a Flag for Poplar*. London: Liberation Films, 138 Fordwich Rd, NW2, 91–4.

Pankhurst, E. S. 1931: *The Suffrage Movement: An Intimate Account of Persons and Ideals*. London: Longman's and Green.

PBMA, various years, Poplar Borough Municipal Alliance, archive, Tower Hamlets Local History Library, 277 Bancroft Rd, London E1.

PMBC, Poplar Metropolitan Borough Council meeting minutes, 1919–1929: Tower Hamlets Local History Library, 277 Bancroft Rd, London E1.

PMBC, Poplar Metropolitan Borough Council Housing Committee meeting minutes, 1919–1929: Tower Hamlets Local History Library, 277 Bancroft Rd, London E1.

PMBC, Poplar Metropolitan Borough Council Works Committee meeting minutes, 1919–1929: Tower Hamlets Local History Library, 277 Bancroft Rd, London E1.

PMBC, 1925: *The Work of Six Years, 1919–25*. Poplar Metropolitan Borough Council, London; Tower Hamlets Local History Library, 277 Bancroft Rd, London E1.

Pollitt, H. 1940: *Serving My Time: An Apprenticeship in Politics*. London: Lawrence and Wishart.

Poplar Labour Party 1922: 'Guilty and proud of it'. Tower Hamlets Local History Library, 277 Bancroft Rd, London E1.

Postgate, R. 1951: *The Life of George Lansbury*. London: Longman.

PRO, various years, files of Ministries of Health (MH), Housing and Local Government (HLG), and Labour (LAB), Public Record Office, Ruskin Ave., Kew, Richmond.

Rees, G. 1985: 'Introduction: class, locality and ideology'. In Rees, G., Burja, J., Littlewood, P., Newby, H. and Rees, T. L. (eds), *Political Action and Social Identity: Class, Locality and Ideology*. London: Macmillan, 1–15.

Rees, G., Bujra, J., Littlewood, P., Newby, H. and Rees, T. L. 1985: *Political Action and Social Identity: Class, Locality and Ideology*. London: Macmillan.

Rees, G., Williamson, H. and Winckler, V. 1987: 'Employers' recruitment/training strategies and local economic restructuring', paper presented at the Urban Change and Conflict conference, Canterbury; copy obtainable from Dr Rees, Department of Sociology, University College, Cardiff.

Representation of the People Act, 1918: *Public General Acts – George V* 7 and 8, chapter 64. London: HMSO.

Robson, W. 1925: *The District Auditor: An Old Menace in a New Disguise*. London: Fabian Society.

Robson, W. 1931: *The Development of Local Government*. London: George Allen and Unwin.

Rose, M. 1957: *The East End of London*. London: Cresset.

Saunders, P. 1984: 'Rethinking local politics'. In Boddy, M. and Fudge C. (eds), *Local Socialism? Labour Councils and New Left Alternatives*. London: Macmillan, 22–48.

Savage, M. 1987: 'Understanding political alignments in contemporary Britain: do localities matter?' *Political Geography Quarterly* 6, 53–76.

Scurr, J. 1924: editorial, *Socialist Review* September issue, page 85.

Stone, D. 1985: 'Municipal socialism 1880–1982: an introduction and bibliography', WP-44, Urban and Regional Studies, University of Sussex, Brighton.

Toynbee Hall, 1922: *Unemployed in East London*. London: P. S. King.

Urry, J. 1981: *The Anatomy of Capitalist Societies: The Economy, Civil Society and the State*. London: Macmillan.

Walters, C. Ensor 1927: foreword in Lax, W., *Lax of Poplar*. London: Epworth Press, 8–13.

Ward, S. V. 1986: 'Implementation versus planmaking: the example of list Q and the depressed areas 1922–39'. *Planning Perspectives* 1, 3–26.
Warde, A. 1986: 'Space, class and voting in Britain'. In Hoggart, K. and Kofman, E. (eds), *Politics, Geography and Social Stratification*. Beckenham, Kent: Croom Helm, 33–61.
Williams, R. 1963: *Culture and Society 1780–1950*. Harmondsworth, Middx: Penguin Books.
Workers' Dreadnought 1917–24: Tower Hamlets Local History Library, 277 Bancroft Rd, London E1.

12 Cindi Katz, 'Sow What You Know: The Struggle for Social Reproduction in Rural Sudan'

Excerpts from: *Annals of the Association of American Geographers* **81**, 488–514 (1991)

A child is perched precariously atop a donkey laden with a sack of sorghum seeds, a digging stick, and a couple of hoes. She steadies herself and rides surely along the canals as the sun rises. It takes nearly an hour to reach her family's field. She and her brother and sister will spend the morning planting part of their family's farm tenancy in sorghum while their father clears the irrigation ditches nearby. Punctuating their work at irregular – and to their father annoying – intervals, this girl and her brother will set up a home-made net trap in an unsuccessful attempt to ensnare some of the birds that descend on the area during the rainy season. On other days they may succeed in trapping a dozen or more small birds which the boy will kill, following Islamic practice, and the children's mother or older sisters will cook for a family meal. School is out during the rains, mainly to allow teachers to return home from rural areas which become inaccessible and difficult to live in during this time. Partly by design, but mostly by coincidence, this schedule also allows all children, including students, to assist their households with the heavy burdens of agricultural work. As the children make their way to the fields, they cross paths with a number of herdboys leading flocks of small animals out to pastures just turning green with the arrival of the rains.

The survival of agricultural production systems over time turns, at least in part, on what children learn about the natural environment, and how they use this knowledge and its attendant skills during their childhood and as they come of age. This narrative drawn from a year of geographic field research in rural Sudan suggests both the importance of and the variation in children's environmental interactions in an agrarian community. What children learn about the environment and how they use this knowledge in their work and play are fundamental cultural forms and practices, shared in a social matrix and bearing a specific relationship to the prevailing social relations of production and reproduction in the area. Inherent in them are the contradictory possibilities of replication, reformulation, and resistance. While this contradictory relationship is increasingly recognized in social theory, there have been few studies in any field and almost none in geography that examine its possibilities and practices on the ground.[1] This piece invokes one such investigation to argue that the cultural forms and practices of social reproduction, such as those which were its focus – the production, exchange, and use of environmental knowledge – have the potential to disrupt, subvert and even reconstitute the accumulation of capital and its attendant social relations of production.

In an agriculturally based economy, learning about the environment – about farming, animal husbandry, and the use of local resources – is an aspect of socialization essential to maintaining and reproducing society. Moreover . . . children's work often is fundamental to the daily maintenance of their households, and thus the community as a whole. My study approached the question of children's environmental interactions as central not only to the activities of production but as fundamental to the daily reproduction of their households, and their environmental learning and knowledge as crucial to the long-term maintenance of the socioeconomic system itself.

My research on these questions enabled me to examine the entwined processes of socioeconomic transition and cultural–ecological change from a perspective little addressed in the development literature in geography, sociology, anthropology, or economics, and to cast new light on the practical response of Third World populations experiencing pronounced shifts in the nature of their articulation with the relations of capitalist production. This study links the human-environment tradition in geography, and in particular its concern with how people come to know the environment and act to transform it, with the larger project of cultural studies which seeks to understand the connections between particular aspects of culture, such as art, ideology, consciousness, dreams and fantasies, literature, knowledge, and everyday life, and social and economic structures and practices (e.g. Thompson, 1963; Gramsci, 1971; Genovese, 1972; Aronowitz, 1973; Bourdieu, 1977; Williams, 1977; Willis, 1977; Said, 1978; Johnson, 1979; Samuel and Jones, 1982; de Certeau, 1984; Spivak, 1987). Its central argument is that under circumstances of socioeconomic transformation, what children learn about the environment and how they acquire and use that knowledge can have contradictory effects not only upon the children as they come of age but on the outcome of the social change itself.

The goals of my research were threefold: (1) to discover and document both the activities through which children acquired and used environmental knowledge, and the content and organization of that knowledge; (2) to examine these as cultural forms and practices, that is, in dynamic articulation with the labor process and the social relations that underlie it; and (3) to describe, and if possible analyze, how children's environmental learning, knowledge, and interactions appeared to be changing in relation to the larger social, economic, and environmental changes taking place in rural Sudan.

In recent years geographers have addressed the complex relationships between human agency and the social and ecnomic structures of society. Their work represents an admirable effort to enrich our understanding of the central problematics of human geography – spatial relationships, the making of place and human–environment relations – by articulating rigorous historical analyses of socioeconomic, spatial, and political structures, and the means by which people create, transform, and respond to them (e.g. Gregory, 1982; Christopherson, 1983; Thrift, 1983; Watts, 1983; Clark and Dear, 1984; Pred, 1984a, b; Smith, 1984; Gregory and Urry, 1985; Warf, 1988; Harvey, 1989, 1990; Soja, 1989). It is difficult, however, to move analytically between structure and substance, and most studies end up concentrating on socioeconomic and political structures without any real analysis or understanding of the practical activities of the people who create and are constituted by these structures. Alternatively, they focus on human behavior and consciousness without locating these in a mutually determining sociospatial and political–economic field. All too often, people are recognized as 'making their own history', but this history is rather thin and pale, a narrative of socioeconomic change that points to but glosses over the material social practices of everyday life to focus only on their temporal and spatial outcomes. Equally problematic, the fact is elided that people do not make their own history out of circumstances of their own choosing, and the material social practices of everyday life are presented as if unbounded by any social or economic structures. In this way, history and geography become voluntaristic free-for-alls.

These are political decisions and omissions. Part of the problem is that few of the analyses claiming to address structure and agency empirically are grounded in theory that can analyze or explain the mutual determinations between the two and their material outcomes over time and space. Moreover, few of these studies draw even partially on an ethnographic approach to address the multiplicity of forms and practices that constitute social activity. We are often left with only the traces of human action that are measurable and lose the texts and textures of that action.

Serious consideration of feminist theory and research over the past two decades would improve this work substantially. Although it is not clear from reading geographic work in what has come to be known as the structurationist school, much of the ground linking human agency to social and economic structure was broken by feminist theorists who turned to 'everyday life' to find subversive power in the voices that had been systematically excluded from history (e.g. Dalla Costa, 1972, 1977; Caulfield, 1974; Ehrenreich and English, 1975; James, 1975; Conference of Socialist Economists, 1977; 'Development and the sexual division of labor', 1981; Moraga and Anzaldua, 1981; Sargent, 1981).[2] In addition to its influence on the direction of research in social science, feminist theory is at the root of much poststructuralist literary theory which analyzes texts as, *inter alia*, expressions of socioeconomic and political struggle (e.g. de Lauretis, 1982; Haraway, 1984; Spivak, 1987). In broadly defining the 'text' and analyzing it against the grain of sociocultural and political–economic structuring, this work, though rarely acknowledged by geographers, is at the root of the social science project linking agency and structure.

As a geographer, I sought as part of my project in Sudan to address a single place – the village of Howa – at a particular historical moment – ten years after its incorporation in a state-sponsored capitalist agricultural project – as one such 'text', constituting at once the ground (literally) for a particular set of material social practices and a repository which expressed in material form the outcomes of these practices. These were understood as mutually constituting. The scale was local, the historical subjects small, but the material social practices associated with their acquisition and use of environmental knowledge are articulated fundamentally with the profound socioeconomic and cultural–ecological changes under way there and in the rest of Sudan. In addressing these, my intent was to analyze theoretically and practically the relationship between the social relations of production and social reproduction under conditions of socioeconomic transformation.

Background

During 1980–81 I conducted a year-long ethnographic study in a farming village I call Howa in central eastern Sudan. My study focused primarily upon the environmental learning, knowledge, and interactions of ten-year-old children addressed in relation to household production and the reproduction of the social relations of production. I chose to undertake this study in an area undergoing profound socioeconomic change in an attempt to locate instances of contestation between the reproduction of what Bernstein (1982, after Marx, 1967, vol. 3) calls the natural economy and the reproduction of capitalist relations of production.[3] The latter had been imposed on the area most decisively by incorporation in the government-sponsored Suki Agricultural Development Project in 1971, ten years prior to my study and the year the children who are its focus were born.

Study site

Howa, a village on the banks of the ephemeral Dinder River east of Sennar in central eastern Sudan (Figure 12.1), was settled in the late nineteenth century by pastoralists. Until 1971, its people cultivated sorghum on a subsistence basis, supplemented by sesame which they sold to passing traders in order to meet their limited needs for cash. Animals were an important source of subsistence and savings for the population. According to local residents and historical accounts of the area, most families kept at least a few goats and sheep for milk and occasional meat, and many households kept some cattle or camels along with flocks of ten to fifty small animals (Tothill, 1948; al Tayib, 1970; Ahmad, 1974; O'Brien, 1978; Gruenbaum, 1979; Duffield, 1981). Many families maintained close ties with relatives who remained exclusively pastoralist, not only having their herds travel with the pastoralists for part of the year, but renewing and maintaining familial relations by intermarriage.

During the 1950s the village was further integrated with the national cash

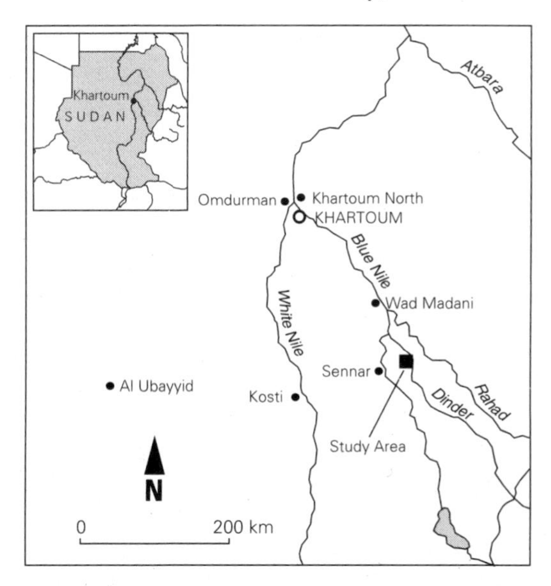

Figure 12.1 Sudan, showing provincial boundaries, main rivers and location of study site (courtesy of the author)

economy through the private cotton schemes along the Blue Nile approximately fifty kilometers away. Men from the village worked as agricultural wage laborers and a few managed to become tenants in these private pump schemes. These tenants apparently began to accumulate capital when cotton fetched record prices during the 1950s and early 1960s. Apart from involvement in these schemes, most of the population of Howa remained subsistence cultivators until 1971.

That year, Howa was incorporated in the State-sponsored Suki Agricultural Development Project. Not only was the basis of the socioeconomic formation of Howa altered from the subsistence production of food crops to the irrigated cultivation of cotton and groundnuts as cash crops, but, more importantly, local control over production and reproduction was undermined and the relation between them interrupted (cf. Barnett, 1975, 1977; Duffield, 1981 for other examples in Sudan). Under the Project, agents of State capitalism – the project authorities – determined not only which crops would be grown by tenants, but also the production schedules and acceptable agricultural tools and practices for their cultivation. The peasant farmers of Howa actively resisted each of these externally imposed changes on their farming practices.

At the root of this disruption in the rhythm of daily and annual work cycles was an altered relationship between the local population and productive resources, primarily the land. Land in the vicinity of Howa historically had been held in common and allocated through customary right by local sheikhs to villagers for dryland cultivation. The local economy was geared primarily to the production or extraction of use-values, i.e. goods to satisfy its own subsistence needs, from a combination of agriculture, animal husbandry and forestry. Under these socioeconomic conditions, constraints on production were determined more by the physical and economic limits of available labor and other resources than by external political–economic relations such as those associated with the Suki Project. It not only circumscribed the land available for cultivation, but determined who had access to it under what conditions.

It would be erroneous to assume that State intervention and economic penetration by merchant capital are the sole sources of transformation in this or any socioeconomic formation. Indeed, my conceptualization of the problematic was one of active negotiation between social, economic, cultural, political, and ecological relations and practices. This study addresses some of its manifestations on the premise that people are not dupes, but rather are active subjects making their own histories and geographies within and against the determinations of historically and geographically specific structures. In Howa these practices were a nexus of struggle between and among the local population and fractions therein, including agents of State capitalism such as schoolteachers and Project authorities, and local religious leaders, particularly of mystical and fundamentalist orders of Islam (cf. el Hassan, 1980). It is noteworthy that these struggles, which are fundamentally over capitalist hegemony, were carried out between, and even within, real people. They are trivialized if they are constructed as roles played out in a structuralist drama that pits the agents of state capitalism against the poor peasants (cf. Thrift, 1983, p. 35).

Many of the changes brought about by inclusion in the agricultural project were welcomed by the local population. For years, many had sought increased incomes as agricultural laborers and, when possible, tenants in the private cottom schemes along the Blue Nile. They saw the Suki Project as offering the opportunity for increased income at home. But under the Project the local community lost much autonomous control over production, and it was over these relations that much of the ensuing struggle turned. This shift in the social relations of production and reproduction – what Watts (1987) calls the 'multi-directional and episodic' nature of capitalist development – marks the larger context for my study of children's acquisition and use of local environmental knowledge.

Methodology

This study was primarily ethnographic. I worked intensively with a small number of children and their families with constant reference to the sociocultural and political–economic dynamic of the village as a whole. The children were from households of tenants and non-tenants of each socioeconomic status. They themselves were a diverse group: boys and girls, students and non-students, of each birth-order position. This population was selected on the basis of a full enumeration and survey of the entire village. This, along with the diversity of the methodology I developed, the standards I maintained in working with the children, and comparisons of my results with those of the few existing relevant studies, gave me confidence in the validity of the information produced. Nevertheless, it was not my intent to produce findings that could be analyzed statistically with any reliability; the complexity of the information and the small sample size raise questions about even the simplest of statistical analyses. Rather, my agenda was to discover, document, and describe the range of children's environmental interactions and the forms and content of their environmental knowledge in all their complexity; and to analyze these in relation to the larger context in which they occurred.

The experience of living in the area for a year and working closely with the children and their families produced extremely rich data. These data were discovered and grounded in a particular temporal and geographic context which I shared. It is important to remember that in this sort of research, distantiation is impossible (cf. Koptiuch, 1985; Probyn, 1990). Unlike most positivist approaches to empirical research, ethnographic research does not claim to be objective. Indeed, its essence is that it is not. Its goal, cultural description, is reached in a wholly subjective manner even though a range of scientific methods are employed. This text is the result of my experiences during a particular year in Howa. Not only my research interests, but my background and personality framed my experience and filtered both what I saw and the way I interpreted it. I do not claim, then, to speak for 'the other' (cf. Clifford and Marcus, 1986; Marcus and Fischer, 1986; Spivak, 1988), the children and adults who participated in this study, but only for myself. To claim otherwise would be dishonest intellectually.

The changing roles of children during socioeconomic and cultural–ecological transformation

The general hypothesis that guided this work from the outset was that a change in the production system, such as was caused by incorporation in the Suki Agricultural Project, would alter the settings and activities for reproducing that production system. My interest in this question was rooted in the perspective that these settings and activities were not simply the outcome of economic factors, but rather were themselves constructed by the material social practices of living historical subjects. My goal was to examine the articulation between the two, for in it lie the multitudinous possibilities of reproduction, reaction, resistance, and reformulation. The household, the peer group and the school or other sites of formal training were the settings of particular concern, and children's work, play, and formal learning were the activities of particular interest. Following Marx (1967), reproduction was understood as physical – encompassing both biological reproduction and the appropriation and/or production of means of subsistence adequate to ensure the daily maintenance of the population – and sociocultural – reproducing the conditions of life and labor in which the skills and knowledge associated with social reproduction figure prominently. My research addressed children's roles in production and the provision of the means of existence, *and* their acquisition of the knowledge and skills necessary to maintain the system of production and reproduction over time. These were examined as lived experiences and contested practices.

Although there is a generally recursive nature to these material social practices addressed analytically in the literature on everyday life (e.g. Lefebvre, 1984), these practices and their relationships to the social relations of production and reproduction are always and everywhere in flux. Everyday life is significant as a critical concept, not as a descriptive notion for the mundane and unspectacular practices by which we construct ourselves and reproduce society, but because inherent in these is the potential for rupture, breakdown and transformation (cf. Kaplan and Ross, 1987; Katz, 1989). At particular historical junctures, the potential for these is heightened. The altered relations of production imposed by inclusion in the Suki Agricultural Project created such a moment in Howa. The acquisition, deployment, content and organization of children's environmental knowledge were addressed to examine the means by which the social formation of Howa was being reproduced and to locate discontinuities in these cultural forms and practices.

By the time of my research in 1981, the Suki Agricultural Development Project, established in 1971, had already had a significant effect in transforming not only the social relations of production and reproduction in Howa, but the local ecology as well. These transformations had led to conditions which altered, among other things, the nature of children's interactions with the environment as well as the means and content of their environmental learning. My initial hypothesis was that as a result of the socioeconomic changes under way, more children would attend school and thus participate less in the work of their households. By extension, I thought that the decline in children's work would begin to erode the traditional relationship between work and play, and, in turn, the experience of the two as means of environmental learning. While a year of field research did not allow sufficient time to analyze definitely how these activities were changing either in themselves or in relation to one another, analysis of the sample population of children and return trips to Howa in 1983 and 1984 made clear that this hypothesis was off the mark.

One of the contradictory effects of incorporation in the agricultural project was that rather than increasing school enrolment and children's play time as I had hypothesized originally, it appeared that the changes wrought by inclusion in the Project demanded more labor from the children. According to teachers in the village school, enrolment had not increased in the years since Howa was included in the Suki Project. Moreover, it was their impression, although they did not keep figures, that a greater proportion of children in the village primary school left before completion than had done so prior to inclusion in the Project. The reasons for this were threefold: (1) the higher labor demands associated with the irrigated cultivation of cotton and groundnuts compared with the rainfed cultivation of sorghum and sesame; (2) the environmental changes brought about by the irrigation project; and (3) the increased need for cash engendered by incorporation in the agricultural development project and the global cash economy which it represented. These are discussed in turn.

In interviews concerning the changes taking place in Howa, adults frequently indicated that the demands of cultivating cotton and groundnuts as cash crops exceed those associated with the rainfed cultivation of sorghum and sesame that had prevailed before 1971. In addition to the labor demands of irrigation and the use of imported fertilizers, pesticides, and herbicides, Project authorities required four weedings of each crop with a short-handled hoe that was more difficult and tiresome to use than the customary long-handled or very short-handled variety.[4] In most households, these labor demands were met with the increased use of family labor, especially children. Many tenant households, particularly those of lower socioeconomic status, reported that children were kept from school enrolment or forced to withdraw after a couple of years because they were needed to help with the full range of agricultural tasks, many of which took place during the school year. Some studies that address children's contribution to household labor suggest that when children are needed for labor-intensive agricultural or other tasks, they are kept home from school (e.g. Landy, 1959; Tienda, 1979). My experience suggests that when this was the case for specific seasonal tasks, households managed without the inputs of their children. But when children's labour was important to a range of tasks such as agriculture, or for an essential year-round activity such as herding, either they were never enrolled or they left school prematurely.

Another reason that children's work in Howa appeared to have increased was the deforestation that had resulted from the shifts in land-use associated with the irrigation project. The traditional system of mixed land-use, combining subsistence dryland agriculture and the raising of small animals, left wooded areas adequate to meet local needs. With the establishment of the Project in 1970–71, the amount of woodland in the vicinity of the village was severely curtailed. By 1981 there were few trees apart from ornamentals within a half-hour's walk from the settlement, and most of the remaining trees within an hour's walk were less than five centimeters in diameter. Adults in the village noted that this situation had increased the time required to procure adequate household fuel supplies. In Howa, where many aspects of fuel provision were the responsibility of children, these changes led to substantial increases in children's work time. Not only did they have to go further afield to collect or cut fuelwood, but the poor quality and small size of most of what was available increased the number of trips per week necessary to provide sufficient wood for domestic consumption.

Finally, children's environmental and other work had increased because of the larger economic changes associated with the introduction of the agricultural project. Not only did the Project bring about enormous changes in the relations of production in Howa, but it heightened the integration of the village into the national cash economy. With the establishment of the Project, access to many goods that had been commonly held or freely available was restricted, for example, wood products. The issue of forestry resources again provides an example of particular relevance to the question of children's labor. As wood products became more difficult to procure, some wealthier households began to purchase them rather than increase the demands on children or other household members. Members of poorer households in the village, including children, began to provide wood for sale. In this way a freely held good gradually becomes a commodity. As more freely held goods become commodities, the need for cash increases. By 1981 this process was well under way, compounded by an explosion in the number of merchants and traders in the village since 1971. Their presence introduced an increasingly wide array of consumer goods to the village.

Commoditization is tied to socioeconomic differentiation. In Howa, the ascendance of the cash economy led to increases in children's workload, because in many households they were needed to earn money to help meet the growing needs for cash. Twelve of the seventeen children in the sample group earned cash that helped provide household subsistence. The twinned processes of commoditization and differentiation particularly affected two resources that children were significant in procuring: wood and water. When marginalized families sought new means to earn cash, it often fell to children to fetch water or cut wood for sale. Ten-year-old children also helped to produce and scavenge charcoal for sale in nearby towns, harvested vegetables from family garden plots and hawked these in the village, and worked as hired help in the tenancies of other village households. In these ways, the increased need for cash, created by the agricultural project and fanned by the exigencies of the monetized economy of which it was a part, increased the work of children in Howa.

These shifts in children's work had consequences for the relationship between work and play, and the nature of these as means for the acquisition and use of environmental knowledge (cf. Katz, 1986a, b), as well as for school attendance. The gains in school enrolment I expected may have been limited, at least in part, by the increased demand for children's labor in Howa. In 1981, 42 per cent of the boys and 4 per cent of the girls aged seven to twelve years old were enrolled in the village school. During 1979 in the largely rural Blue Nile Province as a whole, 53 per cent of the boys and 25 per cent of the girls between these ages were enrolled in primary school (Sudan, Ministry of Education and Guidance, 1981). Whatever the cause, the implications of low school enrolment are serious both for the children as they come of age and for the socioeconomic formation of Sudan as a whole.

My research indicated that most ten-year-old children in Howa were learning the knowledge, skills, and values necessary to reproduce the social relations and practices of production that characterized their community in 1981. In their work and play, children learned the environmental knowledge and practices associated with maintaining a farming community. But for a combination of reasons, it was apparent that relatively few of those ten years old in 1981 would have access to a farm tenancy when they come of age in the 1990s. Displacement of the farming population results from three interrelated phenomena: (1) the static land tenure relationships associated with the agricultural project – with 250 tenancies allocated to the village in 1970, there were from the outset fewer tenancies than the number of households (335 by 1981); (2) the size of the average household in Howa – with a fixed number of tenancies, and an average household size of 5.7, most children do not stand to inherit access to their family's tenancy; and (3) the frequent proximity in age between children and their parents, i.e., the parents of many ten-year-old children were still in their twenties or early thirties and thus unlikely to turn over their tenancies until well after their children reached adulthood. In the absence of further agricultural development (and none was planned for the area), much of what children played and worked at during their childhoods will not exist for them in the world of their adulthood. This eventuality points to a serious disjuncture in the course of social reproduction which could lead to profound shifts in the children's lives as they reach adulthood.

The outcomes differ for boys and girls. My research suggests that while most of the boys were learning to be farmers, they will not be. Rather they will be marginalized as agricultural wage laborers or forced to seek non-agricultural work in the Project headquarters nearby, in regional towns, or in urban centers further away. By contrast, girls were being socialized largely to assume their mothers' social and work roles. While women's roles were likely to be stable for a longer period than men's, it appeared likely that as men increasingly migrate from the area in search of labor, women would assume a larger role in agricultural production. While girls participated extensively in agricultural tasks such as planting and harvesting, unlike boys they were not taught to organize the full range of agricultural operations, had little practical experience in some of these, including clearing and weeding, and had not mastered the use of most tools. Moreover, their participation in all agricultural tasks except harvesting tapered off as they reached puberty. When these girls come of age, they are likely to need knowledge they will not have acquired fully in their childhoods.

This discontinuity between childhood learning and adult opportunities is a part of a process that, in effect, results in the wholesale deskilling and marginalization of rural populations such as those in the area of Howa. Given the general lack of formal schooling which might offer the chance for 'reskilling,' there was little likelihood of most children finding employment in their adulthoods except at the lowest skill requirements. These discontinuities were not the same for all children. The dramatic differences in the rates of school enrolment between girls and boys in Howa (4 per cent and 42 per cent respectively) suggest that girls will be even less capable than boys of undertaking non-farm or non-domestic work. Just as girls lacked experience with the full range of agricultural tasks and were not instructed in their overall organization, many boys in Howa did not have the opportunity to acquire particular agricultural skills and knowledge because their households lacked land in the agricultural project. Children from non-tenant households lost much of the opportunity shared by tenants' children to acquire and use agricultural knowledge in the course of their everyday lives. Thus, certain children, by virtue of their parents' socioeconomic or occupational status, will be less skilled in agriculture than many of their contemporaries, and possibly at a disadvantage in obtaining and holding positions as agricultural wage laborers. Under prevailing socioeconomic conditions, these were among the only viable jobs available. In this way, the process of socioeconomic differentiation engenders further differentiation. My research revealed some of the mechanisms of this process in children's experience.

For the socioeconomic formation of Howa, the intertwined processes of deskilling and differentiation may lead in the short run to declines in rural productivity and, in the long run, to an adult population unable or unwilling to carry out the tasks of production associated with the Suki Project. Rural productivity may decline as the changing relations of production and the ensuing redefinition of vocational skill dislocate the household as the center of production and reproduction (cf. Dalla Costa, 1977). These dual dislocations have resulted historically in the depopulation of the household, first as waged work is found outside of the home and ultimately as children leave in growing numbers to attend school. As a result, less labor is available to the rural household to maintain previous levels of agricultural productivity. The process of labor migration was just beginning in Howa in 1981.

Finally, the increases in children's work in Howa had begun to alter the traditional relationship between work and play, the balance between the two as means for the acquisition and use of environmental knowledge. This process appeared likely to continue whether the work responsibilities of the children continued to increase or greater numbers began to attend school. When play and work are separated, play becomes trivialized as 'childish' activity in the eyes of adults. Severed from work, play remains a central means by which children are socialized in terms of both mastering particular kinds of knowledge or skills and internalizing cultural values and other principles. However important its socializing role may be, play divorced from work is less grounded in the general experience of the community as a whole and thus comes to be viewed as inconsequential. When play becomes isolated and trivialized as an activity *only* for children or as something adults do *only* in their time off from work, the peer group becomes less integral to the larger society because it is no longer a social setting in which work and play are united. This represents the loss of an important aspect of autonomous traditional culture and is thus a means by which capitalism 'colonizes' experience.

Conclusion

This study has examined some of the ways that capitalism is articulated on the landscape of everyday life. Its central logic, the accumulation of capital, at once engenders and is achieved in the colonization of experience and the transformation of time, space, place, and nature. In Howa it was possible to witness these entwined processes taking place. My research on the practices of social reproduction, and in particular children's learning, their use of knowledge, and the organization and content of their knowledge concerning the environment, was spurred, in part, by the desire to find spaces that will not be colonized or to locate instances of experiential decolonization, glimmers of oppositional practice. These ends and those of the children are connected. Engaged myself in the production of knowledge that opposes received notions about the relationship between production and reproduction, I have written this piece at once to point to the subversive power that inheres in the everyday practices by which knowledge is produced, deployed and exchanged, and to reformulate or subvert that knowledge in a way that recognizes the oppositional potential of these very practices of social reproduction.

What I have shown – that children shared a rich and intricate knowledge of a resource use complex under erasure; that their autonomous practices of work and play were being transformed and sundered as unified means for the acquisition and use of particular kinds of knowledge; and that in the wake of capitalism's local impress, children were not being prepared for the world they are likely to face as adults – does not occasion optimism about the potential for opposition. But in examining the acquisition and use of local knowledge as practices of social reproduction in articulation with the political–economic relationships that at once structure and are structured by them, this piece not only has demonstrated the breadth and intensity of children's environmental interactions, but, in constituting these within the critical construction of everyday life, has pointed to their contradictory potential to alter the trajectory of socioeconomic change.

Most of the evidence presented indicates signs of rupture in the means by which children acquired and used environmental knowledge, and breakdown in the relationship between production and social reproduction. These disjunctures appeared to serve the advance of capitalism in Howa. The transformation of the local production system and its attendant processes of socioeconomic differentiation, environmental degradation, commoditization and deskilling are emblematic of the erosion of the countryside as a viable arena of noncapitalist relations of production and reproduction. Capitalist hegemony is neither achieved nor maintained without a struggle, and the community as a whole was responding to the changes imposed upon it in a range of ways. The State regulation of production practices, for example, was successfully limited by the tenants' union when they gained the right to grow sorghum on Project land. The shifting contents of what constitutes adequate socialization for adulthood was recognized by the villagers, and met in the mid-1980s with a village self-help project to construct a girls' school. While the extension of formal schooling *may* further capitalist hegemony in the area, inherent in the nature of education is the possibility that it may not, but, in fact, may enable a population to become conscious of their position in the larger society and resist such changes actively rather than reactively (Freire, 1970). These are but two examples of the local response to the socioeconomic and structural changes imposed in Howa since 1970. The great vitality and variation I discovered in the children's work, play, and learning, and the contradictions that inhere in them as material social practices of production and reproduction, suggest another possible arena by which these changes might be opposed, resisted or subverted. In order for this to be accomplished, children and their elders would have to consciously appropriate and link to political struggle the strength inherent in the production and exchange of knowledge, returning it to their own interests – to steady themselves and ride surely along the canals as the sun rises.

Notes

1 Notable exceptions include Willis (1977) in sociology, Taussig (1980) and Scott (1985) in anthropology, and Johnson (1977) in geography.

2 The literature on everyday life associated with the French social theorist Henri Lefebvre (1984) is germane here as well. In recent years it has been getting increasing attention in North American geography and social science in general. In geography it has been drawn on largely by those interested in the articulation between social and spatial relations (e.g. Soja, 1989; Harvey, 1989), rather than in the reproduction of particular human–environment relations. Its crucial insight is its construction of everyday life as a critical concept, that is, one in which the possibilities of its own transformation are immanent to the very practices of reproduction that constitute it (cf. Kaplan and Ross, 1987). However, this formulation, with the associated literature, is one that I came upon after completing the present project. In tracing my path to it, feminist theory has been most influential and offers emancipatory insights that geographers interested in social change frequently ignore.

3 According to Marx, the natural economy is based in agriculture complemented by domestic handicraft and manufacturing. Its central characteristic is that 'a very insignificant portion' of the product enters into the process of circulation (1967, vol. 3, pp. 786–7). In other words, 'the conditions of the economy are either wholly or for the overwhelming part produced by the economy itself, directly replaced and reproduced out of its gross product' (p. 795). The abstraction natural economy describes a particular relationship to the production of the means of existence and the relative lack of surplus and thus circulation. It should not invoke notions of classlessness; indeed, socioeconomic differentiation is often a characteristic of so-called natural economies. I follow Bernstein (1982) in using Marx's category of natural economy as an abstraction to suggest a social formation in which the production of use-values predominates, although there is an exchange of surpluses at a basic level. This is an apt characterization of the socioeconomic formation of Howa prior to 1971. Since my purpose is an analysis of the relationship between production and reproduction as material social practices in Howa, and not one of the historical transformation of the village as a socioeconomic system, an abstraction such as natural economy is useful as a means to locate the village theoretically.

4 This information was provided in the course of open-ended interviews I conducted with nine couples, eight of whom were parents of children participating in the research. The remaining couples were the grandparents of two children in the sample population. More significantly, a similar perspective was revealed in the open-ended discussions I had with adults in most village households while I was completing the village-wide census at the start of my project. Also, it was reinforced consistently during the informal discussions with a range of adults in Howa throughout the study period.

Selected references

Ahmad, Abd al Ghaffar M. 1974: *Shaykhs and followers: Political struggle in the Rufa'a al-Hoi Nazirate in the Sudan*. Khartoum: Khartoum University Press.

Aronowitz, S. 1973: *False promises: The shaping of American working class consciousness*. New York: McGraw-Hill.

Barnett, T. 1975: The Gezira scheme: production of cotton and the reproduction of underdevelopment. In Okaal, I., Barnett, T. and Booth, D. (eds), *Beyond the sociology of development: Economy and society in Latin America and Africa.* London: Routledge & Kegan Paul, 183–207.

Barnett, T. 1977: *The Gezira scheme: An illusion of development.* London: Frank Cass.

Bernstein, H. 1982: Notes on capital and the peasantry. In Harriss, J. (ed.), *Rural development.* London: Hutchinson University Library, 160–77.

Bourdieu, P. 1977: *Outline of a theory of practice.* Cambridge: Cambridge University Press.

Caldwell, J. C. (ed.) 1977: The persistence of high fertility in the Third World. Canberra: University of Canberra.

Caulfield, M. D. 1974: Imperialism, the family and cultures of resistance. *Socialist Revolution* 20.

Christopherson, S. 1983: The household and class formation: determinants of residential location in Ciudad Juárez. *Environment and Planning D: Society and Space* 1, 323–38.

Clark, G. and Dear, M. 1984: *State apparatus: Structures and language of legitimacy.* Boston: Allen & Unwin.

Clifford, J. and Marcus, G. E. 1986: *Writing culture: The poetics and politics of ethnology.* Berkeley: University of California Press.

Conference of Socialist Economists 1977: *On the political economy of women.* CSE Pamphlet 2. London: CSE.

Dalla Costa, M. 1972: Women and the subversion of the community. In Dalla Costa, M. and James, S. (eds), *The power of women and the subversion of the community.* Bristol: Falling Wall Press, 19–54.

Dalla Costa, M. 1977: Riproduzione e emigrazione. In Serafini, A. *et al.* (eds), *L'operaio multinazionale in Europa*, 2nd edn. Milan: Feltrinelli, 207–41.

de Certeau, M. 1984: *The practice of everyday life.* Berkeley: University of California Press.

de Lauretis, T. 1982: *Alice doesn't: Feminism, semiotics, cinema.* Bloomington: Indiana University Press.

Development and the sexual division of labor. 1981: *Signs: Journal of Women in Culture and Society* Special Issue 7(2), 265–512.

Duffield, M. R. 1981: *Maiurno: Capitalism and rural life in Sudan.* London: Ithaca Press.

Ehrenreich, B. and English, D. 1975: The manufacture of housework. *Socialist Revolution* 26.

Freire, P. 1970: *Pedagogy of the oppressed.* New York: Seabury Press.

Genovese, E. 1972: *Roll, Jordan, roll: The world the slaves made.* New York: Pantheon Books.

Gramsci, A. 1971: *Selections from the prison notebooks*, ed. and trans. by Hoare, Q. and Smith, G. N. New York: International Publishers.

Gregory, D. 1982: *Regional transformation and industrial revolution: A geography of the Yorkshire woollen industry.* Minneapolis: University of Minnesota Press.

Gregory, D. and Urry, J. (eds) 1985: *Social relations and spatial structures.* New York: St Martin's Press.

Gruenbaum, E. 1979: *Patterns of family living: A case study of two villages on the Rahad river.* Monograph 12. Khartoum: Khartoum University, Development Studies and Research Centre.

Haraway, D. 1984: Teddy bear patriarchy: taxidermy in the Garden of Eden, New York City, 1908–1936. *Social Text* 11, 20–64.

Harvey, D. 1989: *The condition of postmodernity.* Oxford: Basil Blackwell.

Harvey, D. 1990: Between space and time: reflections on the geographical imagination. *Annals of the Association of American Geographers* 80, 418–34.

el Hassan, I. S. 1980: On ideology: the case of religion in northern Sudan. PhD dissertation, University of Connecticut.

James, S. 1975: Sex, race and working class power. In James, S. (ed.), *Sex, race and class*. London: Falling Wall Press and Race Today Publications, 9–19.

Johnson, K. 1977: 'Do as the land bids': A study of Otomi resource use on the eve of irrigation. PhD dissertation, Graduate School of Geography, Clark University.

Johnson, R. 1979: Three problematics: elements of a theory of working class culture. In Clark, J., Crichter, C. and Johnson, R. (eds), *Working class culture: Studies in history and theory*. London: Routledge & Kegan Paul, 201–37.

Kaplan, A. and Ross, K. 1987: Introduction. *Yale French Studies* 73, 1–4. Special issue: Everyday life.

Katz, C. 1986a: Children and the environment: work, play and learning in rural Sudan. *Children's Environments Quarterly* 3(4), 43–51.

Katz, C. 1986b: 'If there weren't kids, there wouldn't be fields': children's environmental learning, knowledge and interactions in a changing socioeconomic context in rural Sudan. PhD dissertation, Graduate School of Geography, Clark University.

Katz, C. 1989: 'You can't drive a Chevy through post-Fordist landscape': everyday cultural practices of resistance and reproduction among youth in New York City. Paper presented at the Marxism Now: Traditions and Differences Conference, Amherst, MA.

Koptiuch, K. 1985: Fieldwork in the postmodern world: notes on ethnography in an expanded field. Paper presented at the 84th annual meeting of the American Anthropological Association, Washington, DC.

Landy, D. 1959: *Tropical childhood*. Chapel Hill: NC: University of North Carolina Press.

Lefebvre, H. 1984: *Everyday life in the modern world*. New Brunswick, NJ: Transaction Books.

Marcus, G. E. and Fischer, M. J. 1986: *Anthropology as cultural critique: An experimental moment in the human sciences*. Chicago: University of Chicago Press.

Marx, K. 1967: *Capital*, 3 vols, ed. Engels, F. Trans. Moore, S. and Aveling, E. New York: International Publishers.

Moraga, C. and Anzaldua, G. 1981: *This bridge called my back: Writings by radical women of color*. Watertown, MA: Persephone Press.

O'Brien, J. 1978: How traditional is traditional agriculture? *Sudan Journal of Economic and Social Studies* 2, 1–10.

Pred, A. 1984a: Structuration, biography formation and knowledge: observations on port growth during the late mercantile period. *Environment and Planning D: Society and Space* 2, 251–75.

Pred, A. 1984b: Space as historically contingent process: structuration and the time-geography of becoming places. *Annals of the Association of American Geographers* 74, 279–97.

Probyn, E. 1990: Travels in the postmodern: making sense of the local. In Nicholson, L. J. (ed.), *Feminism/postmodernism*. New York: Routledge, 176–89.

Said, E. 1978: *Orientalism*. New York: Pantheon.

Samuel, R. and Jones, G. S. 1982: *Culture, ideology and politics*. London: Routledge & Kegan Paul.

Sargent, L. (ed.) 1981: *Women and revolution: A discussion of the unhappy marriage of Marxism and feminism*. Boston, MA: South End Press.

Schildkrout, E. 1981: The employment of children in Kano (Nigeria). In Rodgers, G.

and Standing G. (eds), *Child work, poverty and underdevelopment*. Geneva: International Labour Office, 81–112.
Scott, J. C. 1985: *Weapons of the weak: Everyday forms of peasant resistance*. New Haven, CT: Yale University Press.
Smith, N. 1984: *Uneven development*. Oxford: Basil Blackwell.
Soja, E. W. 1989: *Postmodern geographies*. London: Verso.
Spivak, G. C. 1987: *In other worlds: Essays in cultural politics*. New York: Methuen.
Spivak, G. C. 1988: Can the subaltern speak? In Nelson, C. and Grossberg, L. (eds), *Marxism and the interpretation of culture*. Urbana: University of Illinois Press, 271–313.
Sudan, Democratic Republic. Ministry of Education and Guidance 1981: Education in the Sudan: Sector review paper. Prepared for UNICEF Preview Meeting, 22 October 1980.
Taussig, M. 1980: *The devil and commodity fetishism in South America*. Chapel Hill, NC: University of North Carolina Press.
al Tayib, G. el D. 1970: The southeastern Funj area: a geographical survey. Khartoum: University of Khartoum, Sudan Research Unit. Funj Project Paper 1.
Thompson, E. P. 1963: *The making of the English working class*. New York: Vintage Books.
Thrift, N. 1983: On the determination of social action in space and time. *Environment and Planning D: Society and Space* 1, 23–57.
Tienda, M. 1979: Economic activity of children in Peru: labor force behavior in rural and urban contexts. *Rural Sociology* 44, 370–91.
Tothill, J. D. (ed.) 1948: *Agriculture in the Sudan*. London: Oxford University Press.
Warf, B. 1988: Regional transformation, everyday life, and Pacific Northwest lumber production. *Annals of the Association of American Geographers* 78, 326–46.
Watts, M. 1983: *Silent violence*. Berkeley: University of California Press.
Watts, M. 1987: Powers of production – geographers among the peasants. *Environment and Planning D: Society and Space* 5, 215–30.
Williams, R. 1977: *Marxism and literature*. Oxford: Oxford University Press.
Willis, P. 1977: *Learning to labor: How working class kids get working class jobs*. New York: Columbia University Press.

13 Andrew Charlesworth,
'Contesting Places of Memory: The Case of Auschwitz'

Excerpts from: *Society and Space* **12,** 579–93 (1994)

As Young (1993, page 2) has pointed out, as part of a nation's rites or the objects of a people's national pilgrimage, memorials are invested with national soul and memory. Nations have a landscape of historical collective memory constructed out of memorials recalling ennobling events, triumphs over barbarism, or martyrdom when citizens gave their lives in the struggle for national existence. In France in the 1980s, for example, there came into being an ambitious and well-publicised project called *Les Lieux de Mémoire* documenting historical monuments and places and material sites which give rise to the idea of the national patrimony (Nora, 1989). Yet, as Boyarin (1991, page 11) has noted, such projects can be regarded as politically ambiguous. For the very act of memorialisation, of capturing memory so that we do not forget, can by its exclusivity push aside the claims of others for their own collective rights and identities. As Nora has explained, the place of memory is 'created by a play of memory and history'. 'The *lieux de mémoire* only exist because of their capacity for metamorphosis, an endless recycling of their meaning and an unpredictable proliferation of their ramifications' (page 19). Memorial sites are thus both open to different interpretations and malleable to the needs of state power and religious forces. The memorialisation of the Holocaust has not been exempt from these shaping forces.

1995 marks the fiftieth anniversary of the liberation of the concentration and death camps that were to become in the collective social memory of the victors of the Second World War the symbols of the Nazi Holocaust: Dachau, Buchenwald, Bergen–Belsen (liberated by the Western allies), and Auschwitz (liberated by the Red Army, the West's ally, the Soviet Union). Despite its special significance both to Jews and to the international community as symbol of Nazi terror, the landscapes of memorial at Auschwitz have not gone uncontested. In this paper I set out to examine that contestation over symbolic space with particular regard to the explicit attempts at the Catholicising of Auschwitz from the 1970s onwards. In order to do this we need to see how the history of the camp as presented to a postwar audience was refracted through the realpolitik of the Cold War and how this meant, in Young's words, that an icon of remembrance was turned into an idol of remembrance, a process that sought both to de-Judaise Auschwitz and to allow only a limited and controlled expression of Polish national sentiment (Young, 1993, page 14).

The choice of Auschwitz as a site of commemoration of the Nazi terror

The question of how Auschwitz came to be chosen as a site of memorialisation of Nazi terror has not been fully explored.[1] In the postliberation era Auschwitz was not chosen by Jews to be the sign and symbol of Nazi terror, Nazi genocide, the 'Final Solution of the Jewish Question'. Auschwitz as a symbol of Nazi terror predates by at least fifteen years the widespread use of the term 'Holocaust'.[2] Despite Polish prejudices to the contrary, the Jews, reduced in number to 250,000 by the end of the war, had no substantive political influence within the new Polish Communist state. Nor was Auschwitz chosen by the international community. The closing of the Iron Curtain had ensured that. Auschwitz was chosen in 1947 by a Polish Parliament by then dominated by a Stalinist, Moscow-led Communist Party, whose leadership's main task was to suppress Polish national sentiment and stress solidarity between Poland and the Soviet Union.[3] The victory over fascism could be most powerfully commemorated at sites that recalled the horror and terror of the Nazi occupation from which the Soviet Union had liberated the Poles. Auschwitz was, however, not the only such site.

The six death camps whose primary function was the extermination of European Jewry were all possible commemorative sites (Figure 13.1). Of these Chelmno, Belzec, Sobibor, and Treblinka could be deemed inappropriate in that they had been destroyed by the Nazis, leaving little or no trace remaining. This was also very convenient for those who wished to ignore the specificity of Jewish suffering, as these were wholly death camps for Jewish extermination. This left Auschwitz and Majdanek.[4] These two camps had large portions intact on liberation and were both death and concentration camp complexes. They both were camps where Poles, Sinti, Romani, Soviet prisoners of war, and Jews had suffered terribly at the hands of the Nazis. It could be argued that the numbers reported in 1945 as having been killed at each of these camps was the deciding factor. At Auschwitz it was reckoned that 4 million people had perished; at Majdanek it was estimated that 360,000 had been killed.[5] I would argue that Auschwitz was given added significance over Majdanek because of the way that the Polish–Soviet relationship could be refracted through the memorialisation of each camp.

In terms of the victim profile, Majdanek had a better claim to be the place of memorialisation of the Nazi occupation of Poland than Auschwitz. At least 40 per cent of the victims at Majdanek were non-Jewish Poles and only 30 per cent were Jewish. At Auschwitz 87 per cent of the victims were Jewish, of

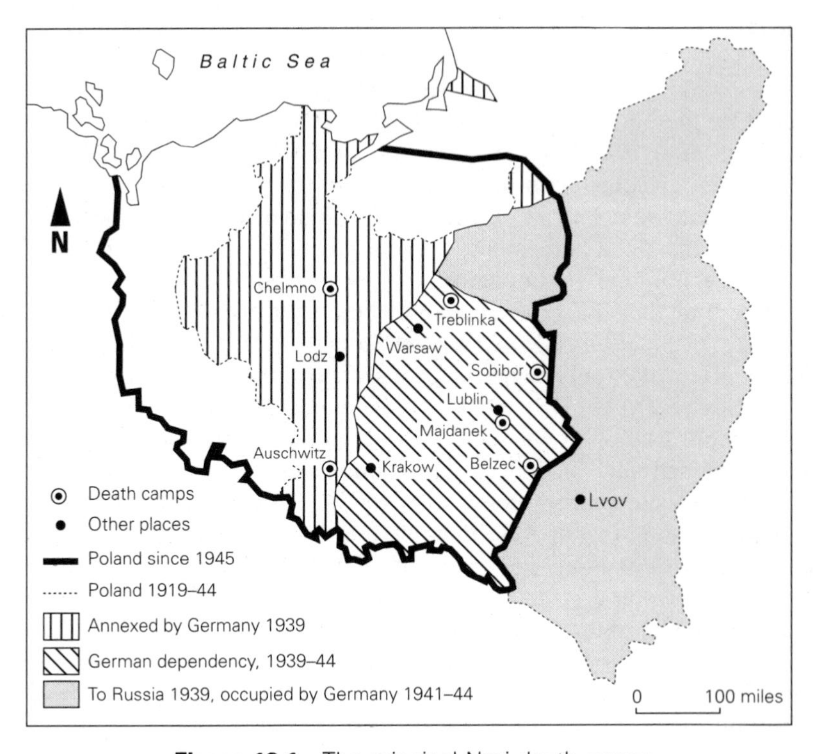

Figure 13.1 The principal Nazi death camps

whom less than a third were Polish Jews and only 7 per cent of total victims were non-Jewish Poles (Piper, 1991; Rajca and Wisniewska, 1983, page 11).

If Majdanek had been chosen it could, however, have undermined rather than cemented Polish–Soviet solidarity and fuelled Polish nationalist sentiment. First, as the Nazi death/concentration camp was taken over with most of its terror apparatus intact, the NKVD (the Soviet secret police) took the opportunity to incorporate it within their system of terror and repression immediately on liberation. Hundreds and then later thousands of members of the Polish Home Army, the noncommunist resistance movement, found themselves incarcerated in the camp (Kersten, 1991, page 95). This fact is still not acknowledged at the camp. Second, on liberation, in 1944 the city of Lublin, on whose outskirts Majdanek was sited, became the seat of the Soviet–Polish puppet government set up to rival the official Polish government in exile. Moreover, the Royal Castle in Lublin which had been a Gestapo prison was also taken over by the NKVD. In April 1945 approximately 8,000 prisoners were held there. A plaque at the castle gateway commemorating that fact publicly has only been placed there since the ending of communist rule in Poland (Kersten, 1991, page 130). The communist authorities did not want to present their Polish compatriots with a focus for remembrance of such historical facts within the Lublin vicinity. Third, Majdanek's location could only serve to remind Poles of their subordination and victimisation by the Soviet Union both between 1939 and 1941 and after 1944. It was close to the eastern territory of prewar Poland which was occupied by the Soviets between 1939 and 1941 and where a Stalinist terror had reigned during that period (Figure 13.1). Two million Poles had been taken to the Gulags, of whom half were dead within a year of arrest. At Katyn 4,500 Polish officers had been murdered by the NKVD (Davies, 1984, page 67).

Moreover, after 1945 Majdanek was less than 90 km from the new Polish–Soviet border, and thus able to act as a place of remembrance for Poles of their latest traumatic loss of territory, as Poland had literally been moved westward by the Yalta agreement.[6] Poland has seen itself since the late 18th century as geopolitical victim, being first partitioned in 1772 and 1793 into a rump state and then in 1795 and 1815 being partitioned completely. The Nazi–Soviet occupation followed by the Soviet Union's political domination of Poland after 1945 only underlined this continuing geopolitical victimisation. The Soviet and the Polish Communist authorities were right to fear that public memorialisation at Majdanek could become a rallying point for anti-Soviet Polish nationalism. Indeed it was only in 1969 that a competition for the first large-scale public monuments to the Holocaust at Majdanek was initiated, postdating that at Auschwitz by ten years (Young, 1993, chapter 5).

Auschwitz with its more western location was much less problematical and could be portrayed as a symbol solely of fascist aggression. First, this fitted in with a Communist geopolitical model, where the fascists had fled westward and could potentially strike eastward again. Auschwitz could be 'orientated' westward, with Germany as the past and potential aggressor. Poland felt particularly vulnerable here because it now occupied former German territory. The Auschwitz guidebooks used to end on this potential threat (Smolen,

1981, pages 113–114).[7] Second, Auschwitz had been the place where Jews from many nations had been brought to be killed. By emphasising its international character and ignoring the fact the victims were Jewish, the Communists linked Poland through the memorialisation of Auschwitz to the other Warsaw Pact countries, both as past and potential victims of German aggression and as present beneficiaries of their liberation by the Red Army and of their continuing defence by the Soviet Union. The film shown to visitors on their arrival at the Auschwitz museum was even up until 1993 one showing a staging of the liberation of the camp by the Red Army. Many of the various exhibitions at Auschwitz were given over to the commemoration of the Nazi aggression in different countries, hence stressing this international element. Communist brotherhood dictated that the German Democratic Republic (East Germany) was allowed an exhibition whereas the Federal Republic of Germany (West Germany) was not. So Auschwitz with its museum, its archives department, its provision for visitors with restaurants and a hotel on site became *the* site within Poland to commemorate the Nazi terror.

In that act of memorialisation throughout the museum and on the memorials and in the literature on the camp, the emphasis was always on 'people', not Jews or Poles, as victims. At Birkenau (Auschwitz II), the Jewish death camp, there were nineteen tablets inscribed in nineteen different languages, commemorative of the '4 million people' who 'suffered and died here at the hands of Nazi murderers'. This was reinforced by the guided tours round Auschwitz whose guides only talked of 'victims' and 'people', making no reference to ethnic origins or religious affiliations. This all not only highlighted the international element but also it succeeded to a large extent in de-Polonising as well as almost totally de-Judaising Auschwitz.[8]

A Polish nationalist Auschwitz

Even so a Polish nationalist Auschwitz could still be discerned by Poles, but one done again at the expense of the Jewish specificity of Auschwitz, thus allowing some controlled expression of Polish nationalism. The original remit for the establishment of the Auschwitz museum in 1947 ran: 'a monument of the martyrdom of the Polish nation and of other nations is to be erected' (Smolen, 1981, page 114). Guidebooks to the museum always opened with the statement that the first prisoners at Auschwitz were Polish prisoners – victims of the Nazi terror campaign against the Polish national leadership (Smolen, 1981, page 7).

Second, during the Nazi Occupation Auschwitz was not in the rump territory of Poland, that of the General Government, but in territory incorporated immediately in 1939 into the Third Reich (Figure 13.1). That is, Auschwitz as a place of martyrdom could be interpreted as a sign to Poles of the continued partitioning of their nation throughout history.

Third, a history of partition was underlined by the fact the camp of Auschwitz I occupied buildings originally constructed as military barracks by the Austrians at the turn of the century, for this part of Poland had been Austrian from 1793 to 1918. The buildings still look from outside like Central

European 19th-century military barracks. Thus a Polish visitor who knew Poland's history, could begin to 'read' the landscape of Auschwitz as a multilayered symbol of Poland's geopolitical victimisation.

The Catholicising of Auschwitz

The memorialisation of Auschwitz in terms of Soviet-Communism ideology, in which some recognition of Polish nationalist sentiment was granted, predominated until the early 1970s. Then there began to emerge a Polish Catholic Auschwitz. This process of Catholicising Auschwitz began at a crucial point in Church–state relations. After a period of spiritual renewal culminating in the celebrations for the one thousand years of Christianity in Poland in 1966, the Poles' sense of unity as Catholics had been significantly strengthened. Following the turmoil surrounding the discrediting of the Gomulka government in 1970 the Church used the opportunity of the new Gierek regime to assert itself by making declarations that inextricably linked the Catholic faith and Polish national self-determination and culture (Tomsky, 1982, pages 10–13; for a fuller account see Weigel, 1992). As with the Communists, Auschwitz would become a symbol of oppression but this time the emphasis would be on Polish Catholic martyrdom. Certainly the de-Judaising of Auschwitz by the Communist authorities allowed the Catholic Church to reclaim Auschwitz more easily. By the 1970s, however, Auschwitz had come to be recognised as *the* symbol of Nazi terror, and, more significantly, of the Holocaust. Catholicising Auschwitz would thus inevitably bring conflict with the Jewish world in particular, but this was to come surprisingly late and was to focus almost exclusively on the presence of the Carmelite nuns in a building adjacent to Auschwitz I.

The history of the active involvement of the Catholic Church in securing a Catholic presence in Auschwitz coincided with the time when the present Pope John Paul II, as Cardinal Karol Wojtyla, was Archbishop of Krakow in whose diocese Auschwitz is located. Cardinal Wojtyla understood clearly the significance of the Auschwitz complex in terms of Polish Catholic and Catholic martyrdom.[9] That significance rested particularly on the campaign to have Father Maximilian Kolbe, the most famous of modern Polish Catholic martyrs, first beatified (1971) and then canonised (1982). As Archbishop of Krakow the future Pope had taken an early interest in Father Kolbe and visited the latter's cell in Block 11 in Auschwitz I on a number of occasions. In a sermon he preached at Auschwitz in 1972, Cardinal Wojtyla made clear that since the beginning of the postwar period 'the Church of Poland' had seen 'the necessity for such a site of sacrifice, of an altar and a sanctuary – precisely in Auschwitz. The beatification of Father' . . . Kolbe made this even more necessary. Cardinal Wojtyla proposed that a church should be erected as Christian tradition dictated for the examples of martyrdom and sainthood down the ages (Minerbi, 1989). Thus the beatification of Father Kolbe in 1971 can be seen as the first part of the Catholicising of Auschwitz.

That the process was going hand in hand with the continued de-Judaising of Auschwitz is clear from this early date. In that 1972 sermon Cardinal Wojtyla made reference to Father Kolbe's actions favouring the conversion of Jews (Minerbi, 1989). What underlines the point is that the mass and sermon were given as part of a mass celebrated at Birkenau. As was to be the case with the Pope's visit to Auschwitz in 1979 the mass took place on the new ramp at Birkenau.

In 1972 and 1979 the locating of masses at Birkenau had symbolic implications for the reclaiming of the Jewish death camp from the Jews. It could be argued that this was the only available open and usable space in the Auschwitz complex, but that type of rationalisation only underscores a lack of sensitivity by the Catholic Church to Jewish sensibilities.[10] The siting of the mass, the religious furniture used, and the Pope's cocelebrants of the mass point directly to the Catholicising of Birkenau and hence the whole Auschwitz complex (Figure 13.2). Here is the description by Catholic commentators of the 1979 mass:

> The mass, which the Pope celebrates in Birkenau, together with 200 priests – former prisoners of concentration camps – has a great significance. The sight of the lookout and barbed wire fencing reminds one of those former days. The altar was built upon a railway loading [*sic*] platform, at the place where, during the German occupation, those chosen for extermination would arrive. Above the altar there ascends a cross, over whose shoulders a cloth is folded in the colours of the camp's

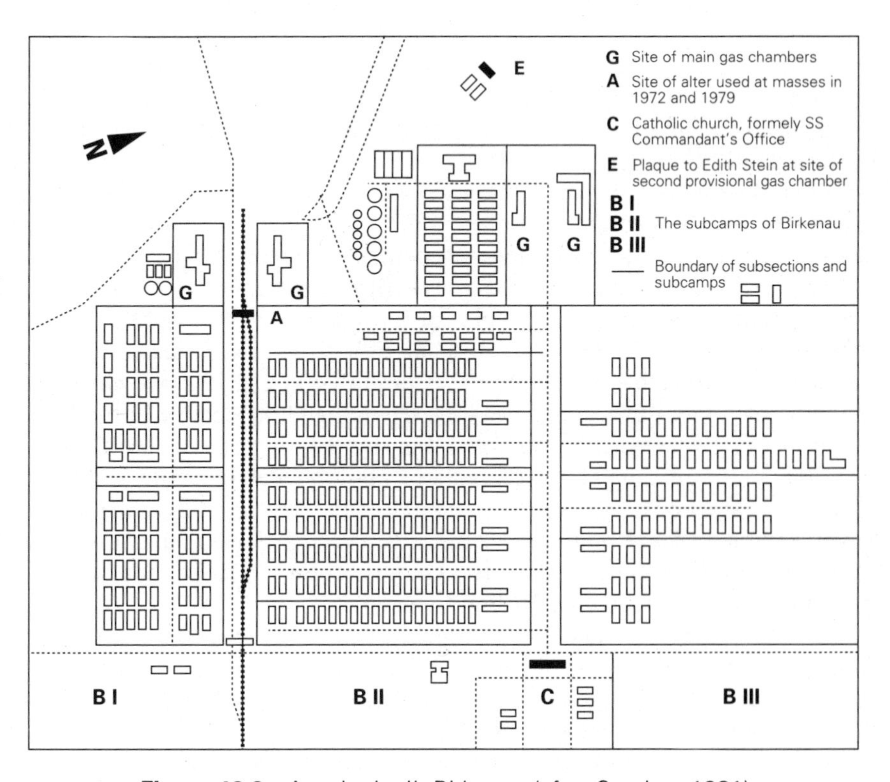

Figure 13.2 Auschwitz II: Birkenau (after Smolen, 1981)

uniform. On this cloth, the letter P is marked – a prison insignia – and the number 16,670 – the number assigned to Father Kolbe (Le Corre and Sobotka, 1984, pages 81–83).

The altar was sited on the ramp where from spring 1944 Jews were unloaded and at the very place where the selection of Jews for gassing or 'life' in the camp took place. The cross towered over Birkenau. It was hung with the prison garb that relatively few Jews ever had a chance to wear, the vast majority going straight to the gas chambers, garb marked with a red triangle, not a star of David, and with the number of the prisoner who in Polish Catholic eyes is *the* martyr of Auschwitz. On the cross was a crown of barbed wire, the Christian symbolism of which was unmistakable.

It should be noted that in all the above commentary there is no direct reference to Jews. Just as the mass sought to layer on top of the Jewish tragedy the Catholic martyrdom, so the commentators implicitly obscure the former tragedy. The Pope in his address uses the same obfuscation. He never mentions the word Jews but refers instead to 'the people who were given by God the commandment "thou shalt not Kill" ', 'a nation whose sons and daughters were destined for total extermination' (Le Corre and Sobotka, 1984, page 83). He never gives a Jewish death count but gives the number of Poles who were killed in the last war (Jewish and non-Jewish Poles) and refers to them only as 'Poles'. In his references to Birkenau as a graveyard, the Pope only says that the graves are 'mostly nameless – like a huge Tomb of the Unknown Soldier' (page 83). This cannot be a slip; it is a conscious attempt not to mention Jews directly. Just as at visits to Majdanek in 1987 and at Mauthausen in 1988 the Pope made no reference to Jews (Rittner and Roth, 1991, page 42).[11]

To the Pope and to Polish Catholics Auschwitz was to have a place in the constellation of sites of Catholic martyrdom and Polish nationhood. The Pope referred to Auschwitz as 'a settling of accounts' by Poles with the 'conscience of mankind', 'another stage of the struggle of this nation, of my nation for its fundamental rights among the peoples of Europe . . . for the right of its own place on the map of Europe' (Le Corre and Sobotka, 1984, pages 83–84). The places he visited on that first pontifical return to Poland in 1979 were meant to underline the unity between the Catholic Church and the destiny of the Polish nation. On that visit he chose to go to Warsaw the political capital; Gniezno, 'the first capital of Poland', 'the cradle of both the Polish nation and Christianity'; Czestochowa, 'the spiritual capital of Poland'; Krakow, 'the ancient Royal capital', 'the Rome of Poland'; Kalwaria, 'Jerusalem's Golgotha, transferred in [*sic*] the neighbourhood of Cracow'; Skalka, the place of martyrdom of the Cracowian Bishop Stanislaus;[12] and Auschwitz (Le Corre and Sobotka, 1984). In his first pontifical sermon preached in Warsaw, the Pope said, 'It is impossible without Christ to understand the history of the Polish nation'. He reiterated this point further on in the sermon when he said 'It is impossible to understand the history of Poland – from St Stanislaus at Skalka to Maximilian Kolbe at Auschwitz – unless we apply to them that single and fundamental criterion, which is Jesus Christ' (Le Corre and Sobotka,

1984, pages 27–28). Thus the Pope established the historical trajectory of martyrdom in Polish Catholic history. Auschwitz was now to be included firmly in that trajectory. And by Auschwitz, reference is being made to the whole Auschwitz complex.

To many people the establishment of a Carmelite convent just the other side of the inner perimeter fence at Auschwitz I in 1984 was the first overt attempt at Catholicising Auschwitz. As we have seen this was part of a larger process. Indeed, all the controversy over the Carmelite convent at Auschwitz I including the most infamous incident, Rabbi Weiss's violent demonstration at the convent, has deflected attention away from the attempt to Catholicise *all* of Auschwitz.[13] For that controversy has always suggested one straight topographic solution. Polish Catholics should be allowed Auschwitz I as their place of martyrdom. The Block of Death, Block 11, where Father Kolbe met his death, is there. Adjacent to this is the Wall of Death, where many Polish prisoners and Catholic priests were shot. The Carmelite convent has occupied a building immediately adjacent to these locations, though outside the inner perimeter fence of the camp. Birkenau could then be left to the Jewish people, for there 98 per cent of the victims killed were Jews. This, however, fails to recognise the Catholicising of both Auschwitz I and Birkenau.

There is in fact a visible signed link between Birkenau and the Carmelite convent at Auschwitz I. The 23-ft high cross in the garden of the convent, which can clearly be seen from inside what was the camp, is that same cross used at the pontifical mass at Birkenau in 1979 (Rittner and Roth, 1991, page 30). It stands as a reminder of that visit and the continuing interest the Pope has taken in the whole Auschwitz complex as a site of Catholic martyrdom. Indeed, in 1983 the Pope's wish expressed eleven years earlier for a church at the site became granted with the conversion of a building at Birkenau that had been the SS commandant's office into a Catholic church (page 20). So this church just outside the barbed wire, with its cross atop the building, overshadows the death camp. Moreover, in the church itself an image of Sister Teresa Benedicta of the Cross (Edith Stein) is venerated. Edith Stein was the other Catholic martyr that the Pope had referred to in his 1979 sermon. She was a converted Jew who had become a Carmelite nun and was gassed at Birkenau along with other Jews brought from Holland (Langbein, 1991, page 96). She was gassed, however, as a Jew, not as a Catholic nun. Nevertheless, in 1987 she was beatified. There is also now a plaque to her on the site of the second provisional gas chamber at Birkenau – the 'white house'. The plaque makes no mention of the fact that she was a Jew.

By 1989 there were reports of Polish Catholic groups, led by priests, marking the fourteen stations of the Cross amongst the remains of Birkenau. When challenged that this was a site of Jewish genocide, one parish priest defended such religious acts by Polish Catholics by reference to Pope John Paul II's description of Auschwitz as 'the Golgotha of our time' and that therefore they had no choice but to pray there (Firestone, 1989).

Auschwitz, Catholic Poland and the Catholic world

Young has argued implicitly against the case I have developed in the essay so far. He has written that 'the problem is not that Poles deliberately displace Jewish memory of the Holocaust with their own, but that as a country bereft of Jews, the memorials can do little but cultivate Polish memory . . . Polish Catholics will remember as Polish Catholics, even when they remember Jewish victims' (1993, page 117). Young, however, fails to recognise that the Catholic Church and Pope John Paul II have long understood the importance of successful contestation over symbolic spaces, contestations that symbolise the very heart of spiritual struggle. In 1970 when he was Archbishop of Krakow the Pope successfully orchestrated a campaign to have a church built opposite the Lenin Steelworks at Nowa Huta – the very symbol of the Communist proletariat (Bloch, 1982, page 69).[14] Such was the powerful symbolism of that church that the Communist authorities in 1979 forbade the Pope from holding a mass in the near vicinity of the church (Martin, 1990, chapter 3).

This should show us that the reason for the steady Catholicising process of Auschwitz may well lie rooted in deeper issues. The Catholicising of Auschwitz has become important, because a Catholicised Auschwitz must stand as a symbol for Poland's rule in Catholicising Europe, in the past, now, and in the future. This is why the layering over of a Jewish Auschwitz is taking place. It is perceived by the Catholic Church that there must be a clear vision of a Polish Catholic Auschwitz. This is why the Pope can call Auschwitz 'the Golgotha of our times', can raise up the host in Birkenau whilst ignoring the offence he is causing to Jewish people and other critics of the Catholicising/de-Judaising of Auschwitz. The link between a Catholic Auschwitz and a Catholic Poland must be made at all costs, for two reasons.

First, Poland stands as a symbol of Catholic martyrdom from actual or threatened religious geopolitical aggression – German Protestant from the West and Russian Orthodox from the East during the partitioning of Poland from the 1790s, then in this century Nazi and Soviet atheistic aggression. Here Father Kolbe's death at Auschwitz is of great symbolic importance. Father Kolbe has become the symbol of the sacred union of the Catholic Church and the Polish nation, of the church sacrificing itself for Poland as he had willingly stepped forward to offer his life to save another's. The Church becomes Poland, Poland becomes the Church, each prepared to be martyred for the other and in that suffering, like Christ, enabled to bring the suffering of others to God the Father. At the 1979 mass at Birkenau the Pope explicitly made reference to the martyrdom of Catholics at Auschwitz, stating that they achieved redemption through it (Bartoszewski, 1990, page 7). Thus Catholic Poland can see itself through its historical suffering to be Christ amongst the nations.[15]

Second, such a role enables Poland to stand as a spiritual fortress from which to launch Catholic renewal and evangelism. Auschwitz is important in this respect in two ways. First, there is the continued controversial line taken on the conversion of Jews. We have noted references to the conversion of

Jews with respect to Father Kolbe and Edith Stein. This is one of the reasons the Carmelites are praying at Auschwitz for 'our strayed brothers and sisters'. Second, there were the terms of the gift of the Carmelite convent to the Pontiff. It was given to the Pope by a Belgian Catholic charity, whose function was to aid Catholic churches behind the Iron Curtain including the territory of the Russian Orthodox Church. One of the original remits of the gift to the Pope was that the convent 'will become a spiritual fortress' (Bartoszewski, 1990, pages 6–9). So it is then but a short step to link through symbolism the convent as spiritual fortress, Auschwitz as spiritual fortress, Poland as spiritual fortress. The Christ amongst nations, standing at the very crossroads of Europe, at the very heart of Europe, is now ready to take on again its religious destiny. Having defeated atheism in its own nation, it can go on to extend the influence of the Catholic Church in the world.[16]

Thus Auschwitz, to many *the* symbol of the Holocaust, has for long taken on different meanings; hence the contestation over its space. We need to be vigilant of this capacity for metamorphosis of *lieux de mémoire* or, if vigilance is too optimistically a liberal ideal, we must be conscious when that capacity is realised. With the fall of communism in Poland this is already happening. One of the first changes at the Auschwitz Museum after the fall of communism was the dismantling of the Bulgarian national exhibition. It could be claimed that the reason for this was that no Bulgarians had been brought to be killed to Auschwitz. Then why has the Danish exhibition been allowed to stand? No Danish Jews were transported to Auschwitz. Both Bulgaria and Denmark are internationally recognised for their successful attempts to rescue Jews. The Bulgarian exhibition was an abstract conception but so again was the Italian one. The Bulgarian national exhibition would seem to have been picked out as a way of the Poles distancing themselves from their former Warsaw Pact colleagues and just as significantly also putting a distance between Catholic Poland and Orthodox Bulgaria.

Despite the new Polish–Jewish Commission on the future of Auschwitz, contestation at places of memorialisation such as Auschwitz does not augur well for the New World Order. It is certain to intensify and should remind political geographers that the battle for people's souls may be just as important as that of nations obtaining what they see as their rightful place in the sun.

Notes

1 The best exploration so far can be found in Bartoszewski (1990, chapter 2).

2 Eley (1983) notes the early 1960s as the point at which the destruction of European Jewry by the Nazis was coded into the term 'The Holocaust'.

3 For the takeover of the Polish state by the communists see Kersten (1991, especially chapter 8). See Davies (1984, pages 106–107) on the Sovietization of the history of the war and liberation of Poland. He exaggerates the Jewish emphasis of that history particularly with regard to Auschwitz and the Warsaw Ghetto; on the latter see also Young (1989).

4 Auschwitz was made up of three principal camps: Auschwitz I (the administrative headquarters and the base camp largely functioning as a concentration camp though

with one gas chamber/crematorium); Auschwitz II – called Birkenau – the death camp built for the extermination of Jews but with a concentration camp attached; and Auschwitz III – Monowitz, a camp housing prisoners who worked at the I. G. Farben Buna industrial plant. There were other subcamps in Auschwitz 'interest zone', a 15 square mile special zone around Auschwitz I. [For shorthand in this paper I will call Auschwitz I and Auschwitz II (Birkenau) when taken together 'Auschwitz'. It is the contestation over this symbolic space that has been most intense. For the problematic over what constitutes Auschwitz, see Webber (1992, page 96, footnote 3).]

5 This figure is now estimated to be incorrect (see Piper, 1991). In the region of 1.1 million men, women, and children are now thought to have been killed, of whom about 960,000 were Jewish. The Majdanek figure comes from Rajca and Wisniewska (1983, page 29).

6 The depth of the trauma relating to the loss is discussed by Davies (1984, pages 102–103).

7 This guide book was on sale as late as 1990.

8 This is taken up in Young (1993, chapter 5). In early 1990 the inscriptions on the tablets were erased on the orders of the Auschwitz Museum director. On a visit to Auschwitz in April 1990 the reason given to me by Dr F. Piper for this was that the number 4 million was incorrect, not that the word 'people' masked the Jewish specificity of Birkenau (see note 5). The wording of the new inscription is still under discussion at the time of writing. See Webber (1992, page 100, footnote 30).

9 Trasatti and Grieco (1982) chart the Pope's involvement in the process of beatification and canonisation of Father Kolbe (see also Reuters, 1987).

10 The site of the BIII subcamp at Birkenau is a large enough open space but is now derelict land and used by locals as a refuse tip.

11 The Pope did make amends the next day but only after a protest by the president of the Austrian Jewish community (Irwin-Zarecka, 1989, page 170, footnote 15).

12 St Stanislaus was the 'Polish Thomas à Becket', a bishop of Cracow murdered on the orders of the King. The parallel with Father Kolbe is meant to be drawn – the Church protecting the Polish people and nation against tyrannical rulers even if it means death to the protector.

13 The best account of the controversy is Bartoszewski (1990). See also Rittner and Roth (1991).

14 For the significance of this church and the second church built in Nowa Huta – that one being the St Maximilian Kolbe Church – in the Church's resistance to the Communist State see Weigel (1992, chapter 5). At the latter church the priest and chaplain of the Lenin Steelworks said 'We tried to give people back their memory' (quoted in Weigel, 1992, page 151).

15 These are themes taken up by Davies (1984). He makes the point about Father Kolbe on page 108.

16 Note how Davies writes that the choice of title for his history of Poland, *The Heart of Europe*, 'happily coincided with the Catholic symbol of the Sacred Heart of Jesus, which is widely revered in Poland . . .' (1984, page x).

References

Bartoszewski, W. T. 1990: *The Convent at Auschwitz*. London: Bowerdean Press.

Bloch, A. 1982: 'Poland in the present'. In Bloch, A. (ed.), *The Real Poland*. New York: Continuum, 53–70.

Boyarin, J. 1991: 'Space, time and the politics of memory'. WP 122, Centre for Studies of Social Change, New School of Research, New York.

Davies, N. 1984: *The Heart of Europe*. Oxford: Clarendon Press.

Eley, G. 1983: 'Holocaust history'. *London Review of Books* 3–17 March, pages 6–9.

Firestone, D. 1989: 'Claims to Auschwitz'. *Newsday* 5 September, page 1.

Gilbert, M. 1986: *The Holocaust*. London: Collins.

Irwin-Zarecka, I. 1989: *Neutralizing Memory: The Jew in Contemporary Poland*. New Brunswick, NJ: Transaction Books.

Kersten, K. 1991: *The Establishment of Communist Rule in Poland 1943–1948*. Berkeley, CA: University of California Press.

Langbein, H. 1991: 'The controversy over the convent at Auschwitz'. In Rittner, C. and Roth, J. K. (eds), *Memory Offended: The Auschwitz Convent Controversy*. New York: Praeger, 95–98.

Le Corre, D. and Sobotka, M. 1984: *John Paul II in Poland*. Bagnolet: Le Corre.

Martin, M. 1990: *The Keys of this Blood*. New York: Simon and Schuster.

Minerbi, S. I. 1989: 'The kidnapping of the Holocaust'. *Jerusalem Post* 25 August, page 1.

Nora, P. 1989: 'Between memory and history: *Les Lieux de Mémoire*'. *Representations* 26(1), 7–25.

Piper, F. 1991: 'Estimating the number of deportees to and victims of the Auschwitz–Birkenau camp'. *Yad Vashem Studies* 21, 49–103.

Rajca, C. and Wisniewska, A. 1983: *Majdanek Concentration Camp*. Lublin: Krajowa Agencja Wydawnicza.

Reuters, 1987: 'Pope cites Auschwitz death-camp in West German TV sermon'. *Reuters North European Service* 26 April.

Rittner, C. and Roth, J. K. (eds) 1991: *Memory Offended: The Auschwitz Convent Controversy*. New York: Praeger.

Smolen, K. 1981: *Auschwitz 1940–1945: Guide Book through the Museum*. Katowice: Krajowa Agencja Wydawnicza.

Tomsky, A. 1982: *Catholic Poland*. Keston: Keston College.

Trasatti, S. and Grieco, G. 1982: *P. Maximilian Kolbe*. St Ottlien: EOS Verlag.

Webber, J. 1992: 'The future of Auschwitz: some personal reflections'. *Religion, State and Society* 20, 81–100.

Weigel, G. 1992: *The Final Revolution*. Oxford: Oxford University Press.

Young, J. E. 1989: 'The biography of a memorial icon: Nathan Rapoport's Warsaw Ghetto monument'. *Representations* 26(1), 69–106.

Young, J. E. 1993: *The Texture of Memory: Holocaust Memorials and Meaning*. London: Yale University Press.

SECTION FOUR

TOWARDS A PROGRESSIVE SOCIAL GEOGRAPHY

Editor's introduction

In which direction(s) should social geography be moving? Clearly, there is no one answer to this question, as it will depend on the social and political values of individual social geographers. It is possible, however, to identify a number of directions which emerge from the literature. First, there is a strong and continuing focus on the importance of work on poverty and inequality and various forms of marginalization and it is still important for social geographers to continue work on the analysis of poverty and deprivation in all its forms. I disagree with David Harvey that the 'mapping of man's inhumanity to man' is both unnecessary and counter-revolutionary as it detracts attention from the processes which generate inequality. I would argue, on the contrary, that it is only by continuously documenting and analysing poverty and deprivation in all its various forms, that it is kept on the political agenda and makes change possible. To do otherwise is to allow it to be conveniently forgotten or swept under the carpet (Philo, 1995).

Secondly, and following from this, I would argue that it is as important today as it was 25 years ago to analyse the geography of the changing structures of social provision and access to facilities such as education, health and housing. The growing trend to privatization of provision and the rolling back of the welfare state in a number of western countries (Thompson, 1990; Edgell and Duke, 1991; Kelsey, 1993; Kelsey and O'Brien, 1995) have made this an important priority for concerned social geographers. Unfortunately, I think that, to some extent, the growing interest in culture, representation and difference may have acted to reduce interest in issues of deprivation and social inequality. As cultural politics, with a small 'p', has become more important, so the interest in social politics with a large 'P' appears to have waned. The rise of postmodernism may, inadvertently, have helped serve the interests of capital and the state in distracting interest away from questions of social inequality and deprivation to a greater concern with personal identity and difference. As Clarke (1991) points out:

> The enthusiasm with which New times has greeted . . . 'new' subjects contrasts strikingly with its silence about what might be termed the 'old' sub-

> jects. Such groups are invisible both to the new Thatcherite ethos of radical individualism and to the New times ethos of 'cultural politics'. Doubly excluded, they are 'invisible subjects', the constituencies and potential constituencies of old politics. The old, the state dependants, the low-paid, and the unorganized stand outside the charmed circle of the new. They are formed in the old ways with limited access to the pleasures of consumption (p. 167).

Clarke goes on to argue that he finds the disappearance of the old subjects particularly puzzling, given that during the Thatcher decade, the patterns of economic and social inequality have deepened and that the webs of gender, class and race are deeply woven together within them. He is concerned with its emphasis on lifestyle choice and identity and suggests that New times 'consistently overreads the degree of volatility and plasticity of identities and cultural forms and, as a result, overestimates the pace of cultural change while underestimating the place and weight of "sedimented" forms, practices and conceptions' (p. 172).

While it is clearly important for social geographers to continue their work on issues of identity, difference, culture and representation, it is important that, in so doing, they do not overlook the persistence of long-standing issues of inequality and deprivation and that, in search to give voice to others, they do not become bogged down in the swamp of relativism which leads to a form of political paralysis in both teaching and research (McDowell, 1994a).

In 'Social justice, postmodernism and the city', Harvey (Chapter 14, this volume) consciously tries to bridge the issues he raised in 1973 in *Social Justice and the City* and some of the debates around postmodernism he took up in 1989 in his book *The Condition of Postmodernity*. Specifically, he attempts to assess to what extent it is possible, and indeed desirable, to link contemporary debates on notions of 'difference', the 'Other', cultural relativism and the like with a set of more traditional concerns with social justice and equality. Working from a discussion of the class, racial and gender struggles over space in the case of Tompkins Square in New York, he points to the problems associated with what Marion Young calls 'openness to unassimilated otherness', or what was traditionally termed tolerance of differences. He then links this to his attempt in the early 1970s to differentiate the various types of argument put forward in support of, or opposition to, a proposed urban freeway in Baltimore, and the multiplicity of different conceptions of social justice that exist. From this he suggests that the concept of justice has to be understood in the way it is embedded in a particular social context. Harvey makes two points from this. The first is that the critique of social rationality and of conceptions such as social justice as policy tools was so ruthlessly pursued by the 'left' in the 1960s that 'it began to generate radical doubt throughout civil society as to the veracity of all universal claims'. Harvey says that it was a short step from this to the unwarranted conclusion made by many postmodernists that 'all forms of metatheory are either misplaced or illegitimate'. His second point is that if the conception of justice varies not only with time and place, but also with the persons concerned, then it is important to look at the way in which a particular society produces such variations in concepts. In other words, it is important 'to look at the material basis for the production of difference, in particular at the production of those radically different experiential worlds out of which divergent language games about social rationality and social justice could arise' (pp. 301–2). Harvey argues that the concepts of justice and rationality have not disappeared over the last 20 years, but their definition and use have changed. He concludes by pointing to Marion Young's (1990) work on justice and the politics of difference, and he then argues that she 'redefines the question of justice away from the purely redistributive model of welfare state capitalism and focuses on what she calls the "five faces" of oppression' (p. 304). Harvey summarizes each of these in turn and advances a series of propositions for 'just planning and policy practices'. His position is clear. As he points out:

> Justice and rationality take on different meanings across space and time and persons, yet the existence of everyday meanings to which people do attach importance and which to them appear unproblematic, gives the terms a political and mobilizing power that can never be neglected. Right and wrong are words that power revolutionary changes and no amount of negative deconstruction of such terms can deny that' (p. 304).

Merrifield (Chapter 16, this volume) takes up some of the issues raised by Harvey, particularly the question of situated knowledge and its relation to questions of social justice. Merrifield argues that epistemologies of situated knowledge offer a basis from which to call into question all privileged knowledge claims, and undercut conceptions of objectivity and neutrality. Rejection of universal truth claims thus potentially gives voice to marginalized 'others'. These insights, says Merrifield, have recently percolated down into human geography where they have generated major debate and led to 'something of a reappraisal and realignment of the left critical geographical program'. But he argues that the recognition of situated knowledge does not mean that equality of positioning is a denial of responsibility and critical enquiry. On the contrary, he says, quoting Haraway (1991), 'Relativism is the perfect mirror twin of totalisation in the ideologies of objectivity'. He argues that as an understanding of reality is accountable and responsible for an *enabling* political practice, the realm of politics conditions what may count as *true* knowledge.

Most of Merrifield's paper is taken up with an account of the geographical expedition movement developed by William Bunge in Detroit in the late 1960s. Bunge argued that it was necessary to 'bring political problems down to earth, to the scale of people's normal lives', and that a geographer's *raison d'être* was an attempt to be useful. Consequently, Bunge rejected most campus geography which severed theory from practice and prioritized 'citing' rather than 'sighting'. Instead, he argued for expeditions which set up base-camps in the inner city, or other deprived areas. The terminology was deliberate in an attempt to subvert the exploration practices of the nineteenth century. The new expeditions were designed to help, rather than destroy, humanity and to undertake field research on issues of money flows, child mortality rates, transportation and employment problems, etc. Merrifield states that one particularly illuminating insight of the expeditions was acknowledgement of the importance of reproduction. Workplace struggles had previously had a privileged position in left research agendas, but the dynamics of class, gender and race in the home, the community and the neighbourhood were crucial in conditioning social relations in the workplace. As Bunge put it: 'the geography of the working class is overwhelmingly at the point of reproduction not the point of production' (1977, p. 60). Merrifield argues that another important aspect of the expedition tradition was an insistence on deepening knowledge and understanding of the practices of everyday life. As Bunge put it:

> if you sit on a front step and listen to people they often talk about the urban geography of the city but seldom about anything that the census is measuring. A child was almost hit . . . by a speeding teenager, a landlord refused to fix the plumbing, the park is becoming a hangout for teenage narcotic users, a lady had her purse snatched on a nearby street.

Merrifield states that Bunge was for a geography, just as Claes Oldenburg was for an art, that 'embroils itself with the everyday crap and still comes out on top'. But, contra both Harvey (1972) and Pile and Rose (1992), mapping became a strategic tool for Bunge to represent the patterns and flows that support misery and alienation. Maps are not just the neutral constructions of a bourgeois male gaze, they can also enable 'others' to assert and represent their differences. They do not necessarily make claims to objective world views or institutionalize a privileged understanding of space as Pile and Rose would have us believe. But, as Merrifield points out, there are assumptions in the expedition concept that raise the difficult question of the extent to which those involved in expeditions can situate themselves in an impoverished area without becoming victim to a form of social and intellectual voyeurism. Are expeditions just another way for liberal middle-class intellectuals to feel that they are doing something good without really changing anything? The view Merrifield reaches is, after Spivak, that 'calling the place of the investigator into question [is] a meaningless piety', and that the 'intellectual's solution is not to abstain from representation'. He concludes by challengingly stating that:

> At a time when left geography is in grave danger of being rendered anodyne though a heady prioritization of discourse and textual politics, there is much to learn from the legacy of practical expeditions into the world of the exploited and oppressed outside the academy: it might at least ensure that critical theory is truly critical (p. 342).

In a related vein, Corbridge (1993) raises a number of important ethical and moral issues regarding the 'claims of distant strangers' by geographers. He points out that development studies have been criticized for being elitist, ethnocentric and dirigiste, that postmodernist and post-colonialist critics have argued that to prescribe development for others is 'to seek once more to represent an Other – to speak for and on behalf of a "Third World" when what is required is a space for the raising of diverse local voices of resistance' (p. 450). Thus, Corbridge states that development studies have reached something of an impasse which has rendered problematic the very notion of progress. But Corbridge argues that while many of the postmodernist and post-colonialist critiques of difference and representation are valid, and that ethnocentrism has undoubtedly characterized much work on development, he suggests that the 'construction of a truly global space-economy makes necessary a wider transformative politics which is sensitive to the needs and rights of distant strangers' (p. 450). Consequently, he asks 'how can we resist the seductive charms of postmodernism? how can we argue for a minimally universalist politics – for a normative development studies – without hitching our stall to an outmoded socialist praxis and without rejecting the essential insights of postmodernism?' (p. 461).

He assesses some of the work which has looked at the boundaries between development studies and moral philosophy, and suggests that the facts of a powerful globalization 'compel us to take seriously new ideas on the nature of moral communities and their boundaries'. He considers a number of other arguments dealing with trans-national justice and suggests that they point to the validity of abstract moral claims in a way which can empower local struggles for development. Thus, universal moral commitments and sensitivity to questions of difference are not incompatible. The postmodern dilemma, says Corbridge, can be avoided if we accept that certain human needs and rights can be taken to be universal, and we learn that in attending to these, we are not dictating to others but to ourselves. Siddaway (1992) and David Smith (1994) have raised a number of similar issues concerning how geographers in developed western countries relate to others elsewhere, and Smith outlines a number of issues arising from the debate on (im)partiality in contemporary moral philosophy. Smith argues that although we may reject the idea of universal responsibility to the whole of humankind, as both a moral and a practical proposition, we should at least consider the 'possibility of contributing to a wider responsibility to the potential "world community" of professional geographers' (p. 366).

He suggests, however, that 'regrettably British geography may be a lost cause', and that competition for money and research ratings is pitting departments against one another in 'a grotesque model of the business world', with macho heads acting more like entrepreneurs than academics. We should try, argues Smith, to reject this model of competitive individualism in favour of trying to build a broader, collaborative structure, and try to 'subvert those forces of darkness turning the practice of geography into an even more extreme expression of hierarchical domination and uneven development' (p. 366). This is a depressing vision. Let us try to ensure it does not come about, and that it is possible to construct a more progressive form of social geography which is concerned not just with understanding the world, but attempting to change it for the better, however that may be defined.

Susan Smith (Chapter 15, this volume) grapples with some of these issues. She notes that in the last 15 years radical geography has advanced a value-committed alternative to the apolitical, quantitative revolution and to the unqualified individualism of classical economics and behavioural science. But she argues that human geography is detached from political theory and practice at a time when the boundaries between civil society and the State are being radically realigned. Her focus is on the increasingly right-wing character of 'new times' and its challenge to geography's old radicalism and she argues that it is important for geography to engage not only in reflective debate about *how* society is structured but in political debates about how, and through what mechanisms societies *should* be structured.

She discusses the challenge of the new right and its critique of collectivism, structuralism and statism which are said to undermine individuality, to foster dependency and lack of initiative. Although few areas of geography seem self-consciously to advocate right-wing politics it is important to look at the implications of new right thinking for social geography, and Smith gives the example of residential differentiation and segregation which can, from this perspective, be interpreted as the outcome of unmediated free choice in the market (S. Smith, 1988). She points out that although neo-conservative sociology has achieved less intellectual credibility and political momentum than its economic counterpart, she suggests that libertarianism, with its insistence on a minimal State, has given a valuable source of legitimacy to market liberalism. This links to a methodological preoccupation with individuals and rejects the underlying assumption of conflict sociology and marxism: 'No social whole is regarded as more than the sum of its parts, and the mechanisms that bind individuals . . . are (like the inequalities that divide them) regarded as natural, inevitable . . . and good' (p. 312). She states that, prescriptively, these views aim to promote a civil society for the new times characterized by what Keane (1988) labels 'self-interest, hard work, flexibility, self-reliance, freedom of choice, private property, the patriarchal family and distrust of state bureaucracy' (p. 7).

In sum, Smith argues that within economics, politics and sociology there is a powerful 'New Right' challenge to the conflict theory which informed radical geography. Although the challenge has not been explicitly incorporated within geography, Smith argues that 'the grounds on which it is erected need to be taken seriously if the discipline wishes to reconstruct its radical tradition'. In particular, she suggests that the 'old' Left formula of collectivism, statism and bureaucratic paternalism is a questionable vehicle to manage the post-Fordist economy and the cultural aspirations of postmodernism. She dismisses the option of advancing or incorporating New Right formulae within geography, rightly pointing out that the freeing up of markets has not prevented social inequality being either systematic or excessive. On the contrary, inequality of income has increased considerably in many western countries in recent years. Smith argues that:

> Any attempt to break new ground in analysing social organization or prescribing social policy must successfully challenge both an uncompromising individualism from the Right and an insensitivity to personal identity on the Left. It should account for the economic and cultural dynamics of the new times, but if it is to signal geographers' greater engagement with political practice, it must also have a normative dimension (p. 313).

She focuses her analysis on the concept of citizenship which she points out has recently re-established itself as a key idea in popular politics though with very different connotations. On the right it emphasizes obligations of citizens rather than their entitlements and duties rather than rights. Smith, however, is concerned with the development of a social democratic notion of citizenship which inspired the modern welfare state. Drawing on the work of Marshall who insisted that the rights of citizenship refer to a spectrum of social rights as well as the political and civil rights embodied in national constitutions – if they exist at all – Smith attempts to outline a new theory of citizenship which has implications for the social and geographical distribution of social rights. As such, it links together questions of class, race, gender and locality with the distribution of rights and points to the locality-based struggles of some groups to achieve their social rights.

14 David Harvey, 'Social Justice, Postmodernism and the City'[1]

Reprinted in full from: *International Journal of Urban and Regional Research* **16**(4), 588–601 (1992)

The title of this essay is a collage of two book titles of mine written nearly 20 years apart, *Social justice and the city* and *The condition of postmodernity*. I here want to consider the relations between them, in part as a way to reflect on the intellectual and political journey many have travelled these last two decades in their attempts to grapple with urban issues, but also to examine how we now might think about urban problems and how by virtue of such thinking we can better position ourselves with respect to solutions. The question of *positionality* is, I shall argue, fundamental to all debates about how to create infrastructures and urban environments for living and working in the twenty-first century.

Justice and the postmodern condition

I begin with a report by John Kifner in the *International Herald Tribune* (1 August 1989) concerning the hotly contested space of Tompkins Square Park in New York City – a space which has been repeatedly fought over, often violently, since the 'police riot' of August 1988. The neighbourhood mix around the park was the primary focus of Kifner's attention. Not only were there nearly 300 homeless people, but there were also:

> Skateboarders, basketball players, mothers with small children, radicals looking like 1960s retreads, spikey-haired punk rockers in torn black, skinheads in heavy working boots looking to beat up the radicals and punks, dreadlocked Rastafarians, heavy-metal bands, chess players, dog walkers – all occupy their spaces in the park, along with professionals carrying their dry-cleaned suits to the renovated 'gentrified' buildings that are changing the character of the neighbourhood.

By night, Kifner notes, the contrasts in the park become even more bizarre:

> The Newcomers Motorcycle Club was having its annual block party at its clubhouse at 12th Street and Avenue B and the street was lined with chromed Harley Davidsons with raised 'apehanger' handlebars and beefy men and hefty women in black leather. A block north a rock concert had spilled out of a 'squat' – an abandoned city-owned building taken over by outlaw renovators, mostly young artists – and the street was filled with young people whose purple hair stood straight up in spikes. At the World Club just off Houston Street near Avenue C, black youths pulled up in the Jeep-type vehicles favored by cash-heavy teen-age crack moguls, high powered speakers blaring. At the corner of Avenue B and Third, considered one of the worst heroin blocks in New York, another concert was going on at an artists' space called The Garage, set in a former gas station walled off by plastic

> bottles and other found objects. The wall formed an enclosed garden looking up at burned-out, abandoned buildings: there was an eerie resemblance to Beirut. The crowd was white and fashionably dressed, and a police sergeant sent to check on the noise shook his head, bemused: 'It's all yuppies'.

This is, of course, the kind of scene that makes New York such a fascinating place, that makes any great city into a stimulating and exciting maelstrom of cultural conflict and change. It is the kind of scene that many a student of urban subcultures would revel in, even seeing in it, as someone like Iain Chambers (1987) does, the origins of that distinctive perspective we now call 'the postmodern':

> Postmodernism, whatever form its intellectualizing might take, has been fundamentally anticipated in the metropolitan cultures of the last twenty years: among the electronic signifiers of cinema, television and video, in recording studios and record players, in fashion and youth styles, in all those sounds, images and diverse histories that are daily mixed, recycled and 'scratched' together on that giant screen that is the contemporary city.

Armed with that insight, we could take the whole paraphernalia of postmodern argumentation and technique and try to 'deconstruct' the seemingly disparate images on that giant screen which is the city. We could dissect and celebrate the fragmentation, the co-presence of multiple discourses – of music, street and body language, dress and technological accoutrements (such as the Harley Davidsons) – and, perhaps, develop sophisticated empathies with the multiple and contradictory codings with which highly differentiated social beings both present themselves to each other and to the world and live out their daily lives. We could affirm or even celebrate the bifurcations in cultural trajectory, the preservation of pre-existing and the creation of entirely new but distinctive 'othernesses' within an otherwise homogenizing world.

On a good day, we could celebrate the scene within the park as a superb example of urban tolerance for difference, an exemplar of what Iris Marion Young calls 'openness to unassimilated otherness'. In a just and civilized society, she argues, the normative ideal of city life:

> instantiates social relations of difference without exclusion. Different groups dwell in the city alongside one another, of necessity interacting in city spaces. If city politics is to be democratic and not dominated by the point of view of one group, it must be a politics that takes account of and provides voice for the different groups that dwell together in the city without forming a community. (Young, 1990: 227)

To the degree that the freedom of city life 'leads to group differentiation, to the formation of affinity groups' (ibid.: 238) of the sort which Kifner identifies in Tompkins Square, so our conception of social justice 'requires not the melting away of differences, but institutions that promote reproduction of and respect for group differences without oppression' (p. 47). We must reject 'the concept of universality as embodied in republican versions of Enlightenment reason' precisely because it sought to 'suppress the popular and linguistic heterogeneity of the urban public' (p. 108). 'In open and accessible public spaces and forums, one should expect to encounter and hear from those who are different, whose social perspectives, experience and affiliations are different.' It then follows, Young argues, that a politics of inclusion 'must promote the ideal of a heterogeneous public, in which persons stand forth with their differences acknowledged and respected, though perhaps not completely understood, by others' (p. 119).

In similar vein, Roberto Unger, the philosophical guru of the critical legal studies movement in the United States, might view the park as a manifestation of a new ideal of community understood as a 'zone of heightened mutual vulnerability, within which people gain a chance to resolve more fully the conflict between the enabling conditions of self-assertion; between their need for attachment and for participation in group life and their fear of subjugation and depersonalization with which such engagement may threaten them' (Unger, 1987: 562). Tompkins Square seems a place where the 'contrast between structure-preserving routine and structure-transforming conflict' softens in such a way as to 'free sociability from its script and to make us available to one another more as the originals we know ourselves to be and less as the placeholders in a system of group contrasts'. The square might even be interpreted as a site of that 'microlevel of cultural-revolutionary defiance and incongruity' which periodically wells upwards into 'the macro-level of institutional innovation' (ibid: 564). Unger is acutely aware, however, that the temptation to 'treat each aspect of cultural revolution as a pretext for endless self-gratification and self-concern' can lead to a failure to 'connect the revolutionary reform of institutional arrangements with the cultural-revolutionary remaking of personal relations'.

So what should the urban policy-maker do in the face of these strictures? The best path is to pull out that well-thumbed copy of Jane Jacobs (1961) and insist that we should both respect and provide for 'spontaneous self-diversification among urban populations' in the formulation of our policies and plans. In so doing we can avoid the critical wrath she directs at city designers, who 'seem neither to recognize this force for self-diversification nor to be attracted by the esthetic problems of expressing it'. Such a strategy can help us live up to expectations of the sort which Young and Unger lay down. We should not, in short, aim to obliterate differences within the park, homogenize it according to some conception of, say, bourgeois taste or social order. We should engage, rather, with an aesthetics which embraces or stimulates that 'spontaneous self-diversification' of which Jacobs speaks. Yet there is an immediate question mark over that suggestion: in what ways, for example, can homelessness be understood as spontaneous self-diversification, and does this mean that we should respond to that problem with designer-style cardboard boxes to make for more jolly and sightly shelters for the homeless? While Jane Jacobs has a point, and one which many urbanists have absorbed these last few years, there is, evidently, much more to the problem than her arguments encompass.

That difficulty is highlighted on a bad day in the park. So-called forces of law and order battle to evict the homeless, erect barriers between violently clashing factions. The park then becomes a locus of exploitation and oppression, an open wound from which bleed the five faces of oppression which Young defines as exploitation, marginalization, powerlessness, cultural imperialism and violence. The potentiality for 'openness to unassimilated otherness' breaks apart and, in much the same way that the cosmopolitan and eminently civilized Beirut of the 1950s suddenly collapsed into an urban maelstrom of warring factions and violent confrontation, so we find sociality collapsing into violence (see Smith, 1989; 1992). This is not unique to New York City but is a condition of urban life in many of our large metropolitan areas – witness events in the *banlieues* of Paris and Lyons, in Brussels, in Liverpool, London and even Oxford in recent times.

In such circumstances Young's pursuit of a vision of justice that is assertive as to difference without reinforcing the forms of oppression gets torn to tatters and Unger's dreams of micro-revolutions in cultural practices which stimulate progressive rather than repressive institutional innovation become just that – dreams. The very best face that we can put upon the whole scene is to recognize that this is how class, ethnic, racial and gender struggle is, as Lefebvre (1991) would put it, being 'inscribed in space'. And what should the planner do? Here is how a subsequent article in the *New York Times* reflected on that dilemma:

> There are neighborhood associations clamoring for the city to close the park and others just as insistent that it remain a refuge for the city's downtrodden. The local Assemblyman, Steven Sanders, yesterday called for a curfew that would effectively evict more than a hundred homeless people camped out in the park. Councilwoman Miriam Friedlander instead recommended that Social Services, like healthcare and drug treatment, be brought directly to the people living in the tent city. 'We do not find the park is being used appropriately,' said Deputy Mayor Barbara J. Fife, 'but we recognise there are various interests.' There is, they go on to say, only one thing that is a consensus, first that there isn't a consensus over what should be done, except that any new plan is likely to provoke more disturbances, more violence.

On 8 June 1991, the question was resolved by evicting everyone from the park and closing it entirely 'for rehabilitation' under a permanent guard of at least 20 police officers. The New York authorities, situated on what Davis (1990: 224) calls 'the bad edge of postmodernity', militarize rather than liberate its public space. In so doing, power is deployed in support of a middle-class quest for 'personal insulation, in residential work, consumption and travel environments, from "unsavory" groups and individuals, even crowds in general'. Genuinely public space is extinguished, militarized or semi-privatized. The heterogeneity of open democracy, the mixing of classes, ethnicities, religions and divergent taste cultures within a common frame of public space, is lost along with the capacity to celebrate unity and community in the midst of diversity. The ultimate irony, as Davis points out, is that 'as the walls have come down in Eastern Europe, they are being erected all over [our cities]'.

And what should the policy-maker and planner do in the face of these conditions? Give up planning and join one of those burgeoning cultural studies progammes which revel in chaotic scenes of the Tompkins Square sort while simultaneously disengaging from any commitment to do something about them? Deploy all the critical powers of deconstruction and semiotics to seek new and engaging interpretations of graffiti which say 'Die, Yuppie Scum'? Should we join revolutionary and anarchist groups and fight for the rights of the poor and the culturally marginalized to express their rights and if necessary make a home for themselves in the park? Or should we throw away that dogeared copy of Jane Jacobs and join with the forces of law and order and help impose some authoritarian solution on the problem?

Decisions of some sort have to be made and actions taken, as about any other facet of urban infrastructure. And while we might all agree that an urban park is a good thing in principle, what are we to make of the fact that the uses turn out to be so conflictual, and that even conceptions as to what the space is for and how it is to be managed diverge radically among competing factions? To hold all the divergent politics of need and desire together within some coherent frame may be a laudable aim, but in practice far too many of the interests are mutually exclusive to allow their mutual accommodation. Even the best shaped compromise (let alone the savagely imposed authoritarian solution) favours one or other factional interest. And that provokes the biggest question of all – what is the *conception* of 'the public' incorporated into the construction of public space?

To answer these questions requires some deeper understanding of the forces at work shaping conflict in the park. Kifner identified drugs and real estate – 'the two most powerful forces in [New York City] today'. Both of them are linked to organized crime and are major pillars of the political economy of contemporary capitalism. We cannot understand events within and around the park or strategize as to its future uses without contextualizing it against a background of the political-economic transformations now occurring in urban life. The problems of Tompkins Square Park have, in short, to be seen in terms of social processes which create homelessness, promote criminal activities of many sorts (from real estate swindles and the crack trade to street muggings), generate hierarchies of power between gentrifiers and the homeless, and facilitate the emergence of deep tensions along the major social fault-lines of class, gender, ethnicity, race and religion, lifestyle and place-bound preferences (see Smith, 1992).

Social justice and modernity

I now leave this very contemporary situation and its associated conundrums and turn to an older story. It turned up when I unearthed from my files a yellowing manuscript, written sometime in the early 1970s, shortly after I finished *Social justice and the city*. I there examined the case of a proposal to put a segment of the Interstate Highway System on an east–west trajectory right through the heart of Baltimore – a proposal first set out in the early 1940s and which has still not been fully resolved. I resurrect this case here in part to show that what we would now often depict as a quintessentially modernist problem was even at that time argued about in ways which contained the seeds, if not the essence, of much of what many now view as a distinctively postmodernist form of argumentation.

My interest in the case at that time, having looked at a lot of the discussion, attended hearings and read a lot of documentation, lay initially in the highly differentiated arguments, articulated by all kinds of different groups, concerning the rights and wrongs of the whole project. There were, I found, seven kinds of arguments being put forward:

1 An *efficiency* argument which concentrated on the relief of traffic congestion and facilitating the easier flow of goods and people throughout the region as well as within the city;
2 An *economic growth* argument which looked to a projected increase (or prevention of loss) in investment and employment opportunities in the city consequent upon improvements in the transport system;
3 An *aesthetic and historical heritage* argument which objected to the way sections of the proposed highway would either destroy or diminish urban environments deemed both attractive and of historical value;
4 A *social and moral order* argument which held that prioritizing highway investment and subsidizing car owners rather than, for example, investing in housing and health care was quite wrong;
5 An *environmentalist/ecological* argument whch considered the impacts of the proposed highway on air quality, noise pollution and the destruction of certain valued environments (such as a river valley park);
6 A *distributive justice* argument which dwelt mainly on the benefits to business and predominantly white middle-class suburban commuters to the detriment of low-income and predominantly African-American inner-city residents;
7 A *neighbourhood and communitarian* argument which considered the way in which close-knit but otherwise fragile and vulnerable communities might be destroyed, divided or disrupted by highway construction.

The arguments were not mutually exclusive, of course, and several of them were merged by proponents of the highway into a common thread – for example, the efficiency of the transport system would stimulate growth and reduce pollution from congestion so as to advantage otherwise disadvantaged inner-city residents. It was also possible to break up each argument into quite distinct parts – the distributive impacts on women with children would be very different from those on male workers.

We would, in these heady postmodern times, be prone to describe these separate arguments as 'discourses', each with its own logic and imperatives. And we would not have to look too closely to see particular 'communities of interest' which articulated a particular discourse as if it was the only one that mattered. The particularistic arguments advanced by such groups proved effective in altering the alignment of the highway but did not stop the highway as a whole. The one group which tried to forge a coalition out of these disparate elements (the *Movement Against Destruction*, otherwise known as *MAD*) and to provide an umbrella for opposition to the highway as a whole turned out to be the least effective in mobilizing people and constituencies even though it was very articulate in its arguments.

The purpose of my own particular enquiry was to see how the arguments (or discourses) for and against the highway worked and if coalitions could be built in principle between seemingly disparate and often highly antagonistic interest groups via the construction of higher-order arguments (discourses) which could provide the basis for consensus. The multiplicity of views and forces has to be set against the fact that either the highway is built or it is not, although in Baltimore, with its wonderful way of doing things, we ended up with a portion of the highway that is called a boulevard (to make us understand that this six-lane two-mile segment of a monster cut through the heart of low-income and predominantly African-American West Baltimore is not what it really is) and another route on a completely different alignment, looping around the city core in such a way as to allay some of the worst political fears of influential communities.

Might there be, then, some higher-order discourse to which everyone could appeal in working out whether or not it made sense to build the highway? A dominant theme in the literature of the 1960s was that it was possible to identify some such higher-order arguments. The phrase that was most frequently used to describe it was *social rationality*. The idea of that did not seem implausible, because each of the seven seemingly distinctive arguments advanced a rational position of some sort and not infrequently appealed to some higher-order rationale to bolster its case. Those arguing on efficiency and growth grounds frequently invoked utilitarian arguments, notions of 'public good' and the greatest benefit to the greatest number, while recognizing (at their best) that individual sacrifices were inevitable and that it was right and proper to offer appropriate compensation for those who would be displaced. Ecologists or communitarians likewise appealed to higher-order arguments – the former to the values inherent in nature and the latter to some higher sense of communitarian values. For all of these reasons, consideration of higher-order arguments over social rationality did not seem unreasonable.

Dahl and Lindblom's *Politics, economics and welfare*, published in 1953, provides a classic statement along these lines. They argue that not only is socialism dead (a conclusion that many would certainly share these days) but also that capitalism is equally dead. What they signal by this is an intellectual tradition which arose out of the experience of the vast market and capitalistic failure of the Great Depression and the second world war and which concluded that some kind of middle ground had to be found between the extremism of a pure and unfettered market economy and the communist vision of an organized and highly centralized economy. They concentrated their theory on the question of rational social action and argued that this required 'processes for both rational *calculation* and effective *control*' (p. 21). Rational calculation and control, as far as they were concerned, depended upon the exercise of rational calculation through price-fixing markets, hierarchy (top-down decision-making), polyarchy (democratic control of leadership) and bargaining (negotiation), and such means should be deployed to achieve the goals of 'freedom, rationality, democracy, subjective equality, security, progress, and appropriate inclusion' (p. 28). There is much that is interesting about Dahl and Lindblom's analysis and it is not too hard to imagine that after the recent highly problematic phase of market triumphalism, particularly in Britain and the United States, there will be some sort of search to resurrect the formulations they proposed. But in so doing it is also useful to remind ourselves of the intense criticism that was levelled during the 1960s and 1970s against their search for some universal prospectus on the socially rational society of the future.

Godelier, for example, in his book *Rationality and irrationality in economics*, savagely attacked the socialist thinking of Oscar Lange for its teleological view of rationality and its presumption that socialism should or could ever be the ultimate achievement of the rational life. Godelier did not attack this notion from the right but from a marxist and historical materialist perspective. His point was that there are different definitions of rationality depending upon the form of social organization and that the rationality embedded in feudalism is different from that of capitalism, which should, presumably, be different again under socialism. Rationality defined from the standpoint of corporate capital is quite different from rationality defined from the standpoint of the working classes. Work of this type helped to fuel the growing radical critique of even the non-teleological and incrementalist thinking of the Dahl and Lindblom sort. This critique suggested that their definition of social rationality was connected to the perpetuation and rational management of a capitalist economic system rather than with the exploration of alternatives. To attack (or deconstruct, as we now would put it) their conception of social rationality was seen by the left at the time as a means to challenge the ideological hegemony of a dominant corporate capitalism. Feminists, those marginalized by racial characteristics, colonized peoples, ethnic and religious minorities echoed that refrain in their work, while adding their own conception of who was the enemy to be challenged and what were the dominant forms of rationality to be contested. The result was to show emphatically that there is no overwhelming and universally acceptable definition of social rationality to which we might appeal, but innumerable different rationalities depending upon social and material circumstances, group identities, and social objectives. Rationality is defined by the nature of the social group and its project rather than the project being dictated by social rationality. The deconstruction of universal claims of social rationality was one of the major achievements and continues to be one of the major legacies of the radical critique of the 1960s and 1970s.

Such a conclusion is, however, more than a little discomforting. It would suggest, to go back to the highway example, that there was no point whatsoever in searching for any higher-order arguments because such arguments simply could not have any purchase upon the political process of decision-making. And it is indeed striking that the one group that tried to build such overall arguments, MAD, was the group that was least successful in actually mobilizing opposition. The fragmented discourses of those who sought to change the alignment of the highway had more effect than the more unified discourse precisely because the former were grounded in the specific and particular local circumstances in which individuals found themselves. Yet the fragmented discourses could never go beyond challenging the alignment of the highway. It did indeed need a more unified discourse, of the sort which MAD sought to articulate, to challenge the concept of the highway in general.

This poses a direct dilemma. If we accept that fragmented discourses are the only authentic discourses and that no unified discourse is possible, then there is no way to challenge the overall qualities of a social system. To mount that more general challenge we need some kind of unified or unifying set of arguments. For this reason, I chose, in this ageing and yellowing manuscript, to take a closer look at the particular question of social justice as a basic ideal that might have more universal appeal.

Social justice

Social justice is but one of the seven criteria I worked with and I evidently hoped that careful investigation of it might rescue the argument from the abyss of formless relativism and infinitely variable discourses and interest grouping. But here too the enquiry proved frustrating. It revealed that there are as many competing theories of social justice as there are ideals of social rationality. Each ideal has its flaws and strengths. Egalitarian views, for example, immediately run into the problem that 'there is nothing more unequal than the equal treatment of unequals' (the modification of doctrines of equality of opportunity in the United States by requirements for affirmative action, for example, recognizes what a significant problem that is). By the time I had thoroughly reviewed positive law theories of justice, utilitarian views (the greatest good of the greatest number), social contract views historically attributed to Rousseau and powerfully revived by John Rawls in his *Theory of justice* in the early 1970s, the various intuitionist, relative deprivation and other interpretations of justice, I found myself in a quandary as to precisely *which* theory of justice is the most just. The theories can, to some degree, be arranged in a hierarchy with respect to each other. The positive law view that justice is a matter of law can be challenged by a utilitarian view which allows us to discriminate between good and bad law on the basis of some greater good, while the social contract and natural rights views suggest that no amount of greater good for a greater number can justify the violation of certain inalienable rights. On the other hand, intuitionist and relative deprivation theories exist in an entirely different dimension.

Yet the basic problem remained. To argue for social justice meant the deployment of some initial criteria to define which theory of social justice was appropriate or more just than another. The infinite regress of higher-order criteria immediately looms, as does, in the other direction, the relative ease of total deconstruction of the notion of justice to the point where it means nothing whatsoever, except whatever people at some particular moment decide they want it to mean. Competing discourses about justice could not be disassociated from competing discourses about positionality in society.

There seemed two ways to go with that argument. The first was to look at how concepts of justice are embedded in language, and that led me to theories of meaning of the sort which Wittgenstein advanced:

> How many kinds of sentence are there? . . . There are *countless* kinds: countless different kinds of use to what we call 'symbols', 'words', 'sentences'. And this multiplicity is not something fixed, given once for all: but new types of language, new language games, as we may say, come into existence and others become obsolete and get forgotten . . . Here the term 'language-*game*' is meant to bring into prominence the fact that the *speaking* of language is part of an activity, or a form of life . . . How did we *learn* the meaning of this word ('good' for instance)? From what sort of examples? in what language games? Then it will be easier for us to see that the word must have a family of meanings. (Wittgenstein, 1967)

From this perspective the concept of justice has to be understood in the way it is embedded in a particular language game. Each language game attaches to the particular social, experiential and perceptual world of the speaker. Justice has no universal meaning, but a whole 'family' of meanings. This finding is completely consistent, of course, with anthropological studies which show that justice among, say, the Nuer, means something completely different from the capitalistic conception of justice. We are back to the point of cultural, linguistic or discourse relativism.

The second path is to admit the relativism of discourses about justice, but to insist that discourses are expressions of social power. In this case the idea of justice has to be set against the formation of certain hegemonic discourses which derive from the power exercised by any ruling class. This is an idea which goes back to Plato, who in the *Republic* has Thrasymachus argue that:

> Each ruling class makes laws that are in its own interest, a democracy democratic laws, a tyranny tyrannical ones and so on; and in making these laws they define as 'right' for their subjects what is in the interest of themselves, the rulers, and if anyone breaks their laws he is punished as a 'wrong-doer'. That is what I mean when I say that 'right' is the same in all states, namely the interest of the established ruling class . . . (Plato, 1965)

Consideration of these two paths brought me to accept a position which is most clearly articulated by Engels in the following terms:

> The stick used to measure what is right and what is not is the most abstract expression of right itself, namely *justice* . . . The development of right for the jurists . . . is nothing more than a striving to bring human conditions, so far as they are expressed in legal terms, ever closer to the ideal of justice, *eternal* justice. And always this justice is but the ideologized, glorified expression of the existing economic relations, now from their conservative and now from their revolutionary angle. The justice of the Greeks and Romans held slavery to be just: the justice of the bourgeois of 1789 demanded the abolition of feudalism on the ground that it was unjust. The conception of eternal justice, therefore, varies not only with time and place, but also with the persons concerned . . . While in everyday life . . . expressions like right, wrong, justice, and sense of right are accepted without

> misunderstanding even with reference to social matters, they create . . . the same hopeless confusion in any scientific investigation of economic relations as would be created, for instance, in modern chemistry if the terminology of the phlogiston theory were to be retained. (Marx and Engels, 1951: 562–4)

It is a short step from this conception to Marx's critique of Proudhon, who, Marx (1967: 88–9) claimed, took his ideal of justice 'from the juridical relations that correspond to the production of commodities' and in so doing was able to present commodity production as 'a form of production as everlasting as justice'. The parallel with Godelier's rebuttal of Lange's (and by extension Dahl and Lindblom's) views on rationality is exact. Taking capitalistic notions of social rationality or of justice, and treating them as universal values to be deployed under socialism, would merely mean the deeper instanciation of capitalist values by way of the socialist project.

The transition from modernist to postmodernist discourses

There are two general points I wish to draw out of the argument so far. First, the critique of social rationality and of conceptions such as social justice as policy tools was something that was originated and so ruthlessly pursued by the 'left' (including marxists) in the 1960s that it began to generate radical doubt throughout civil society as to the veracity of all universal claims. From this it was a short, though as I shall shortly argue, unwarranted, step to conclude, as many postmodernists now do, that all forms of metatheory are either misplaced or illegitimate. Both steps in this process were further reinforced by the emergence of the so-called 'new' social movements – the peace and women's movements, the ecologists, the movements against colonization and racism – each of which came to articulate its own definitions of social justice and rationality. There then seemed to be, as Engels had argued, no philosophical, linguistic or logical way to resolve the resulting divergencies in conceptions of rationality and justice, and thereby to find a way to reconcile competing claims or arbitrate between radically different discourses. The effect was to undermine the legitimacy of state policy, attack all conceptions of bureaucratic rationality and at best place social policy formulation in a quandary and at worst render it powerless except to articulate the ideological and value precepts of those in power. Some of those who participated in the revolutionary movements of the 1970s and 1980s considered that rendering transparent the power and class basis of supposedly universal claims was a necessary prelude to mass revolutionary action.

But there is a second and, I think, more subtle point to be made. If Engels is indeed right to insist that the conception of justice 'varies not only with time and place, but also with the persons concerned', then it seems important to look at the ways in which a particular society produces such variation in concepts. In so doing it seems important, following writers as diverse as Wittgenstein and Marx, to look at the material basis for the production of difference, in particular at the production of those radically different experiential worlds out of which divergent language games about social rationality and social justice could arise. This entails the application of historical-geographical materialist methods and principles to understand the production of those power differentials which in turn produce different conceptions of justice and embed them in a struggle over ideological hegemony between classes, races, ethnic and political groupings as well as across the gender divide. The philosophical, linguistic and logical critique of universal propositions such as justice and of social rationality can be upheld as perfectly correct without necessarily endangering the ontological or epistemological status of a metatheory which confronts the ideological and material functionings and bases of particular discourses. Only in this way can we begin to understand why it is that concepts such as justice which appear as 'hopelessly confused' when examined in abstraction can become such a powerful mobilizing force in everyday life, where, again to quote Engels, 'expressions like right, wrong, justice, and sense of right are accepted without misunderstanding even with reference to social matters'.

From this standpoint we can clearly see that concepts of justice and of rationality have not disappeared from our social and political world these last few years. But their definition and use has changed. The collapse of class compromise in the struggles of the late 1960s and the emergence of the socialist, communist and radical left movements, coinciding as it did with an acute crisis of overaccumulation of capital, posed a serious threat to the stability of the capitalist political-economic system. At the ideological level, the emergence of alternative definitions of both justice and rationality was part of that attack, and it was to this question that my earlier book, *Social justice and the city*, was addressed. But the recession/depression of 1973–5 signalled not only the savage devaluation of capital stock (through the first wave of deindustrialization visited upon the weaker sectors and regions of a world capitalist economy) but the beginning of an attack upon the power of organized labour via widespread unemployment, austerity programmes, restructuring and, eventually, in some instances (such as Britain) institutional reforms.

It was under such conditions that the left penchant for attacking what was interpreted as a capitalist power basis within the welfare state (with its dominant notions of social rationality and just redistributions) connected to an emerging right-wing agenda to defang the power of welfare state capitalism, to get away from any notion whatsoever of a social contract between capital and labour and to abandon political notions of social rationality in favour of market rationality. The important point about this transition, which was phased in over a number of years, though at a quite different pace from country to country (it is only now seriously occurring in Sweden, for example), was that the state was no longer obliged to define rationality and justice, since it was presumed that the market could best do it for us. The idea that just deserts are best arrived at through market behaviours, that a just distribution is whatever the market dictates and that a just organization of social life, of urban investments and of resource allocations (including those usually referred to as environmental) is best arrived at through the market is, of course, relatively old and well-tried. It implies conceptions of justice and rationality of a certain sort, rather than their total abandonment. Indeed, the idea that the market is the best way to achieve the most just and the most rational forms of social organization has become a powerful feature of the hegemonic discourses these last 20 years in both the United States and Britain. The collapse of centrally planned economics throughout much of the world has further boosted a market triumphalism which presumes that the rough justice administered through the market in the course of this transition is not only socially just but also deeply rational. The advantage of this solution, of course, is that there is no need for explicit theoretical, political and social argument over what is or is not socially rational just because it can be presumed that, provided the market functions properly, the outcome is nearly always just and rational. Universal claims about rationality and justice have in no way diminished. They are just as frequently asserted in justification of privatization and of market action as they ever were in support of welfare state capitalism.

The dilemmas inherent in reliance on the market are well known and no one holds to it without some qualification. Problems of market breakdown, of externality effects, the provision of public goods and infrastructures, the clear need for *some* coordination of disparate investment decisions, all of these require some level of government interventionism. Margaret Thatcher may thus have abolished Greater London government, but the business community wants some kind of replacement (though preferably non-elected), because without it city services are disintegrating and London is losing its competitive edge. But there are many voices that go beyond that minimal requirement since free-market capitalism has produced widespread unemployment, radical restructurings and devaluations of capital, slow growth, environmental degradation and a whole host of financial scandals and competitive difficulties, to say nothing of the widening disparities in income distributions in many countries and the social stresses that attach thereto. It is under such conditions that the never quite stilled voice of state regulation, welfare state capitalism, of state management of industrial development, of state planning of environmental quality, land use, transportation systems and physical and social infrastructures, of state incomes and taxation policies which achieve a modicum of redistribution either in kind (via housing, health care, educational services and the like) or through income transfers, is being reasserted. The political questions of social rationality and of social justice over and above that administered through the market are being taken off the back burner and moved to the forefront of the political agenda in many of the advanced capitalist countries. It was exactly in this mode, of course, that Dahl and Lindblom came in back in 1953.

It is here that we have to face up to what Unger calls the 'ideological embarrassment' of the history of politics these last hundred years: its tendency to move merely in repetitive cycles, swinging back and forth between *laissez-faire* and state interventionism without, it seems, finding any way to break out of this binary opposition to turn a spinning wheel of stasis into a spiral of human development. The breakdown of organized communism in eastern Europe and the Soviet Union here provides a major opportunity precisely because of the radical qualities of the break. Yet there are few signs of any similar penchant for ideological and institutional renovation in the advanced capitalist countries, which at best seem to be steering towards another bout of bureaucratic management of capitalism embedded in a general politics of the Dahl and Lindblom sort and at worst to be continuing down the blind ideological track which says that the market always knows best. It is precisely at this political conjuncture that we should remind ourselves of what the radical critique of universal claims of justice and rationality has been all about, without falling into the postmodernist trap of denying the validity of *any* appeal to justice or to rationality as a war cry for political mobilization (even Lyotard, that father figure of postmodern philosophy, hopes for the reassertion of some 'pristine and non-consensual conception of justice' as a means to find a new kind of politics).

For my own part, I think Engels had it right. Justice and rationality take on different meanings across space and time and persons, yet the existence of everyday meanings to which people do attach importance and which to them appear unproblematic, gives the terms a political and mobilizing power that can never be neglected. Right and wrong are words that power revolutionary changes and no amount of negative deconstruction of such terms can deny that. So where, then, have the new social movements and the radical left in general got with their own conception, and how does it challenge both market and corporate welfare capitalism?

Young in her *Justice and the politics of difference* (1990) provides one of the best recent statements. She redefines the question of justice away from the purely redistributive mode of welfare state capitalism and focuses on what she calls the 'five faces' of oppression, and I think each of them is worth thinking about as we consider the struggle to create liveable cities and workable environments for the twenty-first century.

The first face of oppression conjoins the classic notion of exploitation in the workplace with the more recent focus on exploitation of labour in the living place (primarily, of course, that of women working in the domestic sphere). The classic forms of exploitation which Marx described are still omnipresent, though there have been many mutations such that, for example, control over the length of the working day may have been offset by increasing intensity of labour or exposure to more hazardous health conditions not only in blue-collar but also in white-collar occupations. The mitigation of the worst aspects of exploitation has been, to some degree, absorbed into the logic of welfare state capitalism in part through the sheer exercise of class power and trade union muscle. Yet there are still many terrains upon which chronic exploitation can be identified and which will only be addressed to the degree that active struggle raises issues. The conditions of the unemployed, the homeless, the lack of purchasing power for basic needs and services for substantial portions of the population (immigrants, women, children) absolutely have to be addressed. All of which leads to my first proposition: *that just planning and policy practices must confront directly the problem of creating forms of social and political organization and systems of production and consumption which minimize the exploitation of labour power both in the workplace and the living place.*

The second face of oppression arises out of what Young calls *marginalization.* 'Marginals', she writes, 'are people the system of labour cannot or will not use.' This is most typically the case with individuals marked by race, ethnicity, region, gender, immigration status, age, and the like. The consequence is that 'a whole category of people is expelled from useful participation in social life and thus potentially subjected to severe material deprivation and even extermination'. The characteristic response of welfare state capitalism has been either to place such marginal groups under tight surveillance or, at best, to induce a condition of dependency in which state support provides a justification to 'suspend all basic rights to privacy, respect, and individual choice'. The responses among the marginalized have sometimes been both violent and vociferous, in some instances turning their marginalization into a heroic stand against the state and against any form of inclusion into what has for long only ever offered them oppressive surveillance and demeaning subservience. Marginality is one of the crucial problems facing urban life in the twenty-first century and consideration of it leads to the second principle: *that just planning and policy practices must confront the phenomenon of marginalization in a non-paternalistic mode and find ways to organize and militate within the politics of marginalization in such a way as to liberate captive groups from this distinctive form of oppression.*

Powerlessness is, in certain ways, an even more widespread problem than marginality. We are here talking of the ability to express political power as well as to engage in the particular politics of self-expression which we encountered in Tompkins Square Park. The ability to be listened to with respect is strictly circumscribed within welfare state capitalism and failure on this score has played a key role in the collapse of state communism. Professional groups have advantages in this regard which place them in a different category to most others and the temptation always stands, for even the most politicized of us, to speak for others without listening to them. Political inclusion is, if anything, diminished by the decline of trade unionism, of political parties, and of traditional institutions, yet it is at the same time revived by the organization of new social movements. But the increasing scale of international dependency and interdependency makes it harder and harder to offset powerlessness in general. Like the struggle against the Baltimore expressway, the mobilization of political power among the oppressed in society is increasingly a local affair, unable to address the structural characteristics of either market or welfare state capitalism as a whole. This leads to my third proposition: *just planning and policy practices must empower rather than deprive the oppressed of access to political power and the ability to engage in self-expression.*

What Young calls *cultural imperialism* relates to the ways in which 'the dominant meanings of a society render the particular perspective of one's own group invisible at the same time as they stereotype one's group and mark it out as the Other'. Arguments of this sort have been most clearly articulated by feminists and black liberation theorists, but they are also implicit in liberation theology as well as in many domains of cultural theory. This is, in some respects, the most difficult form of oppression to identify clearly, yet there can surely be no doubt that there are many social groups in our societies who find or feel themselves 'defined from the outside, positioned, placed, by a network of dominant meanings they experience as arising from elsewhere, from those with whom they do not identify and who do not identify with them'. The alienation and social unrest to be found in many western European and North American cities (to say nothing of its re-emergence throughout much of eastern Europe) bears all the marks of a reaction to cultural imperialism, and here too, welfare state capitalism has in the past proved both unsympathetic and unmoved. From this comes a fourth proposition: *that just planning and policy practices must be particularly sensitive to issues of cultural imperialism and seek, by a variety of means, to eliminate the imperialist attitude both in the design of urban projects and modes of popular consultation.*

Fifth, there is the issue of *violence*. It is hard to consider urban futures and living environments into the twenty-first century without confronting the problem of burgeoning levels of physical violence. The fear of violence against persons and property, though often exaggerated, has a material grounding in the social conditions of market capitalism and calls for some kind of organized response. There is, furthermore, the intricate problem of the violence of organized crime and its interdigitation with capitalist enterprise and state activities. The problem at the first level is, as Davis points out in his consideration of Los Angeles, that the most characteristic response is to search for defensible urban spaces, to militarize urban space and to create living environments which are more rather than less exclusionary. The difficulty with the second level is that the equivalent of the *mafiosi* in many cities (an emergent problem in the contemporary Soviet Union, for example) has become so powerful in urban governance that it is they, rather than elected officials and state bureaucrats, who hold the true reins of power. No society can function without certain forms of social control and we have to consider what that might be in the face of a Foucauldian insistence that all forms of social control are oppressive, no matter what the level of violence to which they are addressed. Here too there are innumerable dilemmas to be solved, but we surely know enough to advance a fifth proposition: *a just planning and policy practice must seek out non-exclusionary and non-militarized forms of social control to contain the increasing levels of both personal and institutionalized violence without destroying capacities for empowerment and self-expression.*

Finally, I want to add a sixth principle to those which Young advances. This derives from the fact that all social projects are ecological projects and vice versa. While I resist the view that 'nature has rights' or that nature can be 'oppressed', the justice due to future generations and to other inhabitants of the globe requires intense scrutiny of all social projects for assessment of their ecological consequences. Human beings necessarily appropriate and transform the world around them in the course of making their own history, but they do not have to do so with such reckless abandon as to jeopardize the fate of peoples separated from us in either space or time. The final proposition is, then: *that just planning and policy practices will clearly recognize that the necessary ecological consequences of all social projects have impacts on future generations as well as upon distant peoples and take steps to ensure a reasonable mitigation of negative impacts.*

I do not argue that these six principles can or even should be unified, let alone turned into some convenient and formulaic composite strategy. Indeed, the six dimensions of justice here outlined are frequently in conflict with each other as far as their application to individual persons – the exploited male worker may be a cultural imperialist on matters of race and gender while the thoroughly oppressed person may be the bearer of social injustice as violence. On the other hand, I do not believe the principles can be applied in isolation from each other either. Simply to leave matters at the level of a 'non-consensual conception of justice, as someone like Lyotard (1984) would do, is not to confront some central issues of the social processes which produce such a differentiated conception of justice in the first place. This then suggests that social policy and planning has to work at two levels. The different faces of oppression have to be confronted for what they are and as they are manifest in daily life, but in the longer term and at the same time the underlying sources of the different forms of oppression in the heart of the political economy of capitalism must also be confronted, not as the fount of all evil but in terms of capitalism's revolutionary dynamic which transforms, disrupts, deconstructs and reconstructs ways of living, working, relating to each other and to the environment. From such a standpoint the issue is never about whether or not there shall be change, but what sort of change we can anticipate, plan for, and proactively shape in the years to come.

I would hope that consideration of the varieties of justice as well as of this deeper problematic might set the tone for present deliberations. By appeal to them, we might see ways to break with the political, imaginative and institutional constraints which have for too long inhibited the advanced capitalist societies in their developmental path. The critique of universal notions of justice and rationality, no matter whether embedded in the market or in state welfare capitalism, stands still. But it is both valuable and potentially liberating to look at alternative conceptions of both justice and rationality as these have emerged within the new social movements these last two decades. And while it will in the end ever be true, as Marx and Plato observed, that 'between equal rights force decides', the authoritarian imposition of solutions to many of our urban ills these past few years and the inability to listen to alternative conceptions of both justice and rationality are very much a part of the problem. The conceptions I have outlined speak to many of the marginalized, the oppressed and the exploited in this time and place. For many of us, and for many of them, the formulations may well appear obvious, unproblematic and just plain common sense. And it is precisely because of such widely held conceptions that so much welfare-state paternalism and market rhetoric fails. It is, by the same token, precisely out of such conceptions that a genuinely liberatory and transformative politics can be made. 'Seize the time and the place', they would say around Tompkins Square Park, and this does indeed appear an appropriate time and place to do so. If some of the walls are coming down all over eastern Europe, then surely we can set about bringing them down in our own cities as well.

Note

1 This is the text of a plenary paper delivered in Berlin on 9 October 1991 to the European Workshop on the Improvement of the Built Environment and Social Integration in Cities, sponsored by the European Foundation for the Improvement of Living and Working Conditions. It is reproduced here with the permission of the European Foundation.

References

Chambers, I. 1987: Maps for the metropolis: a possible guide to the present. *Cultural Studies* 1, 1–22.
Dahl, R. and Lindblom, C. 1953: *Politics, economics and welfare*. New York: Harper.
Davis, M. 1990: *City of quartz: excavating the future in Los Angeles*. London: Verso.
Godelier, M. 1972: *Rationality and irrationality in economics*. London: New Left Books.
Harvey, D. 1973: *Social justice and the city*. London: Edward Arnold.
—— 1989: *The condition of postmodernity*. Oxford: Blackwell.
Jacobs, J. 1961: *The death and life of great American cities*. New York: Vintage.
Kifner, J. 1989: No miracles in the park: homeless New Yorkers amid drug lords and slumlords. *International Herald Tribune*, 1 August 1989, p. 6.
Lefebvre, H. 1991: *The production of space*. Oxford: Blackwell.
Lyotard, J. 1984: *The postmodern condition*. Manchester: Manchester University Press.
Marx, K. 1967: *Capital*, vol. I. New York: International Publishers.
Marx, K. and Engels, F. 1951: *Selected works*, vol. I. Moscow: Progress Publishers.
Plato 1965: *The republic*. Harmondsworth, Middlesex: Penguin Books.
Rawls, J. 1971: *A theory of justice*. Cambridge, MA: Harvard University Press.
Smith, N. 1989: Tompkins Square: riots, rents and redskins. *Portable Lower East Side* 6, 1–36.
—— 1992: New city, new frontier: the Lower East Side as wild, wild west. In Sorkin, M. (ed.), *Variations on a theme park: the new American city and the end of public space*. New York: Noonday.
Unger, R. 1987: *False necessity: anti-necessitarian social theory in the service of radical democracy*. Cambridge: Cambridge University Press.
Wittgenstein, L. 1967: *Philosophical investigations*. Oxford: Blackwell.
Young, I. M. 1990: *Justice and the politics of difference*. Princeton, NJ: Princeton University Press.

15 Susan J. Smith,

'Society, Space and Citizenship: A Human Geography for the "New Times"?'

Reprinted in full from: *Transactions of the Institute of British Geographers* **14**, 144–56 (1989)

In the last 15 years, geography has advanced a value-committed alternative to the apolitical quantitative 'revolution' and to the unqualified individualism of classical economics and behavioural science. This sociology of conflict has Marxian origins, but grew to embrace the work of Max Weber, Anthony Giddens, Ralf Dahrendorf and many others (Binns, 1977). Among geographers, the impetus came from Harvey (1973) and Peet (1977), and conflict perspectives now run throughout the discipline's critical tradition.[1]

This radical geography drew attention to the power structures that sustain systematic disadvantage and to the collective struggles required to confront inequality. Self-consciously aligned with the political Left, conflict perspectives inspired renewed concern for the poor, and encouraged a continuing tradition of welfare studies (Smith, 1977; Knox, 1988). Additionally, and notwithstanding a preoccupation with class relations and their constitution through territory (Scott and Storper, 1986), this approach could embrace the patriarchal division of urban space (Women and Geography Study Group, 1984) and the racisms that structure residential segregation (Jackson, 1987).

It is no coincidence that the incorporation of conflict perspectives into modern social science coincided, in the 'West', with the strengthening hand of trade unionism and the rise of corporate socialism. It gained momentum from a 'rediscovery' of poverty in the 1960s and was nurtured by political consensus on the immorality of inequality and on the validity of welfare transfers through State intervention (Marquand, 1988). In recent years, however, this consensus has dissolved (N. Johnson, 1987), and with it has gone the authority of the conflict approach. An ideological shift to the Right, as much as post-Fordism's blurred class boundaries or postmodernism's critique of metanarrative, has begun to question the relevance of the intellectual Left.

Despite this political realignment, geography, unlike many other areas of social science (especially politics and economics), has few self-conscious advocates of the so-called 'new Right', and has scarcely begun to participate in reconstructing the Left. Hepple (1988) attributes this to a broader failure within the discipline to engage in political practice. It reflects, he claims, geographers' attraction to foundationalist theories and their preference for an ontological and epistemological 'high ground' where analytical choices are objectivized rather than politicized. Thus, although Dear (1988) has more generally condemned the disengagement of geography from mainstream social science, in my view, the most problematic aspect of this is a detachment from political theory and practice at a time when the boundaries between civil society and the State are being radically realigned (Keane, 1988a, b).

Seeking to fill a strategic omission in the practice of geography, this paper exposes the links between analytical and normative theory, arguing for the importance of engaging not only in reflective debate about how society is structured, but also in political debates about how, and through what mechanisms, societies *should* be structured. In speculating on geography's development beyond the sociology of conflict, I shall not, therefore, directly take up debates surrounding the flexibility or disorganization of capitalism on its path to post-Fordism; nor enter speculation on the associated cultural transformations of postmodernism. These themes are now in the mainstream of geographical debate (see Albertsen, 1988; Cooke, 1988a; Hall, 1988; Harvey, 1987; Lash and Urry, 1987; Schoenberger, 1988). My focus is, rather, on the (intimately related) political element of the new times and on its challenge to geography's 'old' radicalism.

I begin by exploring briefly the implications of the ideas of the intellectual Right which now infuse political practice in parts of Europe and the USA. I question the logic of the analytical claims attached to these positions and I find shortcomings in their prescriptive content when judged by moral criteria. My central aim, therefore, is to find a new outlet for that radical spirit associated with the conflict approach. To this end, I outline a social democratic alternative to the prevailing political orthodoxy, arguing that it has analytical advantages over the theories of both the new Right and the 'old' Left. As a normative framework, moreover, its modified statism may inspire a policy framework for the new times that is more socially just than its market oriented counterpart. While my argument is interdisciplinary in scope, I attempt to show that it speaks directly to the core of human geography.

The challenge of the 'new Right'

While the presuppositions of conflict theory have been questioned from many quarters, its radical mandate is attacked most explicitly in the agenda of the new Right. This contains a range of ideas challenging conflict theory on ontological, prescriptive and political grounds: its collectivism and structuralism is said to undermine the integrity of individuals; its statism is expected to foster dependency and undermine the economy; and its explicit partiality is regarded as intellectually dishonest. Because (for reasons already mentioned) no area of geography seems self-consciously to advance the Right wing project, and because the literatures of economics, sociology and political science are already awash with this theme, I shall not dwell on it in depth. Some comment is required, however, partly because the challenge of the new Right is frequently launched on plausible grounds, and partly because neo-conservatism now shapes the political climate of geographers' 'policy-relevant' research.

The political and popular persuasiveness of the new Right flows from two key areas of social science: neo-liberal and neo-classical economics (which have a bearing on politics through public choice theory) and neo-conservative sociology (see Barry, 1987; Elliot and McCrone, 1987; King, 1987; Levitas, 1986). These disciplines coexist as uneasily in academic discourse as in the political programmes they inspire (Kavanagh, 1987), and, initially at least, the former exerted most impact on political practice.[2]

New Right economics embraces the liberal influence of Friedrich von Hayek as well as the neo-classical tradition associated with Milton Friedman.[3] Analytically, these disciplines are characterized by explicit methodological (and usually ontological) individualism and alleged value-neutrality. The 'structures', markets and institutions of a society are reduced to the uncoordinated actions of individuals, while social and economic order is depicted as the spontaneous accumulation of rational choices. In a free market, observes Barry (1987: p. 161), 'a predictable order will emerge from individual decisions', expressing an inherent tendency to order and justice which does not rest on human altruism, but is a happy consequence of selfish personal preferences (King, 1987). These presuppositions have a direct bearing on the interpretation of social organization, and this can be caricatured with reference to a key theme in human geography – the analysis of residential segregation.

From a neo-liberal perspective, residential differentiation expresses the myriad of rational decisions which individuals (or families) take to translate financial resources into housing and locational preferences. Housing markets will never diminish the inherent variety of the residential mosaic, nor offset the 'moderate' inequalities of income and wealth to which this testifies. However, because individuals theoretically have an equal opportunity to compete in those markets, residential differentiation should never become aligned with excessive or systematic inequality. Any social regularities in residential patterns (the development of 'racial' segregation, for example) are therefore seen as a product, above all, of cultural preference. If segregation is problematic (black people in Britain, for instance, tend to cluster in deprived areas) this is less likely to be attributed to discrimination (an economically inefficient and therefore irrational act) than to a legacy of over-intervention by the state – a legacy expected to diminish as the 'commodification' of housing proceeds (this example is elaborated in Smith, 1989).

While, as I shall suggest later, such arguments are hard to sustain, the assumption that free markets can promote social ends without submitting to collectivist principles provides a moral basis for the new Right's claim to practical relevance in managing the economy (see Barry, 1987). Market liberalism is portrayed as a self-regulating, self-correcting system through which supply, demand, profit and loss solve the economic problem of scarcity better than any known alternative (Hayek, 1946). For public services, observes Jessop (1988: p. 10) this 'involves a mixture of privatization, liberalization, and adopting commercial criteria in the residual State sector', while for the private sector 'it involves deregulation and a new legal and political framework (or Ordnungspolitik) to provide passive support for market solutions'. The aim, then, is to develop societies in which markets are 'free' on the assumption that State intervention has few checks and balances; that public policy rarely achieves its aims; and that high rates of taxation inhibit economic initiative, discourage hard work and constrain individuals' liberty to reap the rewards of personal effort (Gamble, 1987).

Although neo-conservative sociology has achieved less intellectual credibility and political momentum than its counterpart in economics (Morton, 1988), one strand – libertarianism – with its insistence on a minimal State, has given a valuable source of cultural legitimacy to market liberalism. Combined (awkwardly) with neo-authoritarianism, this movement retains a methodological preoccupation with individuals, focusing on their happiness, duty, morality and autonomy in a society grounded in tradition, consensus and the rule of law. The totalizing philosophies and holistic explanatory frameworks assumed to characterize conflict sociology are therefore rejected. No social whole is regarded as more than the sum of its parts, and the mechanisms that bind individuals – markets, instinct and culture – are (like the inequalities that divide them) regarded as natural, inevitable, reasonable and good.

Prescriptively, these views aim to promote a civil society for the new times characterized by 'self interest, hard work, flexibility, self-reliance, freedom of choice, private property, the patriarchal family and distrust of state bureaucracy' (Keane, 1988a: p. 7). Central directives are required to promote law, order, moral norms and national pride, but the responsibility for distributing wealth and resources rests with the market, family, and charity. While the essentials of such a society are contestable (Loney, 1987), this prescriptive formula has sustained the cultural momentum of neo-conservatism throughout the 1980s (Elliott and McCrone, 1987).

What I have tried (too briefly) to illustrate is that within economics, politics and sociology, there is a powerful 'new Right' challenge to the collective radicalism of conflict theory. Although this challenge has not been explicitly incorporated within geography, the grounds on which it is erected need to be taken seriously if the discipline wishes to reconstruct its radical tradition. Analytically, the voice and integrity of individuals *has* often been lost to the identity of interest groups and to the machinations of inaccessible corporate bargaining procedures; and a preoccupation with class conflict does afford low priority to personal aspirations and cultural autonomy. Prescriptively, the egalitarian rhetoric of State intervention has been weakly translated into vaguely defined aims (Ashford, 1986; Hindess, 1987); bureaucratic paternalism has over-ridden the ideals of participatory democracy (Deakin, 1987); and there has often been a lack of economic realism amongst Left wing idealists (King, 1987). The 'old' Left formula is, then, a questionable vehicle through which to analyze or manage the economics of post-Fordism and the cultural aspirations of postmodernism. The struggle to realize an alternative is in an arena where geography could – I think should – be participating.

One option, of course, is actively to advance (or passively incorporate) new Right formulae within the discipline. The literature, though, is scattered with indications that neither neo-liberal economics nor neo-conservative sociology provides an adequate interpretation of the new times. After almost a decade of experiment, the freeing up of markets has not prevented social inequality being either systematic or excessive (Deakin, 1987). Instead, 'the opposition between public-sector welfare agents and private-sector high-wage groups has hardened' (Albertsen, 1988: p. 347). Flexible accumulation signals a polarization of income and wealth (Harvey, 1987), and post-Fordism is 'widening the split between core and periphery in the labour market and the welfare system' (Murray, 1988: p. 12). In this context, the tendency to reduce questions of morality to matters of economics, and to construct moral claims by appeal to the constants of human nature (Barker, 1981), seems inadequate as a basis for normative theory. Welfare cuts do diminish State-dependency and demand greater self reliance. But for those at the lower end of the income scale, this produces new forms of dependency on private charity, and defines a life wherein choices are limited rather than extended by the enterprise culture (Walker, 1988).

Another option, and a way of overcoming the analytical and prescriptive shortcomings of the first, is to build geography's search for political relevance on the moral imperatives of conflict theory. To date, the intellectual Left has been lothe to construct an agenda to manage 'the rise of flexible accumulation and the neo-conservative politics that have thus far helped sustain it' (Scott and Cooke, 1988: p. 243). Although the new Right is the butt of countless critiques, analysts currently have a much clearer sense of what the past has nurtured than of what the future should hold (Hall, 1988). Yet Jessop (1988) indicates that neo-liberalism is just one of several strategies available to guide the transition to post-Fordism, and Deakin (1987) argues that even within the constraints of an internationalized economy, political options are more plentiful than new Right orthodoxy allows. Accordingly, the remainder of this paper charts one route towards a more penetrating critique of social organization and a more satisfying vision of social transformation than is currently offered either by the 'old' Left or the 'new' Right.

From social conflict to human rights

Any attempt to break new ground in analyzing social organization or prescribing social policy must successfully challenge both an uncompromising individualism from the Right and an insensitivity to personal identity on the Left. It should account for the economic and cultural dynamics of the new times, but if it is to signal geographers' greater engagement with political practice, it must also have a normative dimension. One set of ideas appropriate to this task may be wrested from a tradition of research on human rights. Since the area is too vast to explore in detail, I focus on just one theme – a rekindling of interest in the notion of citizenship.

Citizenship has recently re-established itself as a regular if contested buzzword in popular politics. Even as the Left seizes it to reconstruct social democracy the Right claims it to promote popular patriotism and common norms. However, the conservative formulation emphasizes the obligations of citizens rather than their entitlements, duties rather than rights, and allegiance to nation rather than to subnational political collectivities seeking concessions from the State. This citizenship is a normative expectation rather than an analytical tool or a prescription for social change, and it is not the sense in which I use it. This paper is concerned with the development of a social democratic notion of citizenship. This originated in the nineteenth century, inspired the welfare states, and today seems poised to recapture the political imagination.

The idea of citizenship refers to relationships between individuals and the community (or State) which impinges on their lives because of who they are and where they live. Early twentieth-century ideas about the rights associated with residence in (or citizenship of) particular nation states are reviewed by Vincent and Plant (1984).[4] The (then) radical insistence on the social role of the State was taken up by T. H. Marshall (1950) whose influential work insists that the rights of citizenship refer not only to the political and civil rights embedded in national constitutions but also to a spectrum of *social* rights.[5] These statutory entitlements were deemed to range from 'the right to a modicum of economic welfare and security' to 'the right to share to the full in the social heritage and to live the life of a civilized being according to the standards prevailing in the society' (Marshall, 1950: p. 11).

Marshall therefore made an analytical distinction between different kinds of rights and drew attention to the possibility that the practices associated with them might shape social, economic and political organization in different ways (this is considered by Barbalet, 1988). This setting of *individuals* into a *structured* relationship with the State – a relationship which is specified theoretically (in terms of the *de jure* entitlements of the public) but which can be interrogated empirically (to monitor whether, and to whom, such rights are effectively available) – is the platform on which the revitalized concept of citizenship in social democratic theory lays its intellectual credentials.

Now that civil society has escaped those State theories which largely ignored the concept of citizenship, Marshall's ideas are being re-evaluated and reformulated (especially useful accounts are given by Dahrendorf, 1988; Deakin, 1987; R. Harris, 1987; Hindess, 1987; Turner, 1986). This could be the basis for new theories of, and prescriptions for, post-modern, post-Fordist societies. Obviously, such formulations cannot be politically neutral, since they directly challenge the neo-conservatism more usually associated with such regimes. Even as an analytical tool, therefore, citizenship theory is premissed on a series of value judgements: that inequality has material rather than 'natural' origins; that inequality can be structured systematically and is not determined by chance; and that, in the apportionment of resources, collective needs can outweigh individual demand.

Like much of conflict theory, and like the ideas of the New Right, 'new' citizenship theory has both analytical and prescriptive implications. Citizenship as critique regards civil, political and social rights as entitlements whose universality – in a *de jure* as well as *de facto* sense – remains to be realized.[6] It offers a comprehensive vehicle through which to explore systematic discrepancies between the obligations required of, and the rights extended to, members of a nation-state. Enduring variations in the availability of these rights, or in the opportunities to exercise them effectively, can thus be conceptualized as forces shaping or structuring society (this stands in contrast to the more usual view that citizenship rights play an integrative role). Prescriptively, on the other hand, citizenship theory provides a vision for the transformation of society which rests neither on the overthrow of the State nor on the sanctity of the market. It remains to illustrate how, in each of these roles, the social democratic formulation of citizenship can usefully be applied to some of the traditional subject matter of human geography. The analytic and prescriptive elements of this application are considered in turn.

Citizenship as critique

The new times bring changes to the entitlements of residents in different nations and localities which cannot be grasped simply by examining economic opportunities or the emergence of new cultural forms. The late twentieth century *may* be characterized by the reorganization of capital and transmutation of social meaning, but these processes are entwined in a politically inspired restructuring of human rights. Post-Fordism, for instance, is novel not only for the increasing flexibility of economic relations, but also for the changing social role of the State required to manage this (Jessop, 1988). Postmodern pastiche is not limited to the world of art or the structure of the built environment, but weaves through the legislative sphere, infusing modern social problems with a parody of nineteenth-century liberalism and early twentieth-century philanthropy. The new times, then, expose a redrawing of relationships between civil society and the State – relationships which have changed over time, and which vary in space. To illustrate the analytical relevance of the citizenship perspective in exploring this interface between political arrangements and social structure, I shall take examples from the literatures on racism and patriarchy, both of which have received considerable input from human geography in the last few years. These themes illustrate the utility of the State–civil society framework in handling the social fragmentation and political realignments of the new times, and in erecting a framework for comparative analysis within and between nation states.

Citizenship and segregation

Today, the relative neglect of 'race' and racism in classical social theory is roundly condemned by the diversity of social identities which inform political action. The new times are characterized by a decline in the class character of politics, and an increase in the political import of new social movements (Gorz, 1982; Lash and Urry, 1987). As a consequence, social conflict has become more pluralistic and is structured by a much wider variety of interests than early Marxism and even Weberianism allowed (Urry, 1988).

Social geography has a longstanding interest in 'racial' segregation, which goes some way to accommodating these new realities. So far, however, this approach has not explained the reproduction of racism in the fragmented social worlds of postmodernism and post-Fordism. It is true that the subdiscipline has, with other parts of social science, rejected the validity of 'race' as an explanatory variable, focusing instead on racism and the process of race formation (see Jackson, 1987). Nevertheless, its most plausible attempts to do this remain in the domain of traditional Marxism – a conflict perspective – which specifies 'racial' differentiation as an outcome of the labour process in a post-colonial era (C. Harris, 1987; Miles, 1982). 'Races', according to this view, are class fractions, economically part of the working class but ideologically divorced from it by a process of racialization that has 'served to reinforce and maintain the economic stratification of wage labour' (Miles, 1982: p. 171). 'Racial' segregation is also persuasively accounted for in this way (Brown, 1981; Doherty, 1973, 1983), offering a powerful critique of the role of space in the process of social differentiation, stripping all allusions of biological or cultural determinism from explanations of 'racial' inequality and exposing instead its material origins in the organization of the economy and the circulation of capital. Yet, impressive though they are, these works do not, in my view, adequately document residential segregation's testimony to the geography of racism in a postmodern context. A preoccupation with the right to work and to extract a fair wage in return for labour (however crucial this may be) obscures our view of the civil contract from which these – but also many other – entitlements flow or are wrested.

Keane (1988a) sees renewed attention to the theme of citizenship as a welcome broadening out of the wage-capital–labour relationship more usually invoked when documenting recent economic and cultural shifts. In reconceptualizing 'racial' segregation, this broadening out is required for two reasons. First, the racism–migrant labour thesis becomes difficult to sustain in late twentieth-century contexts where racialized minorities can no longer be classed as migrants and are unable to sell their labour (Cross, 1983). Secondly, because this thesis is primarily concerned with the process of race formation amongst groups united by their exclusion from full economic rights, it is forced to invoke the persuasiveness of ideas – of racial ideology (which *is* important) – to isolate black people as a class fraction once their subordinate position is carved out by the economy. This, I suggest, *underdetermines* the material origins of 'race' in any context where States intervene in the allocation of resources or the acquisition of wealth, and it underestimates the reproduction of racism in the restructuring of welfare.

On the other hand, if ideas relating to systematic differences in the apportionment of economic rights could be extended to embrace the broader set of entitlements which residence in a nation-state confers, then it may be possible to arrive at a more comprehensive understanding of how residential differentiation is implicated in the process of racial categorization. Britain's experience is illustrative.

At one level, a form of residential segregation in which black people are statistically over-represented in the worst urban environments *can* be seen as an expression of their marginal position in the British labour market (Ward, 1987). Increasingly, moreover, locational inertia constrains any fuller incorporation into the economy (Cross, 1983). However, the marginalization of black Britons is not only sustained by spatial selectivity in economic restructuring but also by systematic discrimination in the availability of other social, civil and political rights.

Formally, the social rights of citizenship are secured through welfare transfers designed to offset (in cash or kind) the inequalities inherent in capitalist economies. However, as Williams (1987), Cohen (1985) and Gordon (1986) show, social policy in Britain has always been 'racially' divisive. The uneven spread of social rights is, it seems, as implicated in the reproduction of racism as is the reorganization of capital. In housing, for instance, throughout the post-war period, direct and indirect forms of discrimination delayed black people's entry to the public sector, and forced them, as home buyers, to rely on sources of finance ineligible for State subsidies (such as tax relief on mortgage interest). An argument can be made that this *de facto* exclusion from the welfare entitlements guaranteed by the 1948 Nationality Act led directly to the racial inequalities now expressed through residential segregation (Smith, 1988).

This practice of (formally or informally) linking benefit eligibility to immigration status has persisted through the realignment of social policy and the 'recommodification' of welfare. It is putthing the onus on all black people to prove their entitlement to social rights in ways not demanded of whites (Cohen, 1985), and it is creating 'a class of people, mainly black, who do not have the normal welfare rights of other permanent residents' (Gordon, 1986: p. 31). Today, the restructuring of welfare required by a liberal path to post-Fordism means that, despite the theoretical eligibility of black Britons for welfare benefits, and despite their disproportionate contribution to the welfare state (through labour and taxation), their ability to secure State-subsidized services and resources may actually be deteriorating relative to that of whites (M. Johnson, 1987).

From this sketch, it is clear that a group (the working class) that seems united by its place in the division of labour may be divided – 'racially' differentiated – according to eligibility for, and ability to exercise, welfare rights (Rex, 1986). This account of the racist exclusivity of British civil life could, if space allowed, be completed by scrutinizing black people's legal and political rights. We would find that black people are less able than whites to exercise their rights before either the criminal or civil law (Bridges and Bunyan, 1983; Fitzgerald, 1987), and that the ability of racialized minorities to realize their political rights is impaired by an insensitive electoral system (Crewe, 1983).

The key point here is that by analyzing the discriminatory processes of racial categorization in a rights-based framework, we acknowledge that within a nation state, the rights to work, pursue a chosen lifestyle and participate in political decisions are all socially differentiated, and may impinge in different ways on the structuring of society and space. The various material bases of racial inequality can in principle be empirically specified, and their import in particular contexts gauged with reference to other axes of domination (including class oppression and patriarchy). This is one way in which analysts might come to grips with the social and spatial structuring of society amid what Dear (1988) regards as 'postmodern disarray'.

The fraternal social contract

A second example of citizenship as critique concerns its utility as a framework for comparative analysis. The State–civil society perspective advanced by Keane (1988a) recognizes that, even with the advent of global capitalism, nation-states are not simply driven by a world economy, but may shape their own socio-economic environment. Individuals' entitlements must always be analyzed, then, in relation to a national constitution and the specific legislative histories these embody. This does not require us to accept that all (or any) nation-states have a just constitution. Quite the contrary, it offers the opportunity to evaluate particular national agendas relative to some wider ideal of human rights. It therefore also provides a coherent yet comprehensive framework for cross-national comparison – a practice that human geography has found appealing, but one whose theoretical rationale is often weakly specified.

I shall illustrate the relevance of the citizenship perspective to this project using a literature on gender relations. Obviously this is not the place to present a full comparative study, but a flavour of what might be achieved can be sketched by drawing on Australian experience. By international standards, Australia has launched a significant attack on the patriarchal structure of social relations. It is a nation characterized by relative success in narrowing the wages and incomes gap between men and women (Gregory and Ho, 1985; Gregory *et al.*, 1986), and, with the possible exception of Canada, Australia has gone further than any other country in the modern world towards incorporating women's issues into mainstream policy-making (Sawer, forthcoming). Immediately, the context of gender relations in Australia differs from that in Britain, where the income differential is greater and no specific attention to women's needs is built into the legislature (see Carter, 1988). These differences are important for understanding the experience of individual women, the collective struggles of feminists, and the salience of gender in political decision-making at various levels of the State.

Scrutinizing the Australian case more closely, the bases of finer grained comparisons are exposed. It is clear that although obvious gains have been made in relation to women's civil and political rights, and in that area of social rights related to participation in the economy, gender inequalities do persist. These derive from discriminatory controls on the availability of social citizenship rights which determine how women are defined in Australian social policy. A collection edited by Baldock and Cass (1988) shows that a range of State policies, from social security and income tax to industrial and family law, construct women as wives, mothers and carers, regulating their social role and reinforcing women's dependency on men (see also Bryson, 1988). Women who break this mould are largely invisible to the State with respect to entitlements (though not obligations).

To an extent, this subordination of women's social rights can be seen as a general theme in 'western' societies. Pateman (1988a, b) shows that civil society is historically a male construct, implicitly excluding women and devaluing the role assigned to them in the domestic sphere. In this context, women's effective rights of citizenship are not a liberating force but an institutionalization of civil inequality. Nevertheless, the Australian example can be used to illustrate that these general themes must be reproduced by different societies in different ways, with varying emphases and consequences. This is because the elements of citizenship defined by Marshall are not abstract rights, but entitlements bound to the social institutions through which they are exercised. Their meaning is determined, then, by *particular* political systems and *specific* material circumstances.

The subversion of women's rights in Australia is, then, specific to a patriarchal conception of Australian citizenship shaped by that country's distinctive cultural, economic and political history. This is best illustrated by de Lepervanche (1988) who shows how the imagery of Australian citizenship is channelled through the medium of the Anzacs: the 'typical' Australian was for many years regarded as a white male, of British descent, who 'came from the bush, was tall, lean and bronzed and practised mateship' (p. 4). This rough, tough and rural identity (which is discussed by Bean, 1981) underwent a subtle change in the 1960s (MacGregor, 1968; Horne, 1972) when the bush-dwelling pioneer became a middle-class suburbanite. But he retained his gender (and his 'race') – women remained only implicit in Australian identity, banished behind the scenes as reproducers of labour, replacers of those lost at war, and sustainers of Anglo-Saxonism. De Lepervanche's poignant account shows how, by the mid 1940s, this marginal place for women had become entrenched in statute, so that 'women's unpaid domestic labour and childcare were the foundations of welfare, full employment and population policies' (p. 10). The Australian social contract became a vision of liberty, equality and, specifically, *fraternity*, and it produced policies which sustain (and 'naturalize') the distinctions between men as protectors or defenders and women as breeders. The social rights of women in Australia are thus tied in specific ways to a patriarchal vision of citizenship, and this underpins the specific forms of gender inequality we observe.

The more general point here is that in approaching the theme of gender relations via a rights-based framework, the ground is cleared for the boundaries drawn between different groups of men and women, living in different parts of the world, to be assessed relative to the sets of rights and obligations conferred through national constitutions and realized or denied through the operation of markets, institutions and informal social relations. This raises questions which can help establish theoretically where the origins of gender inequality lie, yet which can be investigated empirically to illustrate how the structure of gender relations varies in time and over space.

I have suggested so far that there are systematic differences in the entitlements associated with citizenship – that systematic link between individuals and the State – and that the dispensation of welfare, political and civil rights is as relevant as economic management in structuring the social arrangements of the new times. This, I have argued, is one route to explaining how 'races' are constituted and patriarchal structures sustained, though I could equally have taken examples of the construction of old age, the reproduction of class inequalities or the shaping of new political identities.

Although the rights of citizenship may structure the same kinds of inequalities in many societies, these take forms specific to given national settings (expressing distinctive political histories, particular forms of intervention in the economy, and a range of ideas about the legitimacy of welfare transfers). This means that the relationships between racism, patriarchy and class inequality, and, moreover, the salience of space or territory in the structuring of these relations, also vary. At the moment, the literature is littered with claims about the relative importance of these structured inequalities, and analysts speak as if any one of these axes can be specified as ultimately more fundamental than the rest. The citizenship framework acknowledges, in contrast, that there is dynamic interplay between these sets of social relations: each is constituted *through* the others in ways that vary through time and over space. An appreciation of this is particularly crucial when we move beyond the analytical utility of citizenship theory to consider its prescriptive content.

Citizenship in practice

Like 'old' conflict theory and new Right orthodoxy, citizenship theory is as much the basis of a commentary on what ought to be as a critique of what is. I shall illustrate this with reference to the enigmatic concept of locality, whose meaning, I suggest, is not inherent in the spatial division of labour, in the orientation of public policy or in cultural attachment to territory, but remains *to be realized* through the reorganization of civil society and the mobilization of citizenship rights.

Even the most superficial survey of the last decade indicates that there has been a spatial re-organization of citizens' entitlements and of their capacity to mobilize constitutional and statutory rights. 'The "new times"', observes Hall (1988: p. 29) 'seem to have gone "global" and "local" at the same moment'. The shift toward post-Fordism meant that 'regional production space "imploded" into "localities"' even as 'national productive space "exploded" into a complex global space of interlinked localities' (Albertsen, 1988: p. 347). The globalization of the economy, then, has produced increased inter-urban and inter-regional competition and inequalities which Harvey (1987) refers to as a 'shifting and conflict prone spatialization of class polarities', and whose social implications are documented by Winchester and White (1988). In the cultural sphere, too, while postmodernism is firmly identified with the internationalization of capital (Jameson, 1984), its expression is spatially uneven, juxtaposing 'the centres from which communicative networks of global capitalism and its information systems . . . are controlled' with 'ethnic enclaves, dreadful enclosures . . . and decaying bunkers of collective consumption . . .' (Cooke, 1988a: p. 486).

This spatial reorganization of social life is, though, a political as well as an economic and cultural process. In Britain, it has been secured by centralizing State power, undermining local autonomy and replacing the ideal of local representative democracy with the principles of consumer demand (see Lawless and Raban, 1986). Jessop (1988) points out that the strategies employed have occurred at the expense of local government, are motivated more strongly by political ideology than by economic rationality, and are distinctive for 'the priority they have given to winning the political struggle' at the expense of 'short-term economic crisis-management' (Jessop, 1988: pp. 26–7). The consequence is not a strengthening of national cultures or a revival of local identity, but rather a world in which 'the cultural continuity of all places is seriously threatened by flexible accumulation' (Harvey, 1987: p. 279).

Yet, even as the new times undermine the identity and autonomy of place, they provide an opportunity for the principle of locality to realize new political options. Hall (1988) points out that while the 1980s may be associated with the popularity of the political Right, they have also created scope for renewal on the Left. It is, moreover, the politics of *position* which 'provides people with co-ordinates which are specially important in the face of the enormous globalization and transnational character of many of the processes which now shape their lives' (Hall, 1988: p. 29). Certainly, it is through locality that the realities of globalization are experienced and handled by citizens and this, I shall argue, is a basis on which to advance the normative dimension of citizenship.

Prescriptively, the concept of citizenship aims to realign the principles of social democracy with the practicalities of political and social life in the late twentieth century. There are three key areas where this formula challenges the present orthodoxy: these correspond with three strategies through which the mobilization of locality might advance the Left's new claim to practical relevance.

The strategy of equality

Citizenship theory is critical of the Right's attempt to replace what it sees as an initiative-inhibiting State-sponsored strategy of equality with toleration – even encouragement – of 'moderate' inequality as an incentive to compete in the market and promote economic prosperity. So far, critics argue, there is little to indicate that market inequality is moderate, self-regulating or unaligned (Deakin, 1987). Indeed, 'the economic prosperity of the better-off has been built in large measure on the reduced welfare of the poor' (Walker, 1988: p. 17). There is, moreover, no convincing evidence of the supposed trade-off between social equality and economic efficiency: it is easier to demonstrate that 'a more equal and fairer society can be an efficient and more productive society' (Plant, 1984: p. 24).[7] There is, finally, no reason to believe that market liberalism is the only effective strategy for managing post-Fordist regimes (Jessop, 1988), and no reason to assume that a modified Keynesianism with a strengthened labour policy cannot retain some reverence for market forces in a global economy (King, 1987).

The strength of citizenship theory here is its capacity to recognize that markets can, and do, fail, both technically (as a distributive mechanism) and morally (as a force for social cohesion). The alternative quest for full employment, minimum wages, an egalitarian income distribution and a socially integrated civil life is described by Murray (1988). It should, he argues, 'be based on detailed local plans, decentralized public services and full employment centres', achieved through a network of 'social industrial institutions, decentralized, innovative and entrepreneurial' and a public economy 'made up of a honeycomb of decentralized, yet synthetic institutions' (p. 13). In essence, this vision requires the constitution of locality – of 'causal' groups shaping wider processes through collective action (Lovering, 1987). The relevance of this will be appreciated in considering citizenship's second political imperative.

Liberty and autonomy

Social democratic conceptions of citizenship also question the Right's declared monopoly of interest in the integrity and autonomy of individuals, while rejecting the view that society is no more than a collection of persons. Marquand (1988) argues that neo-liberalism's individualism imposes an unnecessary constraint on the political imagination: it encompasses, he claims, a limited vision of society as 'either a kind of hierarchy, held together because those at the bottom obey those at the top' or 'a kind of market, held together by the calculating self-interest of its members' (pp. 228–9). The conception of citizenship advanced by the social democratic Left, in contrast, seeks explicitly to reconcile individual liberty with social responsibility. Thus R. Harris's (1987) communitarianism is compatible with both social variety and personal autonomy: 'one of its goals is precisely to spread more widely the social foundations of individual flourishing' (p. 149). Likewise, Plant's (1984) concern with a fair distribution of liberty requires collective commitment to greater equality of income, wealth and service provision; while Vincent (1988) insists that 'to formulate the concept of an active ethically motivated citizen implies an active morally motivated community and state'.

Differences between the neo-liberal and social democratic prescriptions do not, then, rest on an opposition between personal development and collective action. They differ rather over the priority they assign to the particular rights individuals need to secure self-determination, and on the extent to which these rights direct individuals towards *social* goals. Citizenship theory is concerned with individuals' personal wellbeing and political potential (secured through the social role of the State) as well as with their economic independence. The aim is not, moreover, to preserve economic rights at any cost, but to balance these against social obligation.

Location – through locality – has a part to play in securing this balance. For, as Mouffe (1988) has argued, the new citizenship is a pluralistic not a universalizing concept. Locality offers a basis from which to mobilize those myriad political identities through which specific needs are articulated and particular entitlements claimed. Locality offers a grounding for the exercise of citizenship which 'can accommodate the specificity and multiplicity of democratic demands and provide a pole of identification for a wide range of democratic forces' (Mouffe, 1988: pp. 30–1). Just *how* this might be achieved is addressed in the next section.

Political transformation and the restructuring of social democracy

The prescriptive content of citizenship theory holds a vision for the transformation of society. Like the vision of neo-liberalism, this involves the empowerment of the public in a context where both corporate bargaining and representative democracy have proved insensitive to popular demand. It attacks neo-liberalism, though, for replacing these procedures with respect for the public as consumers rather than as voters: it opposes the elevation of purchasing power above participatory redistribution as the principle by which goods and services are allocated; and it condemns the consumer 'democracy' which aligns political power with personal wealth. Instead, it recommends an extension of public involvement in the affairs of the State through a more open participatory democracy.

The very idea of citizenship raises questions about the conditions for social participation, and about the kinds of political arrangements which might secure a genuine plurality of life. The answers require a programme to extend the rights of citizenship beyond the marketplace, including poor as well as wealthy constituencies in 'the vitalization of civil society and the democratic reform of state power' (Keane, 1988a: p. 25). The struggle for democratic citizenship is, then, 'an attempt to challenge the undemocratic practices of neo-liberalism by constructing different political identities' (Mouffe, 1988: p. 30). Such a project is approached by Marquand (1988), who proposes a moral politics of mutual education to shatter neo-conservative stereotypes of a passive and irresponsible public. In its place may be set 'a differentiated and pluralistic system of power, wherein decisions of interest to collectivities of various sizes are made autonomously by all their members' (Keane, 1988a: p. 3).

The implications of this fly in the face of current political trends. They require decentralization rather than centralization of fiscal and political power, respect for the exercise of local initiative, and a willingness to formalize social obligations. To achieve all these things, however, the social democratic project also requires the reconstitution of locality. All the authors I have cited identify decentralization and local control as the path to democratic citizenship, in all its facets: political, social and civil. A nationally flourishing *political* community will, urges Marquand (1988: p. 239) 'be a mosaic of smaller collectivities, which act as nurseries for the feelings of mutual loyalty and trust which hold the wider community together, and where the skills of self-government may be learned and practised'. Discussing the *social* rights of citizenship, R. Harris (1987: p. 146) identifies community membership as the fabric of society woven through 'a genuine commitment to infusing welfare services with opportunities for participation in flexible and decentralized structures rooted on local communities'. Concentrating on the *civil* rights of citizenship, Keane (1988a) also argues for the strategic importance of local initiative. Using examples from Sweden (the Meidner Plan), West Germany (the Green Party's employment policy), France (the achievements of the Confédération Démocratique du Travail) and Britain (the struggle for child-care), Keane argues that 'decentralization of power is sometimes the most effective cure for an undue parochialism' and that 'through participation in local organization citizens overcome their localism' (Keane, 1988a: p. 21).

In short, mobilization around the principle of locality could be a vehicle for realizing the normative prescription embedded in the concept of citizenship. But even if the advent of flexible accumulation makes community control through decentralized social democracy more feasible (Harvey, 1987), this political transformation cannot be guaranteed. This goal is not the logical endpoint of some 'natural' process: it is, rather, a contested option which remains to be constructed and won. The plausibility of locality may rest on its contribution to this end. For if we accept Cooke's (1988b) view that locality is constituted through its necessary relationship with nation – through the medium of citizenship – then its meaning arises not only from economic fortunes and cultural tradition, but also through a struggle for political, civil and social rights. Moreover, if locality is distinct from 'community' in the way that Cooke suggests – as a means of understanding change rather than monitoring stability – then it has a unique significance in normative theory as a locus for exercising, extending and restructuring the rights of citizenship. Locality is, from this perspective, significant not for what it is (the encapsulation of flexible accumulation and postmodern culture), but for what it can become when used as a vehicle for political negotiation.

Conclusion

This paper focuses on a small, but I think significant, subplot in the story of social theory. It seeks to develop a geography of critique and engagement able to meet the political as well as analytic challenge posed by the 'new times'. My argument builds from geography's welcome appropriation of conflict sociology in the early 1970s and from a critique of this movement advanced by the new Right. I suggested that the claims of this critique to analytical sophistication and practical relevance are hard to sustain: neo-liberal economics provides only a partial account of post-Fordist reconstruction, and neo-conservative sociology cannot accommodate post-modern culture. The core of the paper therefore moved on to develop a framework able not only to grasp the complexities of economic restructuring and cultural realignment but also to reconstruct the moral project of the 'old' Left.

As critique, this rights-based perspective gives insight into the structuring of social relations through the constitutional and statutory entitlements conferred on (and exercised by) the residents of particular nation-states. It offers a conceptualization of categories like 'race', class, gender and locality which accounts for their variable realization in space and time, without losing sight of the individuals whose lives they permeate. The prescriptive content of this framework revives the social role of the State, and offers a vision for the transformation of society through a shift to participatory democracy effected through the mobilization of locality. Citizenship theory thus has practical as well as analytical relevance. It helps explain the structuring of society, but it also provides some normative principles to guide the restructuring of society. It promises, then, one route towards a human geography for the new times.

Notes

1 Here, conflict theory refers to ideas that are (explicitly or implicitly) broadly structuralist and realist, interpreting capitalist societies as the product of a struggle

for power among competing interest groups, driven by a quest to control the production and distribution of material resources.

2 Elliott and McCrone (1987) point out that as this economic programme becomes increasingly difficult to sustain, the cultural project of Tory authoritarianism may take the leading edge in Conservative politics.

3 Neo-liberalism has been most prominent in Britain through Hayek's links with London's Institute of Economic Affairs.

4 The Victorian philosopher T. H. Green was most influential, linking the rights of citizenship to the pre-conditions for active life, certain elements of which (health, education, livelihood) he believed should be guaranteed by the State.

5 Social rights are only statutory entitlements, and are therefore more readily extended and revoked than their constitutional counterparts.

6 Marshall's original ideas are rather complacent in this respect, assuming that essential rights are secure, universally available, and require no further struggle. His views are also vulnerable to the charges of Anglo-centrism, evolutionism and paternalism (Mann, 1987).

7 New citizenship theorists do not, it must be emphasized, lend unqualified support to the principle of equality of outcome. Rather they seek guidelines for assessing what kinds of inequality are legitimate if higher social aims – such as the equal distribution of the value of liberty – are to be met (Mouffe, 1988; Plant, 1984).

References

Albertsen, N. 1988: 'Postmodernism, post-Fordism, and critical social theory'. *Environ. Plan. D: Soc. and Space* 6, 339–65.

Ashford, D. E. 1986: *The emergence of the welfare states.* Oxford: Basil Blackwell.

Baldock, C. V. and Cass, B. (eds) 1988: *Women, social welfare and the state.* Sydney: Allen and Unwin.

Barbalet, J. M. 1988: *Citizenship.* Milton Keynes: Open University Press.

Barker, M. 1981: *The new racism: conservatives and the ideology of the tribe.* London: Frank Cass.

Barry, N. P. 1987: *The new right.* London: Croom Helm.

Bean, C. E. W. 1981: *The story of Anzac.* Brisbane: University of Queensland Press.

Binns, D. 1977: *Beyond the sociology of conflict.* Basingstoke: Macmillan.

Bridges, L. and Bunyan, T. 1983: 'Britain's new urban policing strategy: the Police and Criminal Evidence Act in context'. *J. of Law and Soc.* 10, 85–107.

Brown, K. 1981: 'Race, class and culture: towards a theorization of the "choice/constraint" concept'. In Jackson, P. and Smith, S. J. (eds), *Social interaction and ethnic segregation.* London: Academic Press.

Bryson, L. 1988: 'Women as welfare recipients: women, poverty and the state'. In Baldock, C. V. and Cass, B. (eds), *Women, social welfare and the state.* Sydney: Allen and Unwin, 134–49.

Carter, A. 1988: *The politics of women's rights.* London: Longman.

Cohen, S. 1985: 'Anti-semitism, immigration controls and the welfare state'. *Critical Social Policy* 5, 73–92.

Cooke, P. 1988a: 'Modernity, postmodernity and the city'. *Theory, Culture and Society* 5, 475–92.

Cooke, P. 1988b: 'Locality, structure and agency: a theoretical analysis'. Paper presented at the annual meeting of the Association of American Geographers, Phoenix, Arizona.

Crewe, I. 1983: 'Representation and the ethnic minorities in Britain'. In Glazer, N. and Young, K. (eds), *Ethnic pluralism and public policy*. London: Heinemann.

Cross, M. 1983: 'Racialised poverty and reservation ideology: blacks and the urban labour market'. Paper presented at the fourth Urban Change and Conflict conference, Clacton-on-Sea.

Dahrendorf, R. 1988: *The modern social conflict: an essay on the politics of liberty*. London: Weidenfeld and Nicolson.

Deakin, N. 1987: *The politics of welfare*. London: Methuen.

Dear, M. 1988: 'The postmodern challenge: reconstructing human geography'. *Trans. Inst. Br. Geogr.* N.S. 13, 262–74.

Doherty, J. 1973: 'Race, class and residential segregation in Britain'. *Antipode* 5, 45–51.

Doherty, J. 1983: 'Racial conflict, industrial change and social control in post-war Britain'. In Anderson, J., Duncan, S. and Hudson, R. (eds), *Redundant spaces in cities and regions*. London: Academic Press.

Elliott, B. and McCrone, D. 1987: 'Class, culture and morality: sociological analysis of neo-conservatism'. *Sociological Rev.* 35, 485–515.

Fitzgerald, M. 1987: *Black people and party politics in Britain*. London: Runnymede Trust.

Gamble, A. 1987: 'The weakening of social democracy'. In Loney, M. (ed.), *The state or the market*. Milton Keynes: Open University Press.

Gordon, P. 1986: 'Racism and social security'. *Critical Social Policy* 6, 23–40.

Gorz, A. 1982: *Farewell to the working class*. London: Pluto.

Gregory, R. G. and Ho, V. 1985: 'Equal pay for comparable worth: what can the U.S. learn from the Australian experience?' Discussion Paper 23, ANU Centre for Economic Policy Research, Canberra.

Gregory, R. G., Daly, A. and Ho, V. 1986: 'A tale of two countries: equal pay for women in Australia and Britain'. Discussion Paper 147, ANU Centre for Economic Policy Research, Canberra.

Hall, S. 1988: 'Brave new world'. *Marxism Today* 32, 24–9.

Harris, C. 1987: 'British capitalism: migration and relative surplus-population: a synopsis'. *Migration* 1, 47–96.

Harris, R. 1987: *Justifying state welfare: the new right versus the old left*. Oxford: Basil Blackwell.

Harvey, D. 1973: *Social justice and the city*. London: Edward Arnold.

Harvey, D. 1987: 'Flexible accumulation through urbanization: reflections on "post-modernism" in the American city'. *Antipode* 19, 260–86.

Hayek, F. A. 1946: *Individualism: true and false*. Oxford: Basil Blackwell.

Hepple, L. 1988: 'Social theory and political practice in human geography'. Paper presented at the Anglo-Austrian seminar on geography and social theory, Zell am Moos, Austria.

Hindess, B. 1987: *Freedom, equality and the market*. London: Tavistock.

Horne, D. 1972: *The Australian people*. Sydney: Angus and Robertson.

Jackson, P. (ed.) 1987: *Race and racism: essays in social geography*. London: Allen and Unwin.

Jameson, F. 1984: 'Postmodernism, or the cultural logic of late capitalism'. *New Left Rev.* 146, 53–94.

Jessop, B. 1988: 'Conservative regimes and the transition to post-Fordism: the cases of Britain and West Germany'. Essex Papers in Politics and Government 47, Dept. of Govern t., Univ. of Essex.

Johnson, M. R. D. 1987: 'Ethnic minorities and racism in welfare provision'. In

Jackson, P. (ed.), *Race and racism: essays in social geography*. London: Allen and Unwin, 238–53.

Johnson, N. 1987: 'The break-up of consensus: competitive politics in a declining economy'. In Loney, M. *et al.* (eds), *The state or the market*. London: Sage.

Kavanagh, D. 1987: *Thatcherism and British politics*. Oxford: Oxford University Press.

Keane, J. 1988a: *Democracy and civil society*. London: Verso.

Keane, J. (ed.) 1988b: *Civil society and the state*. London: Verso.

King, D. S. 1987: *The new right: politics, markets and citizenship*. Basingstoke: Macmillan.

Knox, P. 1988: 'Vulnerable and disadvantaged households in US metropolitan areas'. Paper presented to an IGU commission on *Urban systems in transition*, Melbourne.

Lash, S. and Urry, J. 1987: *The end of organised capitalism*. Cambridge: Polity Press.

Lawless, P. and Raban, C. (eds) 1986: *The contemporary British city*. London: Harper and Row.

de Lepervanche, M. 1988: 'Breeders for Australia: a national identity for women'. Unpubl. paper, Dept. of Anthropology, Univ. of Sydney.

Levitas, R. (ed.) 1986: *The ideology of the new right*. Cambridge: Polity Press.

Loney, M. (ed.) 1987: *The state or the market*. London: Sage.

Lovering, J. 1987: 'Class analysis and locality research'. *Swindon Project Working Paper*. Bristol: SAUS.

MacGregor, C. 1986: *People, politics and pop*. Sydney: Ure Smith.

Mann, M. 1987: 'Ruling class strategies and citizenship'. *Sociology* 21, 339–54.

Marquand, D. 1988: *The unprincipled society*. London: Fontana.

Marshall, T. H. 1950: *Citizenship and social class*. Cambridge: Cambridge University Press.

Miles, R. I. 1982: *Racism and migrant labour*. London: Routledge and Kegan Paul.

Morton, B. 1988: 'Libertarians of the old school'. *Times Higher Education Supplement* 22 June, 13.

Mouffe, C. 1988: 'The civics lesson'. *New Statesman and Society* 1, 28–31.

Murray, R. 1988: 'Life after Henry (Ford)'. *Marxism Today* 32, 8–13.

Pateman, C. 1988a: *The sexual contract*. Cambridge: Polity Press.

Pateman, C. 1988b: 'The fraternal social contract'. In Keane, J. (ed.), *Civil society and the state*. London: Verso.

Peet, R. (ed.) 1977: *Radical geography*. London: Methuen.

Plant, R. 1984: *Equality, markets and the state*. Fabian Tract 494. London: Fabian Society.

Rex, J. 1986: *Race and ethnicity*. Milton Keynes: Open University Press.

Sawer, M. forthcoming: 'The long march through the institutions: women's affairs under Fraser and Hawke'. In Head, B. and Patience, A. (eds), *From Fraser to Hawke*. Oxford: Oxford University Press.

Schoenberger, E. 1988: 'From Fordism to flexible accumulation: technology, competitive strategies, and international location'. *Environ. Plann. D: Soc. and Space* 6, 245–62.

Scott, A. and Cooke, P. 1988: 'The new geography and sociology of production' (guest editorial). *Environ. Plann. D: Soc. and Space* 6, 241–4.

Scott, A. J. and Storper, M. (eds) 1986: *Production, work, territory: the geographical anatomy of industrial capitalism*. Boston: Allen and Unwin.

Smith, D. M. 1977: *Human geography: a welfare approach*.London: Edward Arnold.

Smith, S. J. 1988: 'Political interpretations of "racial segregation"'. *Environ. Plann. D: Soc. and Space* 6, 423–44.

Smith, S. J. 1989: *The politics of ' race' and residence*. Cambridge: Polity Press.
Turner, B. S. 1986: *Citizenship and capitalism*. London: Allen and Unwin.
Urry, J. 1988: 'Disorganised capitalism'. *Marxism Today* 32, 30–3.
Vincent, A. 1988: 'Looking after number one'. *Times Higher Educational Supplement* 25 November.
Vincent, A. and Plant, R. 1984: *Philosophy, politics and citizenship*. Oxford: Basil Blackwell.
Walker, A. 1988: 'Dependent relativities'. *Times Higher Educational Supplement* 22 April.
Ward, R. 1987: 'Race and access to housing'. In Smith, S. J. and Mercer, J. (eds), *New perspectives on race and housing in Britain*. Centre for Housing Research, University of Glasgow.
Williams, F. 1987: 'Racism and the discipline of social policy: a critique of welfare theory'. *Critical Social Policy* 1, 4–29.
Winchester, H. and White, P. 1988: 'The location of marginalised groups in the inner city'. *Environ. Plann. D: Soc. and Space* 6, 37–54.
Women and Geography Study Group 1984: *Geography and gender*. London: Hutchinson.

16 Andy Merrifield,

'Situated Knowledge through Exploration: Reflections on Bunge's "Geographical Expeditions"'

Reprinted in full from: *Antipode* **27**(1): 49–70 (1995)

> The need to lend a voice to suffering is a condition of all truth.
>
> Theodor Adorno (1973)

> [T]o seek to know before we know is as absurd as the wise resolution of Scholasticus, not to venture into the water until he had learned to swim.
>
> G. W. F. Hegel (1931)

> It's a strange world. Some people get rich and others eat shit and die.
>
> Hunter Thompson (1988)

Introduction

Over the last decade or so notions of 'situated knowledge,' 'standpoint theory,' and 'positionality' have received an enormous amount of attention in radical scientific and social scientific circles and in the humanities. At its most simplistic, this state of affairs can be taken as something of a reaction to, and critical engagement with, post-modernist and post-structuralist modes of thought which have gained increasing credence within the academic left.

Within feminist academic and activist inquiry particularly, the concept of 'situated knowledge' has called into question the epistemological basis of the western Enlightenment philosophical tradition and scientific practice (see, e.g., Harding, 1986; Hartsock, 1987, 1989/90; Nicholson, 1990; and Haraway, 1991). At the heart of the issue lies a fundamental insistence on the *contextualized* nature of *all* forms of knowledge, meaning and behavior. There is a further recognition of the partial and partisan edge to inquiry, theory construction, and scholarly (re)presentation, as well as an explicit acknowledgement of the importance of the author's biography in this creative process. These challenging insights have in recent years percolated down to the discipline of human geography (see Christopherson, 1989; McDowell, 1991, 1992; Jackson, 1993). In so doing, they have sparked intense debate and ushered in something of a reappraisal and realignment of the left critical geographical program (cf. Massey, 1991; Harvey, 1992).

This engagement has had both constructive and destructive ramifications, though it is not my intention to discuss these further here. Instead, I will argue that versions of situated knowledge and standpoint theory have had a long and rich legacy within radical academic, intellectual, and political endeavor. In this paper, I propose to explore in more depth just one strand of this tradition within geography: the 'geographical expedition' program, an initiative that was linked with the development of radical approaches to urban phenomena in the late 1960s and early 1970s. I want to argue, firstly, that a critical engagement with the legacy of geographical expeditions can contribute toward the methodological and epistemological discussions now echoing within left geography. Secondly, a contemporary redefinition of expeditions can inform the development of a more sophisticated critical urban geography, one that acknowledges the progressive aspects of postmodernist theory, yet does not jettison the hard-fought insights gained from the expeditions' modernist tradition.[1] Finally, I contend that within the expedition concept there lurks an undeveloped *pedagogic* component that can be teased out and built upon as part of a progressive radicalization strategy for teaching and for localized political activity. In what follows I shall confront these themes in turn. The account, however, will be prefaced with some clarifying remarks about situated knowledge arguments.

Situated knowledges

Epistemologies of situated knowledges offer adherents a conceptual platform from which to call into question all privileged knowledge claims. A contextualized basis for knowledge production and scientific accomplishment implies the denial of meta-narratives and totalizing perspectives, especially those that speak in the name of objectivity and neutrality. Rejection of universal truth claims thus gives voice to marginalized 'Others,' those traditionally oppressed and excluded from Western (white male) intellectual and political practices. Situated knowledge, according to feminist historian of science Haraway (1991), is nothing more than a shorthand term that provides intellectual and political space for hitherto silenced voices to be heard. Hence, situated knowledges are 'particularly powerful tools to produce maps of consciousness for people who have been inscribed within the marked categories of race and sex that have been so exuberantly produced in the histories of masculinist, racist and colonialist dominations.' Haraway (1991: 191) argues that this hegemonic Western cultural narrative produces disembodied, detached, unlocatable and *irresponsible* knowledge claims. For Haraway, furthermore, irresponsibility means 'unable to be called into account' because it purports to 'see everything from nowhere.' It follows from this that a spurious doctrine of scientific objectivity provides an ideological veil – a ruse Haraway calls a 'god-trick' – simultaneously beclouding and reinforcing existing and unequal power relations.

Situated knowledges appeal to the social constructivist argument since, as Nicholson stresses, '[d]escriptions of social reality bear a curious relation to the reality that they are about; in part such descriptions help constitute the reality' (cited in McDowell, 1992: 60). Haraway's (1992) research into primatology illustrates that the detached eye of objective science is an ideological fiction and so it is legitimate to criticize the natural sciences on the level of 'values' as well as that of 'facts.' The upshot is an invocation 'for a politics and epistemologies of location, positioning, and situating, where partiality and not universality is the condition of being heard to make rational knowledge claims' (Haraway, 1991: 195). Under such circumstances, knowledge is always embedded in a particular time and space; it doesn't see everything from nowhere but rather *sees somethings from somewhere* (cf. Hartsock, 1989/90: 29). A situated understanding, therefore, provides a position from which to organize, conceptualize, and judge the world. Yet this is always partial, never finished nor whole; it is always woven imperfectly and holds no justifiable claims to absolute privileged knowledge. There are always different and contrasting ways of knowing the world, equally partial and equally contestable.

But this doesn't mean any viewpoint will suffice. Again Haraway is helpful: '"equality" of positioning is a denial of responsibility and critical enquiry. Relativism is the perfect mirror twin of totalisation in the ideologies of objectivity.' Relativism and absolutism present themselves as commensurate 'god-tricks': both deny the stakes in location, embodiment, and partial perspective, both 'make it impossible to see well' (1991: 191). To this extent a committed and situated knowledge offers a corrective to the god-tricks of positivism and some postmodernism: situatedness implies that an understanding of reality is accountable and responsible for an *enabling* political practice. Ultimately, then, the realm of politics conditions what may count as *true* knowledge.

This fundamental insistence is instructive for my desire to resituate the geographical expeditions phenomenon within an epistemology of situated knowledges and attempt to circumvent the current paralysis within 'strong' postmodern critical theory. Insofar as the situated knowledge conceptualization, framed within the expedition ideal, permits a theoretical and political alternative bold enough not to relinquish some sort of universal, ethical anchoring to scholarly endeavor, yet acknowledges 'otherness' and difference, and recognizes that a partial and partisan perspective is preferable precisely because it can be held accountable. Such an understanding, too, places at a premium knowledge produced from the standpoint of the subjugated; and, as Hartsock (1989/90) tells us, knowledge constructed from the standpoint of the dominated and marginalized holds a claim to present a truer and more adequate account of reality.[2]

Geographical expeditions were implemented precisely to acquire such a subjugated standpoint: Bunge's theoretical and practical invocation to set up 'base camp' in the inner city was, *inter alia*, a search for a situated knowledge. Base camp offered quite literally a critical bivouac from where a partisan, responsible and accountable vision of urban society could be constructed in such a way as to inform action. From there geographers had the opportunity 'to learn how to see faithfully from another's point of view.'[3] Expeditions permitted critical vision consequent upon a 'critical positioning' in urban space. Later, I will explore whether geographers' ability to understand the situation of 'the Other' is restricted by their own individual biographies. In the interim, let me outline more directly the geographical expedition tradition.

Situatedness through exploration

The initial impetus to the geographical expedition 'movement' – if it's possible to call it that – rested of course in the turbulent social and political atmosphere of the 1960s: protests against the imperialist war in Vietnam, the May 1968 insurrection in Paris, and the proliferation of civil right demonstrations and large-scale rioting on the part of oppressed and impoverished urban populations in the United States. These widescale and cataclysmic events provided a *zeitgeist* in which many geographers were compelled to reconsider the conceptual and practical basis of their discipline (see Harvey and Smith, 1984). For those most radicalized by such a state of affairs, this was to be a heart-wrenching process of reevaluation which prompted a necessity for greater social relevancy in their geography, as well as a concomitant rejection of an erstwhile 'nice' or status quo geography (Harvey, 1973 Chapt. 4). This was to involve a deep conviction to an intellectual and political project intent on changing society and fashioning a critique of dominant values, ideology and scholarly practices. These concerns were at odds with the inherently capitalistic and imperialistic geographical establishment, and many radical geographers, like Bunge, were forced out of their jobs, marginalized and ostracized from academic geographical circles (see Horvath, 1971). (Bunge, from his Quebec 'exile,' continues to remain something of a geographical *persona non grata* today.) At the time, the pursuit of intellectual and societal transformation was no place for the career-orientated.[4]

Hitherto, Bunge's research was, like that of many who later turned toward radical reinterpretations of spatial and social phenomena (notably Harvey), rooted in the positivistic tradition. His *Theoretical Geography* (1962) (which he dedicated to Walter Christaller) employed mathematical modeling and mapping techniques to attack, *inter alia*, Hartshorne's insistence on locational uniqueness. This commitment to mapping, it should be noted, remained a dominant motif of Bunge's radical geography. But what had changed was the target and scale of Bunge's focus. In 'Perspective on theoretical geography' (1979a: 170), for example, he passionately describes the way in which a stay in a black ghetto hotel in Chicago for the 1966 Martin Luther King demonstrations taught him 'how you have to "get ready to kill the world" to walk across the street to get a corned beef sandwich; that is, I could make it on "the mean streets" – an indispensable skill for urban exploration in antagonistic systems.' In Detroit, moreover, a young black woman, Gwendolyn Warren (who became Director of the Detroit Geographical Expedition), taught Bunge a further lesson on urban reality: children were starving to death and being killed by automobiles in front of their own homes. For the neophyte Bunge, this experience had a powerful effect on his geography; and as he recounts (1979a: 170), Warren and other ghetto dwellers were 'furiously interpreting the world all around me that I could not see because my life had been spent buried in books . . . [This] caused me to reverse my scale and I wrote a book about one square mile in the middle of black industrial Detroit.'

Yet, the said book, *Fitzgerald: Geography of a Revolution* (1971), didn't start out as a geography text. Nevertheless, the collection of highly charged maps and evocative photographs – in a large, atlas-size format – emphasized the usefulness of academic geography while convincing Bunge of the urgent political necessity to 'bring global problems down to earth, to the scale of people's normal lives' (1979a: 170).[5] That, for Bunge, had to be a geographer's *raison d'etre*: in blunt terms, one had to work at being useful. Unsurprisingly, this fundamental insistence necessitated a categoric rejection of contented 'campus geography' which, according to Bunge, tends to sever theory from practice and prioritizes citing as opposed to *sighting*.[6] Crucial here was the practice of exploration: the construction of a critical vantage point that wasn't an exotic quest of geographical plundering or escapism, but rather a 'contributive' expedition. Indeed, Bunge (n.d.: 48–49) claimed that the seven mile journey from rich suburban Detroit to its poor inner city is a trip half-way around the world in terms of infant mortality rates. And as he deduces, 'If half-way around the world is compressed into just seven miles, only a micro-mapping of it could show even the massive features of its geography.'

Such a geography tended 'to shock because it includes the full range of human experience on the earth's surface; not just the recreation land, but the blighted land; not just the affluent, but the poor; not just the beautiful, but the ugly. In America, since most of the humans live in cities, it implies the exploration of these cities' (Bunge, 1977a: 35). Geographers, Bunge insisted, had to take their geographical knowledge to the poor, and local people were to be incorporated in expeditions as both students and professors. Ironically, Bunge retained the label 'expedition' in an attempt to subvert the exploration practices of the nineteenth century. Now, an expedition was to realize its full potential by helping – rather than destroying – the human species, and hence ensure the collective survival of humankind in a machine age which Bunge saw as intent on threatening itself with annihilation.

For Bunge, *survival* became the fragile thread binding logic, ethics and politics (Bunge, 1973a, 1973b). And he did not pull any punches about where his own political loyalties lay: they were and remain virulently anti-capitalist and anti-racist. Thus, Bunge's geography was informed by a deep commitment to socialism:

> It is an illustration of the nature of the mental labour called geography. Geography has been the overwhelming force in leading me to such a deep 'political' position. Having lived and struggled in this neighbourhood of Detroit called Fitzgerald, from which I write, how could I avoid directing my attention to this region? And in the dialectics of work, the commerce between the labour and the worker, how else could my work not help shape what I think – and, therefore, as a geographer, shape me? (1973a: 320).

There is a certain affinity here with the situated, partial, located and responsible standpoint that Haraway asserts as a means of gaining 'objective knowledge' of the world. To be sure, much of Bunge's (1973a) essay on 'Ethics and logic in geography' anticipates and cuts an epistemological swath for Haraway's more recent radical conviction within the history of science 'to see faithfully from the standpoint of the subjugated.' From such a situated and responsible perspective, the expedition would strive to be a democratic rather than an elitist pursuit. Consequently, the points of view of local people themselves were given a relative priority. Professional geographers worked in unison with 'folk geographers': practically informed lay persons such as members of residents' associations, community activists, socially-responsible citizens of all stripes, as well as taxi drivers who, says Bunge, possess an invaluable and sensitive knowledge of urban environments that should be tapped.[7] An important proviso, however, was that the 'power of the expedition itself, who hires and fires, who writes checks and so forth must be in the hands of the people being explored, risky as that sounds to academics' (Bunge, 1977a: 39).

The prototypical Detroit Geographical Expedition and Institute (DGEI), established in the summer of 1969, incorporated these prerequisites and went on to implement a program of community research and education for the black residents of Detroit. Horvath (1971: 73–74) writes that the main purpose of DGEI is to 'find a way in which geographers could make available educational and planning services to inner city Blacks; it represents an attempt by the black community and some professional geographers to build an institution that would link the university to the needs of the disadvantaged Blacks in the city of Detroit.' The interconnection between research and education was therefore fundamental to the operation of the expedition. In terms of the educational component, professional geographers (explorers) set up free university extension/outreach programs on cartography and geographical aspects of urban planning (in conjunction with the University of Michigan) with the aim that any black person could walk off the street and take 45 hours of university credit courses, and if they attained a C grade or better could transfer with sophomore status to any Michigan university (Horvath, 1971: 73–74). All campus teachers were volunteers, the use of practical case studies was the major teaching mode, and the local community participated in any decisions over structure and content of the courses. Such circumstances enabled a productive commerce between campus explorers and local people: '[b]eyond learning the technical skills of the academics, these folk geographers learn to generalize their experiences to a larger world. In return, the campus explorers gain valuable knowledge and insights into the community' (Stephenson, 1974: 99).

Field research was equally crucial to the expedition concept. Each participant in the expedition was involved in research around issues such as political districting, interregional money flows, transportation problems, cartographic skills, and the geography of child death (see Antipodean Staff Reporters, 1969). This information gave considerable grist to the local community's mill insofar as it enabled them to organize themselves around the issues investigated. In 1971, for example, the DGEI was active in lobbying against the encroachment of Wayne State University into the Trumbull community (particularly into Mattaei Playfield; see *Field Notes*, 1972). Field work data gave sustenance to any political program because it could be presented to city politicians/planners and reports made political lobbying much more effective (Stephenson, 1974). The DGEI produced 'A Report to the Parents of Detroit on School Decentralization' (see *Field Notes*, 1970) that highlighted the very real difficulties low-income blacks faced just showing up for free classes: some would come hungry and others couldn't afford bus fares. Geographers and local community participants responded with a series of maps indicating a more suitable and socially just geographical allocation of educational resources, which gave black community leaders a technical study so effective that Detroit Board of Education was compelled to respond (see Horvath, 1971).

Elsewhere, field work investigated the relationship between children and machines in the context of inner city community spaces (or lack of them) allocated for children's play. Expedition research emphasized how low-income high-rise environments, with their dearth of play space, force children onto the streets where they are vulnerable to speeding traffic. Understanding the complex issue of machine versus human space – especially how it varied between rich suburban kids and their impoverished downtown counterparts – became a pivotal concern of the expedition program, as did the antagonism between community and non-community land use. General environmental quality of urban landscapes was also brought under intense scrutiny: hidden landscapes, private landscapes, landscapes of the powerless, toyless landscapes, rat-bitten baby landscapes were all explored with considerable elan.

Many innovative and imaginative ideas were later deepened and sharpened when Bunge moved from Detroit to Toronto where he helped establish the Canadian–American Geographical Expedition (CAGE) in October 1972 (the results of which were published in *The Canadian Alternative* (Bunge and

Bordessa, 1975; see, too, Stephenson, 1974)). Here, five geographical scales were charted: 1) One square mile of Toronto (the base camp neighborhood of Christie Pits); 2) Toronto itself; 3) Canada; 4) North America; and 5) the world. Exploration focused on the way in which each scale impinged upon three different types of spaces: human-kind, machine-kind, and nature. These five different 'scales of survival' in their mutual interaction reveal the relationship between the unique and the general, especially as it unfolds, concretizes, and impacts upon the daily life of low-income urban populations.

So in both Detroit and Toronto theory and practice were galvanized, and expedition 'manuals' and field data reports (like the *Field Notes* series) were compiled to promote community activism and enhance local empowerment (see Colenutt, 1971). Expedition teams in both cities spent time in local people's homes and professional geographers were taught lessons seldom discussed on university campuses. Many of the ideas propounded by Bunge went beyond the received geographical literature of the time. Impoverished black people brought their desperate and often lurid experiences to the campus classroom where academic geographers would listen, attempt to understand, and henceforward incorporate these insights into intellectual and political endeavor.

Researchers grappled to gain trust and respect in the base camp community (Bunge, 1977a: 39). The geographer studied the area from the point of view of the people who live there, investigating in the process 'what is geographically out of whack' (p. 37). They did so, Bunge says, by 'getting a "feel" of the region. By talking, listening, arguing, befriending, and by making enemies of the humans in the region.' The geographers' fate was the fate of the locals. Accordingly, geographers could be held responsible and accountable simply because they were 'expected to live with the mess they help create.' Such situatedness and positionality meant that the geographer was able to be called into account. This excursion beyond the cloisters of the academy was the route whereby geographers became both rigorous and useful in their inexorable quest for knowledge. But it meant, too, a redefinition of the research problematic and intellectual commitment of the researcher away from a smug campus career to one incorporating a dedicated community perspective which pivots around what Howe (1954) in another context called a 'spirit of iconoclasm.'

Resituating geographical expeditions

A particularly illuminating insight of the expeditions was the explicit acknowledgement of the point of reproduction – the so-called 'second front' of struggle (Bunge, 1977b) – in the organization and perpetuation of capitalism. Workplace struggles and practices had heretofore assumed a relative privilege in left research agendas. Yet class, gender and race dynamics outside the factory – in the home, the community and the neighborhood – are crucial in conditioning workplace social relations. For Bunge (1977b: 60), the 'geography of the working class is overwhelmingly at the point of reproduction not the point of production.' It followed that achieving working class unity required a dual concentration at the point of production and reproduction; the 'full power' of the working class cannot be mobilized unless this unified struggle is achieved (Bunge, 1977b).

Bunge's thought also exhibited a geographical sensitivity to the way in which women – particularly poor black women – struggle to look after their children under the oppressive structure of patriarchal capitalism. The 'hidden landscape' of the home became a legitimate domain of geographical inquiry, as did the worlds of child care and child's play. Exploring these hidden landscapes, noted Bunge, was 'an uncovering of furtive and underground groups relative to public and assertive groups' (*Field Manual* (5), p. 3). And, he added, '[f]inding these groups and establishing their geography, their perception of the space, helps them establish their rightful claim to their turf' (p. 5). Bunge readily admits that much of this awareness of the often desperate plight of women at the point of reproduction was gained from the experiences with the black women folk geographers in Fitzgerald.

Another path-breaking aspect of the expedition tradition was an insistence on deepening our understanding of alienation, oppression and exploitation within the practices of everyday life. Curiously enough, this painstaking desire for the everyday – which is to say, the desire to bring seemingly abstract global problems down to the scale of people's lives – bore all the hallmarks of the libertarian, humanist Marxist principles espoused in continental Europe by Henri Lefebvre (see, e.g., Merrifield, 1993a).[8] Of further interest, too, is that both Bunge's and Lefebvre's sensitivity to lived experience and city space have been deeply affected by periodic stints as taxi drivers (in Toronto and Paris respectively). Thus Bunge, like Lefebvre, berates *remote* sensing and appeals instead for what he calls an *intimate* sensing (Bunge, n.d.: 41). A passage from Bunge splendidly characterizes this micro-sensibility of everyday, lived urban spaces; it could have easily been lifted from Lefebvre's *Critique of Everyday Life* (1991; cf. pp. 57, 97, 134): 'if you sit on a front step and listen to people talk they often talk about the urban geography of the city but seldom about anything that the census is measuring: A child was almost hit at the corner by a speeding teenager, a landlord refused to fix the plumbing, the park is becoming a hangout for teenage narcotic users, a lady had her purse snatched on the nearby commercial street' (*Field Manual* (8), undated, pp. 7–8).[9] Bunge, then, was for a geography, just as Claes Oldenburg (1992: 728) was for an art, 'that embroils itself with the everyday crap and still comes out on top.'

Yet mapping became a crucial strategic tool used by the geographer Bunge to represent the complex dialectical relationship between the thing/process and form/flow nature of reality. In other words, mapping the form of everyday life enabled the instantiation of the misery and alienation that abstract money and capital flows produced when they grounded – or did not ground – themselves in specific places. As Bunge (1979a: 172, 1974b: 86) maintained, the 'simplicity' of descriptive maps makes for better propaganda and agitation. However, Bunge saw that in practice the human misery that money transfers – particularly the 'spatial injustice' conditioned by capital and revenue flowing between the inner city and the suburbs – is difficult to depict. So he tried to resolve this problem by showing both kind of maps: abstract money flows and the concrete descriptive human misery that these flows beget. Here, Bunge had recourse to his own brand of mathematical-theoretical geography; and, on the face of it, Livingstone (1992: 330) is correct to spell out that Bunge's '*analytical* apparatus remained firmly that of the spatial science brigade.'

But it is equally clear that expeditions represented a strategic and politically-charged quantitative mapping exercise. Though this doesn't mean Bunge can be fully exculpated from positivistic overtones. Nevertheless, Bunge was wary that the facade of 'logic' and 'objectivity' in mathematical geography could be used to divert attention away from value judgements, or be deployed as a means to reinforce domination (Bunge, 1973a: 325). As a result, Bunge's maps weren't meant to be disembodied, ostensibly objective representations; they were situated, partial and accountable representations, visualizations constructed from the standpoint of the disenfranchised (cf. Wood, 1993: 186–88). So while, for instance, Pile and Rose (1992: 132) are certainly right to assert that the 'development of cartographic skills went hand in hand with colonial expansion,' they surely go too far in adding that 'it is the powerful bourgeois male gaze that constructs maps. By implicitly employing a rhetoric of neutral description, maps institutionalise a certain understanding of space, certain claims to objective worldviews. The act of mapping assumes a totally transparent society and denies not only difference but also different kinds of difference.' For me, this is somewhat simplistic; it also asphyxiates committed geographical enquiry and reverts to the kind of essentialist logic Pile and Rose are trying to challenge in the first place (cf. Gregory, 1994: 6–7). Maps can, as Bunge affectionately underscored, sometimes be enabling for 'Others' to assert and represent their differences.

True, there are big troubles with mapping and 'computer-print-out geography', especially the 'kinds of information being fed into the computer, or better, the lack of certain needed information' (Bunge, 1973c: 1). It follows that the dilemma for Bunge was scale of scrutiny and geographers' positionality in the exploration of the human environment. Seen in this light, Bunge (1973a, 1979a) believed that certain forms of mathematical geography could be employed in the struggle for a genuine and liberating humanism.[10] He argues, for example, that a computer mapping program was implemented in Detroit to investigate the election situation, unemployment data, and those School Boards most sympathetic to children (Bunge, 1971, 1973a). Furthermore, mathematical geographers were instrumental in resisting the building of an expressway in Philadelphia during the early 1970s (Bunge, 1973a: 324).

The philosophical and methodological trajectory of Bunge's project followed Schaefer's earlier contention that spatial relations rather than mere description are the subject matter of a rigorous geography (see Bunge, 1979b). In Bunge's hands, however, the search for generality and a nomothetic understanding of location served not only an intellectual rejoinder to Hartshorne's neo-Kantian idiographic geography, but also offered considerable scope for political maneuver: the 'generality of locations and humans is the essence of the methodological fight necessary to break into a scientific geography and scientific socialism' (Bunge, 1979a: 173). Yet Bunge's search for the 'generalizability of the story' (as he puts it in the foreword to *Fitzgerald*) was situated and partial; one Bunge called a 'disciplined objectivity': in contradistinction to Hartshorne, locations for Bunge are both unique and general at the same time, namely, they are 'sort of unique' in much the same vein as humans are as similar yet as different as snowflakes (Bunge and Bordessa, 1975: 286).[11]

In saying this, certain assumptions implicit in the expedition concept nevertheless spark a potentially bothersome question: to what degree can the individual biography of the geographer involved in the expedition invalidate the ability to empathize and situate oneself authentically in an impoverished community? This concern about authenticity, to be sure, has a strikingly familiar ring about it, and parallels – if not anticipates – contemporary debates within human geography, anthropology and urban ethnography about whether it is possible to speak meaningfully about 'the Other' or the subaltern (Clifford and Markus, 1984; Jackson, 1985; Harvey, 1992; Katz, 1992; see also Spivak, 1988). Accordingly, the thorny issue of political representation has to be confronted. Bunge (1977a) isn't oblivious to the problematical nature of a researcher investigating, joining, and maybe even representing what he calls a 'foreign-to-his-childhood group.' Indeed, as he affirms (1977a: 37), '[b]ig important gaps will exist' because it is 'not possible to totally undo one's past . . . [and] no matter how empathetic, [the researcher] cannot entirely do it.' Thus, biographical baggage is unquestionably influential and can, unless it is recognized and engaged with, compromise the ability of the geographer to learn *with* and comprehend the way 'the Other' exists in an impoverished inner city community.

Certainly, the task will involve emotional difficulty and an honest political and intellectual commitment to the expedition; nonetheless this can, Bunge (1977a) asserts, be achieved through patience (not patronage) and with 'dogged determination' on the part of the geographer. A sensitive dialogue can, therefore, overcome what Harvey (1992) has recently called 'vulgar conceptions of situatedness.' These, Harvey (p. 303) claims, 'dwell almost entirely on the relevance of individual biographies for the situatedness of knowledge. In so doing they dwell on the separatedness and non-compatibility of language games, discourses, and experiential domains, and treat these diversities as biographically and sometimes even institutionally, socially and geographically determined.' For Harvey, this is a non-dialectical and debilitating rendition of situatedness: it anesthetizes and renders powerless any critical and empathetic impulse because it 'proceeds as if each of us exists as autonomous atoms coursing through history and as if none of us is ever capable of throwing off even some of the shackles of that history or of internalising what the condition of being "the other" is all about' (p. 303).[12]

Spivak, too, in her complex essay 'Can the subaltern speak?' (1988), expresses a similar oppositional stance to a problematic essentialist act of foreclosing. While Spivak is concerned with whether it's possible to speak of – or for – the post-colonial other (notably the subaltern women), her insistence that the colonized subaltern subject is 'irretrievably heterogeneous' (p.

284) suggests that the thesis can (and must) be durable and broad enough to encompass all categories of subaltern groups. Invariably, she cautions, the subaltern cannot speak. Spivak propounds a positionality that 'accentuates the fact that calling the place of the investigator into question remains a meaningless piety' (p. 271). Thus intellectuals always have an 'institutional responsibility' which behooves them to be critical of a complete privileging of subaltern consciousness (p. 284). Indeed, as Spivak (1988: 285) reiterates, for the ' "true" subaltern group, whose identity is its difference, there is no unrepresentable subaltern subject that can know and speak itself; the *intellectual's solution is not to abstain from representation*' [emphasis added].[13]

Bunge's own situatedness and positionality within the expedition program has itself been made available through a critical engagement with his own past. His experiences in Fitzgerald have certainly been formative in throwing off the ideological shackles of what Bunge candidly admits was a privileged childhood. 'Being raised bourgeois,' he avers, 'I always knew my class were thieves. It was the explicitness of the misery this produced, not the process, which I had to discover' (1979a: 172). According to Bunge, his father, William Bunge Sr., was head of the fifth largest mortgage bank in the United States and so had a hand in redlining black neighborhoods in many cities. In *Fitzgerald*, Bunge Jr. admitted a certain existential duality that has been forged out of a 'history of generational tensions. Bunge generations alternate between money-making and cause-serving' (1971: 135). 'From this vantage point,' the book recounts, 'Bunge [Jr.] has had good opportunities to explore the attitudes of the rich and poor towards each other'; all of which seemingly corroborates how subjectivity is inevitably fragmented: the Other is internalized within the self in a manner that complements Spivak's and Harvey's dialectical approach. Because, then, it doesn't shy away from themes of situatedness and representation, the geographical expedition principle embodies within its intellectual and organizational ethos a practical and theoretical route that can perhaps deepen ongoing debates within left geography. Let me explore this contention more closely.

Towards a situated pedagogy and praxis?

Learning 'to see faithfully' from the subjugated standpoint of the oppressed is maybe the most challenging insight to regain from the expedition program. By confronting the dialectical relationship between a researcher subject (geographer) and a researched object, the expedition concept seemingly strives for a genuine 'pedagogy of the oppressed' of the sort that radical Brazilian educationalist Paulo Freire (1972) steadfastly invokes.[14] The geographer herein gains a platform to look faithfully at the world through a *dialogical encounter* with others. Within this interaction, the academic geographer has critical faculties to offer in the exploration of cities, particularly in recognizing the shortcomings of framing things solely in terms of the 'concrete experience' of the oppressed or in a quasi-empiricist manner of 'what actually happens' (cf. Spivak, 1988).[15] As Bunge was keen to observe, 'academic geographers have a sense of scale but no sense, while members of communities have sense but no sense of scale' (interview, 1994). A commerce between local people and campus geographers is thus implied, rather than domination of one group by the other (Bunge, n.d.: 76).

The political practice emerging from this dialogue is formed *with* oppressed urban groups. Such is the nub of Freire's argument. For the geographer involved in an expedition, positionality is simultaneously that of teacher and student; and knowledge of particular urban realities is attained through 'common reflection and action' as researcher and researched alike discover themselves dialectically through 'intentional practice' (Freire, 1972: 44). To the degree that they attempt to negate the formalized and contradictory relationships between the one who teaches and the one who is taught, expeditions could become a powerful teaching instrument – much the way they did in both Detroit and Toronto during the late 1960s and 1970s. This, too, would be consistent with Bunge's behest that geographers should on occasion be put under the tutelage of folk geographers; both of whom henceforward give a coherence and an awareness of function to a social group and thereby represent something akin to Gramsci's (1971: 5–6) *organic intellectual*.

Via this activist route, academic geographers can articulate the 'collective will of a people' (Gramsci's phrase) by gaining access and speaking to power elites, or by giving evidence at public inquiries and the like. Here, geographers participating in the expedition can use their research skills and written pamphlets as vital weapons of resistance for oppressed groups. Stephenson (1974: 101), who was active in the Toronto Geographical Expedition, notes that when '[a]rmed with a comprehensive research report, political lobbying is much more effective, but even failure to implement proposals has some value. People begin to realise their position in society, which may in turn lead to more active agitation for change.' The authenticity of the geographer, then, simply lies in the ability to do committed and accountable urban geography.

This dialogue between oppressed communities and professional geographers is, therefore, an indispensable instrument in the process which Freire (1972) calls 'conscientization.'[16] Under such conditions, expeditions become more than an attempt to learn *about* the impoverished: they become an effort to learn *with* them the oppressive reality that confronts ordinary people in their daily lives.[17] From this standpoint, the academic geographers involved in geographical expeditions come to know through a dialogue of mutual recognition both oppressed people's objective situation in the city and their awareness of that situation (cf. Freire, 1972: 68). Furthermore, the researchers can come to have a recognition of themselves as part of this synergy. It might be possible to press this point further if we interpolate Kojève's (1980) highly influential reading of Hegel which asserts that the self-consciousness of an individual subject can only be achieved through the *recognition* (as an equal) of the Other. Such a dialogical interaction could provide the opportunity both to discover, following Haraway, the significant differences ushered in by global systems of domination and to stimulate awareness (for the geographer as well as oppressed subjects) of these restrictive mechanisms, since they cannot be rendered intelligible solely at the level of 'concrete experience.'

That is why radical geographers have a vital contributing role to play. Through expeditions it is incumbent upon the geographer to become a person of action, a radical problem-raiser, a responsible critical analyst participating *with* the oppressed. That said, the ambit of the geographer's responsibility is always ambiguous. And while expeditions in the 1960s and 1970s were intended to be a mutual learning and consciousness-raising program for researcher and researched alike, academic geographers were not, as Bunge insisted, to organize the local community.[18] Academic geographers can stifle community mobilization or centralize the expedition's organizational structure. According to Bunge, the Vancouver Geographical Expedition was a failure because it lacked a true community base and was never self-critical about its democratic failings (Bunge, n.d.: 81). So, there is always an immanent hazard that the voice heard in the supposed symbiosis between academic geographer and folk geographer is skewed toward the overzealous – though well-meaning – academic geographer. As the voice of the oppressed is muted, the expedition program degenerates into a paternalism reminiscent of 19th century Western missionaries and settlement houses.

In both Detroit and Toronto, however, this enfeebling impulse was avoided through *organic* interaction between academics and the local community. In each case, academic geographers persistently asked the local community about their own priorities. Bunge's *Fitzgerald*, for instance, impresses by the sensitivity expressed in the text and through the emotive photographs. Therein, via the medium of the Detroit Geographical Expedition, it was unquestionably black voices documented in an honest and non-patronizing fashion; *Fitzgerald* evoked a representational (not represented) experience of oppressed black people in an American city. In Detroit, Bunge consolidated the radical intellectual ideal that Marshall Berman (forthcoming) articulates as the conjoining of the 'stacks in the library with the signs in the street.'

Concluding remarks

It's been some twenty years since expeditions were initiated. Difficulties plague any assessment of their efficacy, either theoretically or practically. Yet, critical assessment would appear to be in order. Bunge recently – and rather elusively – confirmed that it was too early to tell whether the Detroit expedition and the accompanying book, *Fitzgerald*, were successful (interview, 1994). Nevertheless, expeditions did for a brief sparkling moment threaten what Harvey (1973: 147–52) labeled status quo understandings of capitalist society. Presumably this was why they came under assault from dominant class and social forces within the academy. Bunge claims he was driven out of his university teaching posts (interview), and Horvath (1971: 84) described how expedition radicals were fired, denied promotions, and refused admittance to graduate school, grimly concluding: '[d]ealing with the poor and powerless transforms the advocate into a marginal man.'

What of the reinstigation of a similar radical venture today? Aside from all else, one factor precluding a reassertion of the expedition principle may be lack of time: with burgeoning teaching and administrative work loads and the competitive stresses of 'publish or perish' in an ever more marketized academic world, finding the time to begin living, working and getting to know a potential base camp locality as an insider is extremely difficult. Such pressures aren't denied by Bunge, though he suggests that they can be partly circumvented by geographers implanting themselves within what he calls the 'cracks' between the academy and broader social life (interview 1994). That way, a potentially creative tension might ensue as the scholar restlessly gravitates between formal academic and community-based duties, whereby one simultaneously informs and enriches the other. The prospect for the insertion of some kind of expedition (nominally or otherwise) might thus blossom within the contact zones of these conflictual realities.

To summarize: I've argued that there is much that is instructive about Bunge's geographical expedition program for radical debates resonating today, especially over situatedness, positionality, representation and the political role of left academics. In this paper, I have tried to sketch out the numerous ways expeditions previously acknowledged that these issues were vital aspects of critical scholarship and knowledge production. While it would be hasty – and foolhardy – to think Bunge had definitive answers to such problematical themes, his expeditions did at least show geographers a possible way *into* these dilemmas; that they did so in such a palpable manner makes them all the more suggestive and radical today. At any rate, exploring more deeply the tradition's successes and failures might illuminate the pursuit of the genuinely accountable and responsible situated knowledge that Haraway *et al.* now invoke. At a time when left geography is in grave danger of being rendered anodyne through a heady prioritization of discourse and textual politics, there is much to learn from the legacy of practical expeditions into the world of the exploited and oppressed outside the academy: it might at least ensure that critical theory is truly critical.

Notes

1 I take expeditions to be part and parcel of the modernist tradition of opposition that Berman (1982) so brilliantly identifies: it is a tradition, for example, that offers a 'celebration of urban vitality, diversity and fullness of life' (p. 316), and is intimately related to a 'shout in the street.'

2 Marx's critique of political economy was a powerful version of this thesis, accepting that there are epistemological and ontological distinctions between Marx and Haraway and Hartsock in terms of their notions of truth and objectivity, inasmuch as Marx's scathing analysis of modern capitalism was established from the subjugated standpoint of the working class. For Marx, *true* knowledge could only be produced *within* the confines of capitalist power relations (see Harvey, 1989) and pretending to be 'outside of' or 'beyond' a position in the world through an appeal to any notion of objective neutrality is either intellectually shoddy – because it fails to confront the *political nature* of knowledge production – or outrightly dishonest. 'Objective truth,' according to Marx in his 'Theses on Feuerbach' (III), 'is not a theoretical but a practical question. It is in praxis that humans must prove the truth.' However, I think it is also important to

bear in mind that my discussion below recognizes that Bunge's expedition concept relied on a Marxist notion of truth and science, and as such differs somewhat from the manner in which Haraway and Hartsock deploy the situated knowledge stance.

3 It is also worthwhile here to underscore that this understanding holds a certain similarity with the work of anthropologist Clifford Geertz, notably his *Local Knowledge* (1983) (see especially the chapter 'From the native's point of view').

4 This said, it is perhaps ironic to look at the respectability of much radical geography and the established academic reputation of radical geographers today, many of whom hold senior ranking positions in Anglo-American geography departments. (For a fierce, though marred, polemic on the contradictions between tenure politics and radical politics, see Jacoby, 1987.) This scholarly respectability of geography can, of course, be witnessed by *Antipode*'s own decision to professionalize in 1986 and prosper from the academic credibility provided by publisher Basil Blackwell (cf. Jacoby's comments, pp. 181–82).

5 See, for example, Lewis's (1973) and Ley's (1973) review of *Fitzgerald*, and Bunge's (1974a) trenchant rejoinder to Lewis.

6 'Armchair geographers of the world arise, you have nothing to lose but your middle-aged flab,' was a flamboyant Bunge clarion call at the time (see Bunge, 1977a). 'The academic geographer,' Bunge colorfully adds, 'needs to get off their camp-ass.' This opinion, it should be added, is one Bunge reinforced today (interview, 1994).

7 Interview, 1994.

8 Bunge's project likewise parallels some of the ideas expounded by the Situationists in continental Europe between 1957 and 1972. The strategies formulated by this group focused on the creation of *situations*, creatively constructed encounters and directly lived experiences that could subvert and transform alienated everyday life within urban settings (see Knabb, 1981). Herein, the active production of situations lay at the core of the Situationists' manifesto of an integrated urbanism (so-called unitary urbanism) inasmuch as these situations provided a critical vantage point from which proponents could understand, contest and agitate against the sterility and oppressive nature of market-driven urban landscapes and practices of everyday life.

9 As I read it, this standpoint also closely resembles Marshall Berman's special notion of modernism and its connections with the 'signs in the street': both Bunge's and Berman's theorizations are thus passionate and partisan ones that are deeply embedded in the everyday life of ordinary people. Berman's (1984) vignettes on the desolation and struggles involved in New York daily life, for example, evoked as a response to what he sees as the 'remoteness' of Perry Anderson's vision of modernity, is a telling recognition of this concern.

10 In the foreword to *Fitzgerald*, the book is ambiguously described as 'science: its data are maps, graphics, photographs, and the words of people. But the book also makes a value judgment – the desirability of human survival – and thus transforms itself into a steel-hard hammer of humanism.' And later on: 'The end product [of *Fitzgerald*], like science and art, hopes to be more real than facts alone.' For a brief discussion of the dissolution of art, science and humanism in Bunge's *Fitzgerald*, see Meinig (1983). Certain situated knowledges, such as those drawing upon the anti-humanism of Foucault's post-structuralism – and it is uncertain as to whether Haraway falls into this camp or not – would doubtless want to distance themselves from Bunge's humanist predilections.

11 Bunge's early explorations here were already keenly sensitive to the problem of geographical scale, a topic that writers such as Smith (1992) and Harvey (1993)

have recently sought to address more directly. To this extent, Bunge's expedition project, in Toronto especially, recognized, as Neil Smith has more recently, that capitalism operated in some sort of 'nested hierarchical space.' While Bunge didn't, of course, have any definitive answers to this dilemma, he did at least pinpoint the theoretical and practical importance of arbitrating and translating between different spatial scales.

12 Buber (1987: 88) provides succinct confirmation here when he points out that 'no [hu]man is pure person and no [hu]man is pure individuality.' And Sennett's (1970) far-sighted debunking of the desire for a 'purified identity' in city life likewise reiterates Buber's concern, though less mystically. 'In order to sense the Other,' Sennett (1990: 148) has written more recently, 'one must do the work of accepting oneself as incomplete.'

13 Hartsock (1989/90) makes a similar claim: '[t]here is a role for intellectuals in making these [situated knowledges] clear [and] in explaining a group to itself, in articulating taken-for-granted understandings.' Hartsock's postulation, of course, has close affinities with Gramsci's (1971) category of 'organic intellectual.' For Gramsci, these comprise intellectuals who are *organically bonded* to a place and to a people, to the degree that they feel 'the elementary passions of the people, understanding them and therefore explaining and justifying them in the particular historical [and geographical] situation and connecting them dialectically to the laws of history' (p. 418).

14 Since the paper was first drafted, it has come to my attention that Gregory (1978: 161–64) had already made allusions with respect to this possibility. Although Gregory's main purpose was to emphasize the parallels between Freire's 'theory of dialogical action' and Habermas's 'theory of communicative competence' for furthering 'committed explanation in geography,' Gregory stresses that: 'when the practical lessons which they [geographical expeditions] contained were translated into theoretical terms the language was Freire's' (p. 162).

15 The problematical nature of 'concrete experience' or 'common sense' understandings of social reality was of course emphasized by Gramsci. Indeed, he wrote (1971: 422) that common sense is 'a chaotic aggregate of disparate conceptions, and one can find there anything that one likes'.

16 Conscientization refers, for instance, 'to learning to perceive social, political, and economic contradictions and to take action against the oppressive element of society' (Freire, 1972: 15).

17 As a shorthand definition, by 'ordinary people' I mean those people who may have the capabilities to intellectualize but don't, as Gramsci identified, have the capacity to *function* as an intellectual.

18 A further point of qualification might also be useful here: I accept, as did Bunge, that the nature of representation *within* respective oppressed communities is problematical; and I am aware that local community leaders' playing a 'vanguard' role is both unavoidable and frequently divisive. Moreover, while I likewise accept Young's (1990) favoring of a 'politics of difference' for checking romantic – and potentially reactionary – notions of community, I also believe it possible, then as now, to speak of 'community action' comprising a group of people bonded by *commonality* and organizing around a common grievance in a way that isn't simply a NIMBY ordeal (see Merrifield, 1993b). And as with any collective action there are leaders and spokepersons. A community's mobilization here is, furthermore, likely to involve a 'militant particularist' component and activate internal as well as external controversy. This might be as much about the *form* of resistance as it is about the actual grievance itself.

References

Adorno, T. 1973: *Negative Dialectics*. London: Routledge.

Antipodean Staff Reporters 1969: 'Where it's at: Two programs in search of geographers II: The Detroit Geographical Expedition.' *Antipode* 1, 45–6.

Berman, M. 1982: *All That Is Solid Melts into Air*. London: Verso.

Berman, M. 1984: The signs in the street: A response to Perry Anderson. *New Left Review* 144, 114–23.

Berman, M. forthcoming: Justice/Just-Us: Rap and social justice in America. In Merrifield, A. and Swyngedouw, E. (eds), *The Urbanization of Injustice*. London: Lawrence and Wishart.

Buber, M. 1987: *I and Thou*. Edinburgh: T. and T. Clark.

Bunge, W. 1962: *Theoretical Geography*. Stockholm: Lund Studies in Geography.

Bunge, W. 1971: *Fitzgerald: Geography of a Revolution*. Cambridge, Masso: Schenkman.

Bunge, W. 1973a: Ethics and logic in geography. In Chorley, R. J. (ed.), *Directions in Geography*. London: Methuen, 317–31.

Bunge, W. 1973b: The geography of human survival. *Annals of the Association of American Geographers* 63, 275–95.

Bunge, W. 1973c: 'Urban stations – The tradition of geographic base camp urbanized.' Unpublished paper.

Bunge, W. 1974a: Fitzgerald from a distance. *Annals of the Association of American Geographers* 64, 485–9.

Bunge, W. 1974b: Simplicity again. *Geographical Analysis* 6, 85–9.

Bunge, W. 1977a: The first years of the Detroit Geographical Expedition: A personal report. In Peet, R. (ed.), *Radical Geography*. London: Methuen, 31–9.

Bunge, W. 1977b: The point of reproduction: A second front. *Antipode* 9, 60–76.

Bunge, W. 1979a: Perspective on theoretical geography. *Annals of the Association of American Geographers* 69, 169–74.

Bunge, W. 1979b: Fred K. Schaefer and the science of geography. *Annals of the Association of American Geographers* 69, 128–32.

Bunge, W. no date: 'The Methodology of Exploring Citites.' Unpublished manuscript.

Bunge, W. and Bordessa, R. 1975: *The Canadian Alternative – Survival, Expeditions and Urban Change*. Ontario: York University.

Christopherson, S. 1989: On being outside 'the project.' *Antipode* 21, 83–9.

Clifford, J. and Markus, G. (eds) 1984: *Writing Culture: The Poetics and Politics of Ethnography*. Berkeley: University of California Press.

Colenutt, R. 1971: Postscript on the Detroit Geographical Expedition. *Antipode* 3, 85.

Field Manual no date: *No. 5: Hidden Landscapes*. Detroit: Detroit Geographical Expedition.

Field Manual no date: *No. 8: Toronto – Detroit: A Tale of Two Countries*. Toronto: Toronto Geographical Expedition.

Field Notes 1970: *No. 2 – A Report to the Parents of Detroit on School Decentralization*. Detroit: Detroit Geographical Expedition and Institute.

Field Notes 1972: *No. 4 – The Trumbull Community*. Detroit: Detroit Geographical Expedition.

Freire, P. 1972: *Pedagogy of the Oppressed*. Harmondsworth: Penguin.

Geertz, C. 1983: *Local Knowledge: Further Essays in Interpretative Anthropology*. New York: Basic Books.

Gramsci, A. 1971: *Selections from Prison Notebooks*. London: Lawrence and Wishart.

Gregory, D. 1978: *Ideology, Science and Human Geography*. London: Hutchinson.

Gregory, D. 1994: *Geographical Imaginations*. Oxford: Basil Blackwell.

Haraway, D. 1991: Situated knowledges: The science question in feminism and the privilege of partial perspective. In Haraway, D., *Simians, Cyborgs and Women*. London: Free Association Press, 183–201.

Haraway, D. 1992: *Primate Visions: Gender, Race and Nature in the World of Modern Science*. London: Verso.

Harding, S. 1986: *The Science Question in Feminism*. Milton Keynes: Open University Press.

Hartsock, N. 1987: Rethinking modernism: Minority versus majority theories. *Cultural Critique* No. 7, 187–206.

Hartsock, N. 1989/90: Postmodernism and political change: Issues for feminist theory. *Cultural Critique* No. 14, 15–33.

Harvey, D. 1973: *Social Justice and the City*. London: Edward Arnold.

Harvey, D. 1989: From models to Marx: Notes on the project to 'remodel contemporary geography.' In MacMillan, W. (ed.), *Remodelling Geography*. Oxford: Basil Blackwell, 211–16.

Harvey, D. 1992: Postmodern morality plays. *Antipode* 24, 300–26.

Harvey, D. 1993: The nature of the environment: The dialectics of social and environmental change. In Panitch, L. and Miliband, R. (eds), *Socialist Register 1993*. London: Merlin Press, 1–51.

Harvey, D. and Smith, N. 1984: Geography: From capitals to capital. In Ollman, B. and Vernoff, E. (eds), *The Left Academy: Marxist Scholarship on American Campuses*, Volume II. New York: Praeger, 99–121.

Hegel, G. W. F. 1931: *The Logic of Hegel* (Part 1 of *The Encyclopedia of the Philosophical Sciences*). London: Oxford University Press.

Horvath, R. 1971: The 'Detroit Geographical Expedition and Institute' experience. *Antipode* 3, 73–85.

Howe, I. 1954: This age of conformity. *Partisan Review* 21, 7–33.

Jackson, P. 1985: Urban ethnography. *Progress in Human Geography* 9, 157–76.

Jackson, P. 1993: Editorial: Visible and voice. *Environment and Planning D: Society and Space* 11, 123–6.

Jacoby, R. 1987: *The Last Intellectuals – American Culture in the Age of Academe*. New York: Noonday Press.

Katz, C. 1992: All the world is staged: Intellectuals and the projects of ethnography. *Environment and Planning D: Society and Space* 10, 495–510.

Knabb, K. 1981: *Situationist Anthology*. Berkeley: Bureau of Public Secrets.

Kojève, A. 1980: *Introduction to the Reading of Hegel: Lectures on the Phenomenology of Spirit*. Ithaca: Cornell University Press.

Lefebvre, H. 1991: *Critique of Everyday Life* – Volume One. London: Verso.

Lewis, J. 1973: Review of William Bunge's 'Fitzgerald.' *Annals of the Association of American Geographers* 63, 131–2.

Ley, D. 1973: Review of William Bunge's 'Fitzgerald.' *Annals of the Association of American Geographers* 63, 133–5.

Livingstone, D. 1992: *The Geographical Tradition*. Oxford: Basil Blackwell.

McDowell, L. 1991: The baby and the bathwater: Deconstruction and feminist theory in geography. *Geoforum* 22, 123–33.

McDowell, L. 1992: Multiple voices: Speaking from inside and outside 'the Project.' *Antipode* 24, 56–71.

Marx, K. 1975: Theses on Feuerbach. In *Early Writings*. Harmondsworth: Penguin, 421–3.

Massey, D. 1991: Flexible sexism. *Environment and Planning D: Society and Space* 9, 31–57.

Meinig, D. 1983: Geography as an art. *Transactions of the Institute of British Geographers* 8, 314–28.

Merrifield, A. 1993a: Place and space: A Lefebvrian reconciliation. *Transactions of the Institute of British Geographers* N.S. 18, 516–31.

Merrifield, A. 1993b: The struggle over place: Redeveloping American Can in south-east Baltimore. *Transactions of the Institute of British Geographers* N.S. 18, 102–21.

Nicholson, L. (ed.) 1990: *Feminism/Postmodernism*. London: Routledge.

Oldenburg, C. 1990: I am for an Art . . . In Harrison, C. and Wood, P. (eds), *Art in Theory 1900–1990*. Oxford: Basil Blackwell, 727–30.

Pile, S. and Rose, G. 1992: All or nothing: Politics and critique in the modernism–postmodernism debate. *Environment and Planning D: Society and Space* 10, 123–36.

Sennett, R. 1970: *The Uses of Disorder: Personal Identity and City Life*. Harmondsworth: Penguin.

Sennett, R. 1990: *The Conscience of the Eye – The Design and Social Life of Cities*. London: Faber and Faber.

Smith, N. 1992: Geography, difference and the politics of scale. In Doherty, J. Graham, E. and Malek, M. (eds), *Postmodernism and the Social Sciences*. London: Macmillan.

Spivak, G. 1988: Can the subaltern speak? In Nelson, C. and Grossberg, L. (eds), *Marxism and the Interpretation of Culture*. London: Macmillan, 271–313.

Stephenson, D. 1974: *The Toronto Geographical Expedition. Antipode* 6, 98–101.

Thompson, H. 1988: *Generation of Swine*. London: Picador.

Wood, D. 1993: *The Power of Maps*. London: Routledge.

Young, I. M. 1990: *Justice and the Politics of Difference*. Princeton, New Jersey: Princeton University Press.

REFERENCES

Adams, J. 1969: Directional bias in intra-urban migration. *Economic Geography* 45, 302–23.

Adams, J. and Gilder, K. A. 1972: Household location and intra-urban migration. In Herbert, D. and Johnston, R. J. (eds), *Geography and the urban environment*, Volume 4. Chichester: John Wiley.

Barnett, C. 1993: Peddling postmodernism: a response to Strohmayer and Hannah's 'Domesticating Postmodernism'. *Antipode* 25(4), 345–58.

Barnett, C. and Low, M. 1996: Speculating on theory: towards a political economy of academic publishing. *Area* 28(1), 13–24.

Barrow, H. H. 1923: Geography as human ecology. *Annals of the Association of American Geographers* 13, 1–14.

Bassett, K. and Short, J. 1980: *Housing and residential structure: alternative approaches*. London: Routledge.

Berman, M. 1984: The signs in the streets: a response to Perry Anderson. *New Left Review* 144, 114–28.

Black, R. 1994: Forced migration and environmental change: the impact of refugees on host environments. *Journal of Environmental Management* 42, 261–77.

Black, R. 1996: Immigration and social justice: towards a progressive European immigration policy. *Transactions of the Institute of British Geographers* 21(1), 64–75.

Blaikie, P., Cannon, R., Davis, L. and Wisner, B. 1994: *At risk: natural hazards, people's vulnerability and disasters*. London: Routledge.

Blaut, J. 1979: A radical critique of cultural geography. *Antipode* 11, 25–9.

Bonvalet, C., Carpenter, J. and White, P. 1995: The residential mobility of ethnic minorities: a longitudinal analysis. *Urban Studies* 32(1), 87–103.

Brown, L. A. and Holmes, J. 1971: Search behaviour in an intra-urban migration context: a spatial perspective. *Environment and Planning A* 3, 307–26.

Brown, L. A. and Moore, E. G. 1970: The intra-urban migration process: a perspective. *Geografiska Annaler, Series B* 52, 1–13.

Bunge, W. 1977: The point of reproduction: a second front. *Antipode* 9, 60–76.

Burgess, E. 1919: The growth of the city. *American Journal of Sociology*

Buttimer, A. 1968: Social geography. In Sills, D. (ed.), *International Encyclopedia of Social Science*, Volume 6. New York, 134–45.

Buttimer, A. 1971: Society and milieu in the French geographic tradition. *Association of American Geographers, Monograph Series 6*.

Carlstein, T., Parkes, D. and Thrift, N. (eds) 1978: *Human activity and time geography*. London: Edward Arnold.

Cater, J. and Jones, T. 1989: *Social geography: an introduction to contemporary issues*. London: Edward Arnold.

Castle, I. M. and Gittus, E. 1957: The distribution of social defects in Liverpool. *Sociological Review* 5, 43–64.

Charlesworth, A. 1983: The spatial diffusion of rural protest: on historical and comparative perspective of rural riots in the nineteenth century. *Society and Space* 1, 251–63.

Clarke, J. 1991: *New times and old enemies: essays on cultural studies and America*. London: HarperCollins Academic.

Coates, B. E., Knox, P. and Johnston, R. 1977: *Geography and inequality*. Oxford: Oxford University Press.

Cohen, A. 1980: Drama and politics in the development of a London carnival. *Man* 15, 65–87.

Cohen, A. 1982: A polyethnic London carnival as a contested cultural performance. *Ethnic and Racial Studies* 5, 23–41.

Cohen, S. 1995: Sounding out the city: music and the sensuous reproduction of place. *Transactions of the Institute of British Geographers* 20(4), 434–46.

Collison, S. 1996: Visa requirements, carrier sanctions, 'safe third countries' and 'readmission': the development of an asylum 'buffer zone' in Europe. *Transactions of the Institute of British Geographers* 21(1), 76–90.

Condon, S. A. and Ogden, P. E. 1991: Afro-Caribbean migrants in France: employment, state policy and the migration process. *Transactions of the Institute of British Geographers* 16, 440–57.

Corbridge, S. 1993: Marxisms, modernities, and moralities: development praxis and the claims of distant strangers. *Society and Space* 11, 449–72.

Cosgrove, D. 1983: Towards a radical cultural geography. *Antipode* 15, 1–11.

Cosgrove, D. 1984: *Social formations and symbolic landscape*. London: Croom Helm.

Cosgrove, D. and Jackson, P. 1987: New directions in cultural geography. *Area* 19(2), 95–101.

Cotterell, M. 1992: St Patrick's Day parades in nineteenth century Toronto: a study of immigrant adjustment and elite control. *Histoire Sociale* 25(49), 57–73.

Crilley, D. 1993: Architecture as advertising: constructing the image of redevelopment. In Kearns, G. and Philo, C. (eds), *Selling places: the city as cultural capital, past and present*. Oxford: Pergamon.

Daniels, S. and Rycroft, S. 1993: Mapping the modern city: Alan Sillitoe's Nottingham novels. *Transactions of the Institute of British Geographers* 18, 460–80.

Davis, M. 1990: *City of quartz*. London: Verso.

Davis, S. and Albaum, M. 1972: Mobility problems of the poor in Indianapolis. In Peet, R. (ed.), *Geographical Perspectives on American Poverty*. Antipode Monographs in Social Geography No. 1. Worcester, MA.

Dear, M. and Gleeson, B. 1991: Community attitudes towards the homeless. *Urban Geography* 12(2), 155–76.

Dear, M. and Scott, A. 1981: *Urbanization and urban planning in capitalist society*. London: Methuen.

Dear, M. and Wolch, J. 1987: *Landscapes of despair: from deinstitutionalization to homelessness*. Princeton, NJ: Princeton University Press.

Derrida, J. 1973: *Speech and phenomena*. Evanston: Northwestern University Press.

Derrida, J. 1978: *Writing and difference*. London: Routledge and Kegan Paul.

Derrida, J. 1981: *Positions*. Chicago: University of Chicago Press.

de Vise, P. 1972: Cook County Hospital: bulwark of Chicago's apartheid health system and prototype of the nation's public hospitals. *Antipode* 3(1), 9–20.
Dickens, P., Duncan, S. S., Goodwin, M. and Gray, F. 1985: *Housing, states and localities*. London: Methuen.
Dingemans, D. 1978: Redlining and mortgage lending in Sacramento, CA. *Annals of the Association of American Geographers* 69, 225–39.
Dow, K. 1992: Exploring differences in our common future: the meaning of vulnerability to global environmental change. *Geoforum* 23(3), 417–36.
Duncan, J. and Duncan, N. 1988: (Re)reading the landscape. *Society and Space* 6, 117–26.
Duncan, J. and Ley, D. 1982: Structural marxism and human geography: a critical assessment. *Annals of the Association of American Geographers* 72, 30–59.
Duncan, J. and Ley, D. (eds) 1993a: *Representing cultural geography*. Andover: Chapman and Hall.
Duncan, J. and Ley, D. 1993b: *Place, culture and representation*. London: Routledge.
Duncan, N. and Sharp, J. P. 1993: Confronting representation(s). *Society and Space* 11, 473–86.
Duncan, S. S. 1976: Research directions in social geography: housing opportunities and constraints. *Transactions of the Institute of British Geographers,* ns 1(1), 10–19.
Duncan, S. 1979: Qualitative change in human geography – an introduction. *Geoforum* 10(1), 1–4.
Edgell, S. and Duke, V. 1991: *A measure of Thatcherism: a sociology of Britain*. London: HarperCollins.
Eyles, J. 1974: Social theory and social geography. *Progress in Geography* 6, 27–87.
Eyles, J. 1981: Why geography cannot be Marxist: towards an understanding of lived experience. *Environment and Planning A* 13, 1371–88.
Eyles, J. 1989: The geography of everyday life. In Gregory, D. and Walford, R. (eds), *Horizons in Human Geography*. London: Macmillan, 102–17.
Fainstein, N. 1993: Race, class and segregation: discourses about African-Americans. *International Journal of Urban and Regional Research* 17(3), 384–403.
Firey, W. 1945: Sentiment and symbolism as sociological variables. *American Sociological Review* 10, 140–8.
Fitzgerald, W. 1946: Correspondence. *Geographical Journal* 107.
Frankenberg, R. 1966: *Communities in Britain*. Harmondsworth: Penguin.
Friedland, W. 1982: The end of rural society and the future of rural sociology. *Rural Sociology* 47, 589–608.
Garmanikow, E. 1978: Introduction. Special issue on women and the city. *International Journal of Urban and Regional Research* 2, 390–403.
Geertz, C. 1973: *The interpretation of cultures*. New York: Basic Books.
Geertz, C. 1983: *Local knowledge*. New York: Basic Books.
Geras, N. 1995: Language, truth and justice. *New Left Review* 209, 110–35.
Giddens, A. 1979: *Central problems in sociological theory: action, structure and contradiction in social analysis*. London: Macmillan.
Giggs, J. A. 1973: The distribution of schizophrenics in Nottingham. *Transactions of the Institute of British Geographers* 59, 55–76.

Gilbert, E. W. and Steel, R. N. 1945: Social geography and its place in Colonial Studies. *Geographical Journal* 105.

Gleeson, B. 1995: A public space for women: the case of charity in colonial Melbourne. *Area* 27(3), 193–207.

Goheen, P. 1993: The ritual of the streets in mid-19th-century Toronto. *Society and Space* 11(2), 127–46.

Golledge, R., Brown, L. and Williamson, F. 1972: Behavioural approaches in geography: an overview. *Australian Geographer* 12(1), 59–79.

Graham, B. 1994: The search for common ground: Estyn Evans's Ireland. *Transactions of the Institute of British Geographers* 19(2), 183–201.

Gray, F. 1975: Non-explanation in urban geography. *Area* 7(4), 228–35.

Gregory, D. 1981: Human agency and human geography. *Transactions of the Institute of British Geographers*, ns 6(1), 1–18.

Gregory, D. and Urry, J. (eds) 1985: *Social relations and spatial structures.* Basingstoke: Macmillan.

Gregson, N. 1987: Structuration theory: some thoughts on the possibilities for empirical research. *Society and Space* 5, 73–91.

Gruffudd, P. 1994: Back to the land: historiography, rurality and the nation in interwar Wales. *Transactions of the Institute of British Geographers* 19(1), 61–77.

Hall, S. and Jefferson, T. (eds) 1976: *Resistance through rituals.* London: Hutchinson.

Hamnett, C. 1972: The social patterning of cities. In *Social geography: new trends in geography.* Milton Keynes: Open University Press, 29–63.

Hamnett, C. 1979: Area based explanations: a critical appraisal. In Herbert, D. T. and Smith, D. M. (eds), *Social problems and the city.* Oxford: Oxford University Press, 244–60.

Hamnett, C. 1984: Gentrification and residential location theory: a review and assessment. In Herbert, D. T. and Johnston, R. J. (eds), *Geography and the urban environment*, Volume 6. London: John Wiley, 283–319.

Hamnett, C. 1991: The blind men and the elephant: the explanation of gentrification. *Transactions of the Institute of British Geographers*, ns 16, 173–89.

Hamnett, C., McDowell, L. and Sarre, P. 1989: *The changing social structure.* London: Sage.

Hannah, M. and Strohmayer, U. 1991: Ornamentalism, geography and the labour of language in structuration theory. *Society and Space* 9, 309–27.

Harvey, D. 1969: Social justice and spatial systems. In Peet, R. (ed.), *Geographical perspectives on American poverty.* Antipode Monographs in Social Geopgraphy No. 1. Worcester, MA.

Harvey, D. 1972: Revolutionary and counter-revolutionary theory in geography and the problem of ghetto formation. *Antipode* 4(2), 1–13.

Harvey, D. 1973: *Social justice and the city.* London: Edward Arnold.

Harvey, D. 1974: Class monopoly rent, finance capital and the urban revolution. *Regional Studies* 8, 239–55.

Harvey, D. 1985: *The urbanization of capital.* Oxford: Basil Blackwell.

Harvey, D. 1989: *The condition of postmodernity.* Oxford: Basil Blackwell.

Herbert, D. T. 1976: The study of delinquency areas: a social geographical approach. *Transactions of the Institute of British Geographers* 1(4), 472–92.

Hoch, C. 1991: Urban homelessness: a case study of Chicago. *Urban Geography* 12, 137–54.

Hoggart, K. 1990: Let's do away with the rural. *Journal of Rural Studies* 6, 245–57.

Holcomb, B. 1986: Geography and urban women. *Urban Geography* 7, 448–56.

Jackson, P. 1988: Street life: the politics of carnival. *Society and Space* 6, 213–27.

Jackson, P. 1992: The politics of the streets: a geography of Caribana. *Political Geography* 11, 130–51.

Johnson, N. C. 1994: Sculpting heroic histories: celebrating the centenary of the 1798 rebellion in Ireland. *Transactions of the Institute of British Geographers* 19(1), 78–93.

Johnson, N. C. 1995: Cast in stone: monuments, geography and nationalism. *Society and Space* 13, 51–65.

Johnston, R. J. 1966: The location of high status residential areas. *Geografiska Annaler* 48B, 23–35.

Johnston, R. J. 1972a: *Urban residential patterns.* London: G. Bell and Sons.

Johnston, R. J. 1972b: Towards a general model of intra-urban residential patterns: some cross-cultural observations. *Progress in Human Geography* 4, 85–124.

Johnston, R. J. 1973: Possible extensions to the factorial ecology method: a note. *Environment and Planning* 5, 719–34.

Johnston, R. J. 1991: *A question of place: exploring the practice of human geography.* Oxford: Basil Blackwell.

Jones, E. 1960: *A social geography of Belfast.* London: Oxford University Press.

Jones, E. 1975: *Readings in social geography.* Oxford: Oxford University Press.

Keane, J. 1988: *Democracy and civil society.* London: Verso.

Kearns, G. and Philo, C. 1993: *Selling places: the city as cultural capital, past and present.* Oxford: Pergamon.

Keith, M. 1987: 'Something happened': the problems of explaining the 1980 and 1981 riots in British cities. In Jackson, P. (ed.), *Race and racism.* London: Allen and Unwin, 275–303.

Kelsey, J. 1993: *Rolling back the state.* Wellington, New Zealand: Bridget Williams Books.

Kelsey, J. and O'Brien, M. 1995: *Setting the record straight: social development in Aotearoa/New Zealand.* Wellington, New Zealand: Association of Non-Government Organisations of Aotearoa.

Knights, M. 1996: Bangladeshi immigrants in Italy: from geopolitics to micropolitics. *Transactions of the Institute of British Geographers* 21(1), 105–23.

Knox, P. 1975: *Social well-being: a spatial perspective.* Oxford: Oxford University Press.

Knox, P. 1978: The intra-urban ecology of primary medical care: patterns of accessibility and their policy implications. *Environment and Planning A* 10, 415–35.

Knox, P. 1987: *Urban social geography,* 2nd edition. London: Longman.

Kong, L. 1995: Music and cultural politics: ideology and resistance in Singapore. *Transactions of the Institute of British Geographers* 20(4), 447–59.

Kuper, H. 1972: The language of sites and the politics of space. *American Anthropologist,* ns 74, 411–25.

Lee, T. R. 1973: Ethnic and social class factors in residential segregation: some implications for dispersal. *Environment and Planning* 5, 477–90.

Ley, D. 1983: Gentrification and the politics of the new middle class. *Society and Space* 12, 53–74.

Ley, D. and Cybriwsky, R. 1974: Urban graffiti as territorial markers. *Annals of the Association of American Geographers* 64, 491–505.

Littlejohn, J. 1963: *Westrigg: the sociology of a Cheviot parish*. London: Routledge and Kegan Paul.

McDowell, L. 1983: Towards an understanding of the gender division of urban space. *Society and Space* 1, 59–72.

McDowell, L. 1989: Gender divisions. In Hamnett, C., McDowell, L. and Sarre, P. (eds), *The changing social structure*. London: Sage, 158–98.

McDowell, L. 1991: Life without father and Ford: the new gender order of post-Fordism. *Transactions of the Institute of British Geographers* 16(4), 400–19.

McDowell, L. 1992: Doing gender: feminism, feminists and research methods in human geography. *Transactions of the Institute of British Geographers* 17(4), 399–416.

McDowell, L. 1994a: Polyphony and pedagogic authority. *Area* 26(3), 241–8.

McDowell, L. 1994b: The transformation of cultural geography. In Gregory, D., Martin, R. and Smith, G. (eds), *Human geography: society, space and social science*. London: Macmillan.

McDowell, L. and Massey, D. 1984: A woman's place. In Massey, D. and Allen, J. (eds), *Geography matters*. Cambridge: Cambridge University Press, 128–47.

McLuhan, M. 1964: *Understanding media: extensions of man*. London: Routledge and Kegan Paul.

MacMaster, N. 1990: The battle for Mousehold Heath, 1857–1884: 'popular politics' and the Victorian public park. *Past and Present* 127, 117–54.

Marston, S. A. 1989: Public rituals and community power: St Patrick's Day parades in Lowell, Mass., 1841–74. *Political Geography Quarterly* 8, 255–69.

Massey, D. 1979: In what sense a regional problem? *Regional Studies* 13, 233–43.

Massey, D. 1984: Introduction: geography matters. In Massey, D. and Allen, J. (eds), *Geography Matters*. Cambridge: Cambridge University Press, 1–11.

Massey, D. 1985: New directions in space. In Gregory, D. and Urry, J. (eds), *Social relations and spatial structures*. London: Macmillan, 9–19.

Mercer, D. 1972: Behavioural geography and the sociology of social action. *Area* 4, 48–51.

Mills, C. 1988: Life on the upslope: the postmodern landscape of gentrification. *Society and Space* 6, 169–89.

Moscovici, S. 1976: *La Psychoanalyse: son image et son public*. Paris: Presses Universitaires de France.

Moscovici, S. 1981: On social representation. In Forgas, J. (ed.), *Social cognition: perspectives on everyday understanding*. London: Academic Press, 181–209.

Munt, I. 1987: Economic restructuring, culture and gentrification: a case study of Battersea, London. *Environment and Planning A* 19, 1175–97.

Murdoch, J. and Pratt, A. C. 1993: Rural studies: modernism, postmodernism and the 'post-rural'. *Journal of Rural Studies* 9(4), 411–27.

Murie, A. 1974: *Housing movement and housing choice, Occasional Paper No. 28*. Birmingham: Centre for Urban and Regional Studies.

Newby, H. 1986: Locality and rurality: the restructuring of rural social relations. *Regional Studies* 20(3), 209–16.

O'Loughlin, J. 1980: The distribution and migration of foreigners in German cities. *Geographical Review* 70, 253–75.

Pacione, M. 1987: *Social geography: progress and prospect.* London: Croom Helm.

Pahl, R. 1965: Class and community in English commuter villages. *Sociologica Ruralis* 5(1).

Pahl, R. 1966: The urban–rural continuum. *Sociologica Ruralis* 6, 299–327.

Pahl, R. 1970a: *Whose city?* London: Longman.

Pahl, R. 1970b: Trends in social geography. In *Whose city?* London: Longman, 125–46.

Pahl, R. 1971: Poverty and the urban system. In Chisholm, M. and Manners, G. (eds), *Spatial policy problems of the British economy.* Cambridge: Cambridge University Press.

Park, R. E. 1904 (1972): The crowd and the public. In Elsner, H. Jr (ed.), *The crowd and the public and other essays.* Chicago: University of Chicago Press.

Park, R. E. 1916: The city: suggestions for the investigation of human behavior in the urban environment. *American Journal of Sociology* 20, 577–612.

Park, R. E. and Burgess, E. 1925: *The city.* Chicago: University of Chicago Press.

Peach, C. 1968: *West Indian migration to Britain: a social geography.* Oxford: Oxford University Press.

Peach, C. (ed.) 1975: *Urban social segregation.* London: Longman.

Peach, C. 1985: Immigrants and the 1981 urban riots in Britain. In White, P. E. and Van der Knapp, G. A. (eds), *Contemporary studies of migration.* Norwich: Geo Books, 146–52.

Peach, C. 1996: Does Britain have ghettoes? *Transactions of the Institute of British Geographers* 21(1), 216–35.

Peach, C. and Byron, M. 1993: Caribbean tenants in council housing: 'race', class and gender. *New Community* 19(3), 407–23.

Peach, C. and Shah, S. 1980: The contribution of council house allocation to West Indian desegregation in London, 1961–71. *Urban Studies* 17, 333–41.

Peet, R. 1970: Some issues in the social geography of American poverty. In Peet, R. (ed.), *Geographical Perspectives on American Poverty.* Antipode Monographs in Social Geography No. 1. Worcester, MA.

Philo, C. 1992: Neglected rural geographies. *Journal of Rural Studies* 8(2), 193–207.

Philo, C. 1995: *The social geography of poverty in the UK.* London: Child Poverty Action Group.

Pile, S. and Rose, G. 1992: All or nothing: politics and critique in the modernism–postmodernism debate. *Society and Space* 10, 123–36.

Potter, J. and Wetherell, M. 1987: *Discourse and social psychology: beyond attitudes and behaviour.* London: Sage.

Pratt, G. 1982: Class analysis and urban domestic property: a critical re-examination. *International Journal of Urban and Regional Research* 6, 481–502.

Pratt, G. 1991: Discourses of locality. *Environment and Planning A* 23, 257–66.

Pred, A. and Palm, R. 1974: *A time-geographic perspective on problems of inequality for women.* Working paper 236. Berkeley: Institute of Urban and Regional Development, University of California.

Pred, A. and Palm, R. 1978: The status of American women: a time-geographic view. In Lanegren, D. and Palm, R. (eds), *An invitation to geography.* New York: McGraw-Hill.

Rawstron, E. and Coates, B. 1966: Opportunity and affluence. *Geography* 53, 1–15.

Robson, B. 1969: *Urban analysis: a study of city structure*. Cambridge: Cambridge University Press.

Rorty, R. 1980: *Philosophy and the mirror of nature*. Oxford: Oxford University Press.

Rorty, R. 1982: *Consequences of pragmatism*. Hemel Hempstead: Harvester.

Rorty, R. 1991: *Objectivity, relativism and truth*. Cambridge: Cambridge University Press.

Rose, D. 1984: Rethinking gentrification: beyond the uneven development of marxist theory. *Society and Space* 2, 47–74.

Rose, G. 1994: The cultural politics of place: local representation and oppositional discourse in two films. *Transactions of the Institute of British Geographers* 19(1), 46–60.

Rose, H. M. 1970: The development of an urban sub-system: the case of the Negro ghetto. *Annals of the Association of American Geographers* 61(1), 1–17.

Rose, H. M. 1972: The spatial development of black residential sub-systems. *Economic Geography* 48(3), 43–66.

Rowe, S. and Wolch, J. 1990: Social networks in time and space: homelessness in Skid Row, LA. *Annals of the Association of American Geographers* 80, 184–204.

Said, E. 1979: *Orientalism*. New York: Vintage Books.

Sarre, P. 1986: Choice and constraint in ethnic minority housing. *Housing Studies* 1(2), 71–86.

Sarre, P. 1987: Realism in practice. *Area* 19(1), 3–10.

Sarre, P., Phillips, D. and Skellington, R. 1989: *Ethnic minority housing: explanations and policies*. Aldershot: Avebury.

Saunders, P. 1979: *Urban politics*. London: Hutchinson.

Saunders, P. 1984: Beyond housing classes: the sociological significance of private property rights in means of consumption. *International Journal of Urban and Regional Research* 8(2), 202–27.

Saunders, P. 1985: Space, the city and urban sociology. In Gregory, D. and Urry, J. (eds), *Social relations and spatial structures*. London: Macmillan, 67–89.

Sayer, A. 1977: Gravity models and spatial autocorrelations or atrophy in urban and regional modelling, *Area* 9(3).

Sayer, A. 1979: Epistemology and conceptions of people in nature in geography, *Geoforum* 10, 19–43.

Sayer, A. 1985: The difference that space makes. In Gregory, D. and Urry, J. (eds), *Social relations and spatial structures*. London: Macmillan, 49–66.

Shields, R. 1991: *Places on the margins: alternative geographies of modernity*. London: Routledge.

Short, J. R. 1978: Residential mobility in the private housing market of Bristol. *Transactions of the Institute of British Geographers*, ns 3(4), 533–47.

Sibley, D. 1981: *Outsiders in urban societies*. Oxford: Basil Blackwell.

Sibley, D. 1988: Purification of space. *Society and Space* 6, 409–21.

Sibley, D. 1994: The sin of transgression. *Area* 26(3), 300–3.

Sibley, D. 1995: *Geographies of exclusion: society and difference in the West*. London: Routledge.

Siddaway, J. D. 1992: In other worlds: on the politics of research by 'First World' geographers in the 'Third World'. *Area* 24, 403–8.

Simmel, G. 1903: The metropolis and mental life, reprinted in Levine, D. N. (ed.), *Georg Simmel on individuality and social forms* (1971), 324–39.

Smith, D. M. 1977: *Human geography: a welfare approach*. London: Edward Arnold.

Smith, D. M. 1994: On professional responsibility to distant others. *Area* 26(4), 359–67.

Smith, D.M. and Eyles, J. 1974: *Qualitative methods in human geography*.

Smith, N. 1979: Toward a theory of gentrification: a back to the city movement by capital not people. *Journal of the American Planning Association* 45, 538–48.

Smith, S. 1988: Political interpretations of 'racial segregation' in Britain. *Society and Space* 8, 423–44.

Smith, S. J. 1993: Bounding the Borders: claiming space and making place in rural Scotland. *Transactions of the Institute of British Geographers* 18(3), 291–309.

Spivak, G. 1988: Can the subaltern speak? Speculations on widow sacrifice. In Nelson, C. and Grossberg, L. (eds), *Marxism and the interpretation of culture*. London: Macmillan, 271–313.

Strohmayer, U. and Hannah, M. 1992: Domesticating postmodernism. *Antipode* 24, 29–55.

Swyngedouw, E. 1989: The heart of the place: resurrection of locality in an age of hyperspace. *Geografiska Annaler* 71B(1), 31–42.

Thompson, G. 1990: *The political economy of the New Right*. London: Pinter.

Tivers, J. 1985: *Women attached: the daily lives of women with young children*. London: Croom Helm.

Valentine, G. 1993: (Hetero)sexing space: lesbian perceptions and experiences of everyday spaces. *Society and Space* 11(4), 375–496.

Vidal de la Blache, P. 1926: *Principles of human geography*. New York.

Visher, S. S. 1929: Social geography. Social Forces 10, 351–4.

Wallis, C. P. and Maliphant, P. 1967: Delinquent areas in the county of London: ecological factors. *British Journal of Criminology* 7, 250–84.

Watts, M. J. and Bohle, H. G. 1993: The space of vulnerability: the causal structure of hunger and famine. *Progress in Human Geography* 17(1), 43–67.

Western, J. 1991: *Outcast Cape Town*. London: Allen and Unwin.

Western, J. 1993: Ambivalent attachments to place in London: twelve Barbadian families. *Society and Space* 11(2), 147–70.

White, P. and Jackson, P. 1995: (Re)theorising Population Geography, *International Journal of Population Geography I*, 111–23.

Williams, P. 1978a: Building societies and the inner city. *Transactions of the Institute of British Geographers*, ns. 3, 23–34.

Williams, P. 1978b: Urban managerialism: a concept of relevance. *Area* 10, 236–40.

Williams, P. 1982: Restructuring urban managerialism: towards a political economy of urban allocation. *Environment and Planning A* 14, 95–105.

Williams, W. 1956: *The sociology of an English village*. London: Routledge and Kegan Paul.

Williams, W. 1963: *A West Country village: Ashworthy*. London: Routledge and Kegan Paul.

Winchester, H. and White, P. E. 1988: The location of marginalised groups in the inner city. *Society and Space* 6, 37–54.

Wirth, L. 1939: Urbanism as a way of life. *American Journal of Sociology* 44, 1–24.

Wolfe, J., Drover, G. and Skelton, I. 1980: Inner city real estate activity in Montreal: institutional characteristics of decline. *Canadian Geographer* 24, 348–67.

Wolpert, J. 1965: Behavioural aspects of the decision to migrate. *Papers of the Regional Science Association* 15, 159–69.

Wolpert, J. 1967: Distance and directional bias in inter-urban migratory streams. *Annals of the Association of American Geographers* 57, 605–16.

Young, I. M. 1990: *Justice and the politics of difference.* Princeton, NJ: Princeton University Press.

Zetter, R. 1988: Refugees and refugee studies: a label and an agenda. *Journal of Refugee Studies* 1(1), 1–6.

INDEX